Benchmark Papers
in Animal Behavior

Series Editor: Martin W. Schein
West Virginia University

Benchmark Papers
in Animal Behavior / 8

A BENCHMARK® Books Series

VERTEBRATE
SOCIAL
ORGANIZATION

Edited by

EDWIN M. BANKS

Univerity of Illinois
Urbana-Champaign

Dowden, Hutchinson
& Ross, Inc.
STROUDSBURG, PENNSYLVANIA

Copyright © 1977 by **Dowden, Hutchinson & Ross, Inc.**
Benchmark Papers in Animal Behavior, Volume 8
Library of Congress Catalog Card Number: 76-26571
ISBN: 0-87933-122-4

79 78 77   1 2 3 4 5
Manufactured in the United States of America.

LIBRARY OF CONGRESS CATALOGING IN PUBLICATION DATA
Main entry under title:
Vertebrate social organization.
   (Benchmark papers in animal behavior ; 8)
   Includes bibliographies and index.
   1. Vertebrates—Behavior—Addresses, essays, lectures.   2.   Social
behavior in animals—Addresses, essays, lectures.   I.   Banks, Edwin M.
[DNLM: 1. Vertebrates. 2. Behavior, Animal. 3. Social behavior.
W1 BE514 v.7 / QL775 V567]
QL775.V47   596'.05   76-26571
ISBN   0-87933-122-4

Exclusive Distributor: **Halsted Press**
A Division of John Wiley & Sons, Inc.
ISBN: 0-470-98950-5

SERIES EDITOR'S FOREWORD

It was not too many years ago that virtually all research publications dealing with animal behavior could be housed within the covers of a very few hard-bound volumes which were easily accessible to the few workers in the field. Times have changed! Present day students of animal behavior have all they can do to keep abreast of developments within their own areas of special interest, let alone in the field as a whole—and of course we have long since given up attempts to maintain more than a superficial awareness of what is happening "in biology," "in psychology," "in sociology," or in any of the broad fields touching upon or encompassing the behavioral sicences.

It was even fewer years ago that those who taught animal behavior courses could easily choose a suitable textbook from among the very few that were available; all "covered" the field, according to the bias of the author. Students working on a special project used *the* text and *the* journal as reference sources, and for the most part successfully covered their assigned topics. Times have changed! The present day teacher of animal behavior is confronted with a bewildering array of books to choose from, some purported to be all-encompassing, others confessing to strictly delimited coverage, and still others being simply collections of recent and profound writings.

In response to the problem of the steadily increasing and overwhelming volume of information in the area, the Benchmark Papers in Animal Behavior was launched as a series of single topic volumes designed to be some things to some people. Each volume contains a collection of what an expert considers to be *the* significant research papers in a given topic area. Each volume, then, serves several purposes. To the teacher, a Benchmark volume serves as a supplement to other written materials assigned to students; it permits in-depth consideration of a particular topic while at the same time confronting the student (often for the first time) with original research papers of outstanding quality. To the researcher, a Benchmark volume serves to save countless hours digging through the various journals to find *the* basic articles in a specified area of interest; often the journals are not easily available. To the student, a Benchmark volume provides a readily accessible set of original papers on the topic, a set that forms the core of the more extensive bibliography that is likely to be compiled; it also permits the student to see at first hand what an "expert" thinks is important in the area, and to react accordingly. Finally, to the librarian, a Benchmark volume represents a collection of important papers from many diverse sources, thus making readily available materials that might otherwise not be economically possible to obtain or physically possible to keep in stock.

The choice of topics to be covered in this series is no small matter. Each of us could come up with a long list of possible topics and then search for potential volume editors. Alternatively, we could draw up long lists of recognized and prominent scholars and try to persuade them to do a volume on topics of their choice. For the most part, I have followed a mix of both approaches: match a distinguished researcher with a desired topic, and the results should be outstanding. And so it is with the present volume.

Dr. Banks was an obvious choice as Editor for the present volume which, together with *Territory* (Volume 2) and *Social Hierarchy and Dominance* (Volume 3) in the Benchmark Series, completes a trilogy on the organization of vertebrate societies. Dr. Banks, one of the last students of the renowned ecologist, W. C. Allee, is himself an eminent and well respected behavioral ecologist, very active in teaching and research; few people are as appropriate as he is to put together the present volume.

MARTIN W. SCHEIN

PREFACE

The purpose of this Benchmark volume is to provide a collection of essays, observations, and experimental studies focusing on the subject of social organization in vertebrate animals. Other Benchmark volumes treat the topics of territoriality (Stokes, 1974), social hierarchy (Schein, 1975), and behavior as an ecological factor (Davis, 1974). The interested student will find that these three volumes plus the present one offer a veritable treasure of information on various aspects of sociobiology.

The approach adopted in the present volume was to select articles of historical importance that emphasized evolutionary perspectives. The difficulty of the task was evident from the outset because, until quite recently, the vast majority of students of vertebrate social organization have not been concerned with an examination of ultimate causal factors. It would, for example, have been relatively simple to assemble a collection of outstanding contributions to our understanding of ecological and evolutionary effects on social structures drawn from the literature of the past five or ten years.

The articles comprising this volume were selected after consideration of a number of factors, many of which reflect my personal biases and experiences as a student, teacher, and researcher of social behavior and organization. It has always struck me as somewhat shameful how reluctant each new wave of students is to examine the ideas and suggestions of past investigators. The emphasis upon the latest dogmas and methodologies needs to be leavened by an appreciation that our scientific forebears had important messages to communicate. In many instances, their ideas and formulations have been proven erroneous. But, often enough, the questions they raised and struggled to answer are still to be resolved.

EDWIN M. BANKS

REFERENCES

Davis, D. E. 1974. *Behavior as an Ecological Factor*. Dowden, Hutchinson & Ross. Stroudsburg, Pa. 390 pp.

Schein, M. W. 1975. *Social Hierarchy and Dominance*. Dowden, Hutchinson & Ross. Stroudsburg, Pa. 401 pp.

Stokes, A. W. 1974. *Territory*. Dowden, Hutchinson & Ross. Stroudsburg, Pa. 398 pp.

CONTENTS

Contents

PART III: COMPARATIVE STUDIES

PART IV: GENETICS AND SOCIAL STRUCTURE

CONTENTS BY AUTHOR

VERTEBRATE SOCIAL ORGANIZATION

INTRODUCTION

Studies dating back to the second and third decades of this century represent a convenient starting point at which to begin an evaluation of the sociobiology of vertebrates. During these years, progress in our understanding of social structures took the form of documentation, species by species, of how grouping tendencies vary around a limited number of common themes. This focus of attention on the variation in the expression of territorial behaviors, flocking and schooling aggregates, and dominance hierarchies produced an ample data base. Associated with these studies were attempts to subject the component behaviors to detailed quantitative analyses. This was essentially a reductionist approach—a search for proximal causes. This behavioral view contributed, in part, to the development of conceptual typologies or stereotypes. The aftereffects of typological thinking in this area are still being combated today. Witness the recent concern about the concept of dominance—in the sense of the relative rank or status between pairs of animals (Rowell, 1974)—where, over time, a term that is most properly used for descriptive convenience has come to carry the burden of explanatory meaning. This situation is now receiving needed correction.

A more recent trend in sociobiology has been to explore social organization in a much broader ecological and evolutionary framework, that is, to ask questions relating to ultimate causal mechanisms. Social systems evolve within definable but changing ecosystems, which are themselves subject to varying pressures of natural selection. Given a particular social system, what are the ecological correlates and constraints associated with it? The goal of such an inquiry must be to so understand fundamental relationships that predictions regarding ecological factors and social organization may emerge. There can be little doubt that this is an exceedingly difficult approach because it requires an understanding of complex interactions at many levels of biological organization.

Not only is it necessary to develop a quantitative data base of behavioral dynamics within an organized group, but a more rigorous characterization of cost–benefit aspects of sociality has become imperative. Why expend energy in defending a breeding territory? In what ways does an alpha male in a dominance hierarchy exploit his position, that is, increase his fitness at the expense of other males? Earlier answers to questions such as these, although plausible enough, need to be reformulated in the light of current ways of thinking that include concern for insight into the time and energy budgets of members of an organized social structure, and an evaluation of the total behavioral–ecological matrix in terms of modern evolutionary theory.

In very few cases do we have the information required for a thorough understanding of cost–benefit ratios of the behavioral components involved in the development and maintenance of a social system. On the input side, one needs to know the specifics regarding dietary components—their nutritional and energetic properties, their distribution and accessibility, their rate of intake and assimilation. The prospect of answering these questions by relying solely on methods of direct observation is obviously quite limited. On the output side, the paucity of information concerning the energetic costs to the animal of engaging in the various components of social behavior limits our ability to conduct very meaningful analyses of the cost–benefit aspects of sociality. These costs would be expected to vary with season and social context; for example, an initial dyadic encounter between adult males, even though highly ritualized, would be expected to require greater energetic expenditure than would a more casual encounter between the same pair after a period of long-established relationship. Once such cost–benefit analyses have been conducted, we may be in a more reasonable position to inquire about the effects of natural selection on these systems of social organization. Our goal should be that of evaluating current social structures in the context of evolutionary strategies and thus of developing a theoretic base for sociobiology which encompasses the broad elements of behavior, ecology, genetic, and evolutionary strategies.

Alexander (1974) has suggested that social behaviors within groups may be expected to evolve for at least three reasons: (1) to enhance the original advantages of group living, e.g., predator confrontation–cooperative hunting; (2) because they reduce the likelihood of disease and parasite transmission, principally by means of the development and spread of resistance to various diseases; and, perhaps most important, (3) because the social behaviors concerned with reproductive competition have considerable impact on both group members and the total

2

species population. Of course, behaviors that evolve under one set of selection constraints may acquire other functions without losing the first, e.g., grooming patterns in primate social units may be related quite closely to the social dynamics and organization of the group, while at the same time playing an important role in curbing parasite infestation. In fact, it may well have been that the role of grooming in parasite control evolved prior to its social role.

In recent years, a great deal of attention has focused on how best to explain the manner in which reproductive competition is related to a particular kind of social organization. The geneticist Fisher (1930) set the stage by developing models of sex-ratio selection in terms of parental expenditures. Robert Trivers (1974) has more recently extended these ideas, which involve parental manipulation of offspring as a reflection of parental care. Referring once again to Richard Alexander's views, three classes of social behavior, singly or in combination, must be considered in any meaningful explanation of how reproductive competition is related to a particular variety of social organization. These classes of social behavior include reciprocity, nepotism, and manipulation of descendants. It should be possible to discover how the social behaviors in each of these classes evolved, given the make-up of initial groups and their different backgrounds of selective values for group living. The information required here includes: (1) the conditions under which each individual will gain, (2) the conditions under which reproductive interests of different individuals will coincide or conflict, and (3) the extent to which various sorts of asymmetry under what conditions will permit one or another individual to succeed in cases of conflict. These conditions have already been defined for reciprocity by Trivers (1971) and for nepotism by Hamilton (1963). The case for parental manipulation is being developed by Trivers (1974). He considered the consequences of increasing parental investment for aspects of sexual selection, sex-ratio selection, and parent-offspring competition.

I would like to examine, in a bit more detail, a few studies that might serve as signposts along the road to placing sociobiology in a proper ecologic and evolutionary perspective.

In 1968, a three-year study of the adaptative nature of social organization in two species of tree squirrels situated in the Cascade Mountains of British Columbia was published by C. C. Smith. Briefly, the social organization of the red and Douglas squirrels consists of year-round territories occupied and defended by single individuals. Both sexes establish and defend these nonoverlapping and contiguous territories except during a brief breeding season. Smith's study provided information regarding the behavioral components of territory defense, vocalizations, breeding behavior, and parental care; i.e., it followed the traditional proximal-cause approach. Smith also included in his study observations and measurements relating to time and energy budgets and an analysis of food resources, including a listing of the various plant species eaten and a breakdown of which plant parts were used for food. Seasonal availability of foods, their caloric and nutritive values, and the time the squirrels spent in eating were also evaluated. These data provided the basis for estimates of individual daily energy requirements. Smith's analysis was carried further by estimations of the energy supply available per territory and the relationship between territory size and food supply. The conclusions of these studies were that (1) nesting sites were not limiting factors; (2) food was frequently in short supply— territory size was found to be inversely proportional to the density of preferred food supply—thus suggesting that the size of the defended area is determined by the amount of food it contains; and (3) food supplies of tree squirrels in the conifer forests are packaged in easily defended parcels, e.g., mushrooms, cones. An analysis such as this permits us to infer a great deal about the social organization displayed by these animals. Conditions are optimized for each animal, male or female, to gather, cache, and defend essential food stores, which are seasonally abundant and without which few are able to survive the winter. Moreover, surprisingly little time is spent in the energetically more costly aspects of territorial defense which is, instead, mediated to a large extent by vocalizations. Selection, in conjunction with the promiscuous mating system of these species, would favor aggressive, resident individuals, those both familiar with the area and more aggressive than vagrants seeking entry into the area.

Are there any areas in the geographic range of these squirrels where the ecological and competitive situation differs from the northwest population studied by Smith? And if so, are there any differences in the social organization that can be related to ecologic factors? A study of the red squirrel in the eastern deciduous forest area revealed reciprocal tolerance and a lack of territoriality. In these areas, the diet of the red squirrel is comprised largely of nuts and other foods also used by such competitors as the eastern gray and the fox squirrel. It has been suggested that defense of a food store in the deciduous forest may be impossible for the red squirrel because of competition provided by gray and fox squirrels. As a consequence, the territorial behavior displayed so vividly in the northwestern populations of red squirrels is not apparent, or at least, is less effective in the eastern populations of this species. What we may derive from this analysis is that territorial behavior operates to insure each territorial individual an adequate supply of seasonally abundant food, in those ecologic circumstances where food is, in fact, a defendable resource. Hence, the answer to the question "Why defend territory?" seems clear enough in this case. Another important notion emerging from these studies is that knowledge of the social organization and ecological concomitants of a species in one region of its geographic range does not necessarily permit us to assume that other populations of the same species situated in other ecologic settings will display the same kind of social organizations. In fact, the accumulating evidence suggests the contrary and provides further support to the idea that studies of social organization that do not include ecologic analyses are apt to be inadequate.

Another approach to the goal of obtaining clearer understanding of ecological and evolutionary effects on social organization may be exemplified by the studies of Brown (1974) on the New World jays. Brown's point of departure has been that of working with a group of animals that are relatively homogeneous in morphology, inhabit a wide geographic range including many habitat types, and comprise species showing a variety of social systems. Behavioral comparisons of such species within the group, in conjunction with a detailed analysis of their population ecology, may lead to an understanding of the evolutionary processes that gave rise to the diversity in social organization.

Sociality in the jay group has evolved into two distinct systems, colonial and communal. Brown's analysis suggests how these two might be interconnected by various intermediates to the classical system of an all-purpose territory. I will articulate the argument as it applies to the probable evolution of the colonial system from the classical all-purpose territory.

The social organization of mainland California populations of the scrub jay is that of the classic all-purpose territory; all conspecifics are excluded by the territorial resident through aggressive behaviors. To date, only one species of New World jay shows the colonial life style, the piñon jay. In brief, colonies of piñon jay are made up of as many as 200 individuals. Nests are built in clumps and foraging is accomplished in large, closely packed flocks. Aggressive behavior is only rarely observed, and the range used by one colony is not defended against other colonies. At the onset of the breeding season, adults leave the colony, court, and build nests; the young are fed and cared for by the mated pair. Helpers at the nest, in the sense of conspecifics other than the parents who contribute to care and feeding of the young, do not constitute an integral part of the piñon jay system.

Certain ecological distinctions between the scrub and piñon jay bear importantly upon the differences in their social organization. For example, the scrub jay defends the entire home range against conspecifics, whereas the piñon jay may share a common home range with hundreds of conspecifics. Population expansion by the scrub jay evidently occurs by means of dispersion and settling new areas. In contrast, population expansion by the piñon jay is accomplished by colony growth.

Of concern to our presentation is the question of whether there exists one or more species of jays exhibiting an intermediate condition between the more or less regular spacing of the scrub jay and the colonial life style of piñon jays. Brown suggests that the population ecology and behavior of the Steller's jay provides an appropriate link. In the Steller's jay, individuals penetrate deeply into the ranges of conspecifics while, at the same time, maintaining an exclusive core area around their nests. These core areas of dominance are not considered to be classic territories by Brown, and are, in fact, restricted to small areas compared to the areal extent of the home range. In no sense, can the Steller's jay be considered colonial; nest sites are spaced rather than clumped, as a reflection of the area of dominance just referred to. In short, the social system of the Stellar's jay represents an apparent compromise. A pair not only forages within its own home range, but may fly some distance to exploit a temporarily rich food resource while at the same time maintaining its core area for nesting and most foraging activities.

Thus, this species gains some of the social advantages of flocking—such as the discovery of rich, temporary food sources and the detection of predators—without at the same time suffering the disadvantages of increased vulnerability to predation that appears to be a concomitant of the closely spaced nests of a colony. The organizational system of the Steller's jay may well be an adaptation to environmental conditions intermediate between those favoring coloniality and all-purpose territories. It seems appropriate to add the caveat that Brown's analysis neither attempts to derive the pattern of one jay species from another, nor does it necessarily establish the evolutionary direction of sociality. These studies do, however, serve to focus attention on the interplay of behavioral and ecologic factors in the evolution of social systems and also serve to emphasize the heuristic value of the comparative approach in dealing with these issues.

Students of primate sociobiology, also, have begun to examine observed variations in social systems in terms of ecologic constraints. Although it is true that a relatively large amount of field-generated information has appeared in the journals during the past 15 years, it is beyond dispute that our knowledge of the salient behavioral and ecological details necessary for the development of a global theory of primate social evolution is still quite spotty and incomplete.

Nevertheless, it is equally clear that there is a substantial amount of intertaxa variation with respect to such basic elements as group size and composition; the spacing arrangement within groups and the manner in which this variable changes with activity (sleeping vs. foraging) and with season; the types of social interactions among group members, competitive and cooperative, as in predator defense; mateship systems, which vary from monogamy of the gibbon to the almost complete promiscuity of the chimpanzee; the various kinds of infant care, which, in some species is the exclusive responsibility of the mother whereas, in other species, it involves a substantial amount of male participation; the priority system, whether stable and linear, composed of multi-individual coalitions, or is determined in part by maternal status; and finally, in this incomplete listing of the elements of primate sociobiology, the existence of classes of specialties within the group organization, such as leaders, defenders, control animals, and the like. To bring some order out of these abundant variations and attempt further to evaluate the role of ecological factors acting on them is a task of monumental proportions.

There have been at least two major recent attempts at such a synthesis. Crook, in 1970, proposed a five-grade system, each grade representing a level of adaptation to varying habitat types, e.g., forest; forest fringe; rich savanna; and arid, somewhat harsh environments. This gradation was said to reflect the overall pattern of social evolution in the primates, with a good deal of parallel radiation occurring in each major primate taxonomic unit. The following trend emerges from this approach; (1) primitive, insectivorous forest prosimians, solitary and nocturnal; leading to (2) social, diurnal fruit and leaf eaters of the forest; which in turn lead to (3) diurnal, omnivorous troop formers exhibiting different social units in open savanna. Crook's social classification scheme is based upon relative size of the group; the ratio of adult males to adult females; mechanisms of dispersion, that is, the presence of defended territories; modes of defense or, in the absence of defended territories, the nature of home range utilization; and whether troop permeability or exchange, particularly of males, is evident. Each grade is also characterized with respect to the major habitat types referred to earlier.

This admittedly tentative scheme was followed in 1972 by another attempt at synthesis by Eisenberg and his colleagues. The range of social organization and feeding ecology for primates offered by these workers differs somewhat from that proposed by Crook in that one feature of primate sociality, that of the reproductive group, is elevated to major status as an organic unit that exhibits stages of growth and decline and that varies with habitat. A new category of social unit, the age-graded-male troop represents another innovation in this analysis.

Thus, the range of social organization visualized by Eisenberg proceeds from the solitary prosimian species to the parental family, the one-male troop, the age-graded male troop, and the multi-male troop. Under each of these structural categories, major taxonomic families may be arranged on the basis of food habits—that is, the insect, fruit, leaf eaters or omnivorous forms—and a further gradation relating to arboreality, semiterrestriality, and true terrestriality. A series of hypothetical schemes are advanced to account for the evolution of the various grades of social structure among the primates.

In all of these arguments, constant care is required to differentiate between similarities that are best u.1derstood as the consequences of close phylogenetic relationships and similarities that are related more clearly to specific adaptations to ecologic contingencies. This precaution holds whatever the phylogenetic status of the group under analysis. In most instances, the gaps in our knowledge need to be remedied before a final, universally acceptable theory of animal social organization can be attained.

One of the reasons for arranging this volume on vertebrate social organization was to determine whether it is yet possible to discover in the ever-growing animal literature some basic concepts whose application might enhance our understanding of the human condition. As a first premise, I will restate the view of most biologists that man, as other animals, is a product of selection. That is to say, all his basic traits have been accumulated essentially because they fostered the best reproductive performance. At the same time, the explanation for why it is that any species exhibits group living, centers on at least these three considerations. The first of these is the reduced susceptibility to predation by virtue of group defense or because group living results in some members of the group being more vulnerable to predation than others, as in fish schools or herds of ungulates. A second important consideration relates to species whose prey are of such a nature that individuals or nuclear families are incompetent, by themselves, in obtaining sufficient nourishment. In these circumstances, group or pack hunting has evolved as the most parsimonious strategy for living. A third consideration relates group living to shortages of critical resources such as breeding sites. This latter case is exemplified by the phenomenon of colonial breeding in marine birds, such as the gannet, where exceedingly large numbers of birds congregate during the breeding season on island outcrops of small areal extent.

These conditions either singly or in combination are the most obvious bases for group living, which would otherwise be disadvantageous because of a reduction in the individual's ability to outcompete, that is, to outreproduce his neighbors.

In the case of man, there seems to be a considerable amount of evidence, at least to this nonspecialist in such matters, that early man subsisted to a greater or lesser degree on animals larger and faster than man himself. Also, it is usually argued that for the bulk of his existence, man lived in small hunter–gatherer groups, and that the hunting persuasion facilitated the rapid increase in brain size, tool making, and tool use. For the sake of providing another viewpoint, I would like to suggest an alternative possibility first called to my attention by Alexander in 1971.

The same data used in the man, the group-hunter, argument could as well be interpreted to suggest that during the long period of protohominid evolution, man was a warlike, cannibalistic animal. He was forced to live in ever-larger groups by the pressure of predators, rather than being, in a sense, forced to live in groups to facilitate hunting prey animals. Predator pressure arose not from animals bigger, stronger, and faster, but rather from bigger and stronger groups of men. There would appear to be no upper limit to group size if its main function were to conduct war on other groups of men. On the other hand, group size should have remained fairly constant if its major function were that of hunting prey animals, particularly with the documented increase in hunting technology.

In terms of reproductive competition, it is not implausible to suggest that war has been the chief *modus operandi* at man's intergroup level, and that it has been a much more potent factor in bringing about the rapid evolution of brain size and intelligence than could be afforded by the much less forceful and indirect competition suggested by the idea that man evolved intelligence in order to hunt prey more effectively.

This hypothetical view emphasizes intraspecific competition and leads to the notion that the most potent environmental change in early stages of man's evolution was "culture," not in the sense of effecting a better means of hunting, but rather as effecting a better means of warfare.

The foregoing is, of course, highly speculative and may, indeed, be completely erroneous. I offer it not in support of the idea that man is inevitably aggressive or warlike, but rather to stress the idea that if we are to gain some measure of understanding about the human condition, we must attempt to deal with the evolutionary background of our species in terms of modern theory. It is in this sense, I believe, that we can make use of the wealth of information available from studies of the sociobiology of other species.

REFERENCES

Alexander, R. D. 1971. The search for an evolutionary philosophy of man. *Proc. Roy. Soc. Victoria* **84**: 99–120.

Alexander, R. D. 1974. The evolution of social behavior. *Ann. Rev. Ecol. Syst.* **5**: 325–383.

Brown, J. L. 1974. Alternate routes to sociality in jays—with a theory for the evolution of altruism and communal breeding. *Amer. Zool.* **14**(1): 63–80.

Crook, J. H. 1970. The socio-ecology of primates. In J. H. Crook (ed.). *Social Behavior in Birds and Mammals.* Academic Press, New York.

Eisenberg, J. F., N. A. Muckenhirn, and R. Rudran. 1972. The relation between ecology and social structure in primates. *Science* **176**: 863–874.

Fisher, R. A. 1930. *The genetical theory of natural selection.* Clarendon Press, Oxford, 1958.

Hamilton, W. D. 1963. The evolution of altruistic behavior. *Amer. Nat.* **97**: 354–356.

Rowell, T. E. 1974. The concept of social dominance. *Behav. Biol.* **11**: 131–154.

Smith, C. C. 1968. Adaptive nature of social organization in the genus of tree squirrels, *Tamiasciurus. Ecol. Monogr.* **38**: 31–63.

Trivers, R. L. 1971. The evolution of reciprocal altruism. *Quart. Rev. Biol.* **46**: 35–57.

Trivers, R. L. 1974. Parent-offspring conflict. *Amer. Zool.* **14**(1): 249–264.

Part I

SOME THEORETICAL AND HISTORICAL CONSIDERATIONS

Editor's Comments
on Papers 1 Through 6

That an excerpt of Kropotkin's *Mutual Aid* has been selected as the opening item of this section reflects my early introduction to sociobiology under the late W. C. Allee. Much of Allee's philosophical orientation to investigations of social organization was influenced by the ideas of Kropotkin. Kropotkin (1842–1921) lived during the period of social and economic ferment of the late nineteenth and early twentieth centuries. Although born to the landed nobility of Russia, he developed early into a social libertarian. As a young man, Kropotkin traveled extensively in eastern Asia and won acclaim for a brilliant theory of the cartography of that area. He was an ardent admirer and student of Darwin's work, and during his travels he actively sought further support of Darwin's theory of evolution by careful observation of Siberian wildlife. These studies led Kropotkin to dismiss what he considered to be a gross overemphasis by Darwin's supporters of the concept that competition was the central factor in the natural selection of organisms. Kropotkin was also influenced by the ideas of a Russian zoologist, Karl Kessler, who in 1880 expressed the kernel of the concept that cooperation, rather than competition or conflict, was the overriding factor in evolution. Kessler died shortly thereafter and Kropotkin adopted and spent many years systematically developing Kessler's "law of mutual aid." The two chapters of Kropotkin's book included below reveal a wealth of natural history observations supporting the idea that mutual cooperation in such activities as hunting, migration, and reproduction play a significant role in natural selection. Of course, Kropotkin provided little insight into the mechanisms by which cooperative behaviors might further the maintenance and evolution of social species, but his comments are nonetheless of substantial historical importance.

William Morton Wheeler (1865-1937) was best known for his studies of insect social organization. In his essay on "Societal Evolution," (Paper 2) Wheeler takes a broader view and presents a discussion of various categories of associations and societies. For the interested reader, the most extensive classification scheme of this type was that developed by a German worker, Deegener, in 1918 (cited by Wheeler). A partial translation of this scheme can be found in Allee, 1927 (cited by Wheeler). Wheeler stressed the importance of the development of the family as the major element in the evolution of sociality. In the social insects, family bonds involve the exchange of food, a process termed *trophallaxis* by Wheeler. The complexity of the social structures in many vertebrates is reflected by Wheeler's attempt to invoke gregarious or herd instincts as the bases for social bonding. Without further explanation, Wheeler presents an interesting and critical review of classical hypotheses regarding the origin of sociality in man. Paper 2 provides a stimulating view of the questions to be answered in our attempts to understand the origin of sociality; many of these same questions are still being asked today.

W. C. Allee (1885-1955) was an eminent American ecologist and student of social organization (see Schein, 1975). In Paper 3, "Concerning the Origin of Sociality," Allee developed a general statement regarding "unconscious cooperation" or "protocooperation" as he later called it. He proposed a very broad definition of sociality as "all integrations of two or more organisms into a supra-individualistic unity on which natural selection can act. . . ." This definition has quite a modern ring to it; it is not unlike the concepts expressed by Wynne-Edwards in 1962, which gave rise to the group-selection hypothesis. It is of some historical interest that the hotly debated group-selection hypothesis was clearly stated by Allee in the present essay. One also finds echoes of Kropotkin's emphasis on the role of cooperation rather than competition as a mechanism of natural selection. Allee's argument involves the interrelationships among optimal population densities, grouping tendencies, learned and unlearned cooperation, and the origin of sociality. Rather than being restricted to higher taxa, sociality is viewed as a widespread biological phenomenon; an inherited, innate tendency toward social cooperation is the mode in nature, according to Allee.

The paper by T. C. Schneirla (1902–1968) on "Problems in the Biopsychology of Social Organization" is included here (Paper 4) for a number of reasons. In it, Schneirla crystalizes his general views regarding the weaknesses inherent in the supraorganism analogy when used to compare social organizations across widely disparate taxa. He reiterates and supports Wheeler's concept of trophallaxis as an integral factor in the origin, maintenance, and evolution of sociality. Schneirla argues pursuasively that trophallaxis, though it may be the cornerstone of social organization in insects and in man, operates in ways that are qualitatively different in widely separated taxa. The outcome of trophallaxis in insect societies is restricted by biological factors; whereas in man and most other mammals, this phenomenon merely sets the stage for far more complex mother–young interactions. Schneirla views trophallactic responses in higher forms as the basis for all learned approach and avoidance responses. Further, in this essay, Schneirla denigrates the utility of the dominance concept as of being inadequate to form the basis of an understanding of vertebrate social organization (cf. Rowell, 1974). Rather, he proposed that various forms of social facilitatory responses, based on earlier trophallactic interactions, represent more meaningful systems to be examined as major unifying factors in the sociality of vertebrates. What is needed, therefore, is more emphasis on the development of these facilitatory responses as the binders that lead to cohesiveness in the social structures of vertebrates. Schneirla's concepts have had a significant impact on investigations into the bases of biosociology, although his style of writing is somewhat difficult to comprehend.

The last two selections in this section are of more recent vintage. The article by Scott (Paper 5), "The Analysis of Social Organization in Animals," was based on a paper presented at a symposium in honor of W. C. Allee at the 1954 meetings of the American Institute of Biological Sciences. In this paper, Scott, a leading student of sociobiology, outlines a general procedure for the study of problems of social organization. He also includes a classification scheme for behavior patterns that he had developed earlier. The scheme has provided a useful systematic framework for students interested in making intertaxa comparisons of behavior. Many, but not all, of the terms and their definitions are in current use. Scott also developed a classification of social relationships, leading from simple aggregations to the more complex relationships displayed by the social vertebrates. Of particular interest is his articulation of the critical-period concept in the context of the process of socialization. This is an area in which Scott has long been a leader.

The essay by McBride (Paper 6), "A General Theory of Social Organization and Behaviour," represents a bold attempt to establish a tentative theoretical framework for studies of social organization. McBride was trained as a geneticist at Edinburgh. He has published extensively in various areas of animal behavior. McBride proposes that the major distinction between aggregated and nonaggregated species is a function of the distance between conspecifics and their arrangements in space. In addition to these two spatial relationships, a third, referred to as the "social dimension," appears to be a basic component. The social dimension is comprised of a wide range of social behaviors, but McBride places special emphasis on social dominance. An extension of this idea leads to the expression of a theory of fields of social forces. Thus, the behavior of animals toward neighbors may result either in centripetal, i.e., mutually attractive, or centrifugal, i.e., repulsive responses. This formulation is patterned after the concept of individual distance of Hediger (1950) and Conder (1949). A social animal conducts itself as though an area immediately in front of its head is its own territory, or social force field. The intensity of a force field can be related to the social dominance of the individual, hence the primary manifestation of a social force field is the relative level of aggressiveness of the individual. McBride suggests that most social organizations are, in fact, concerned with the regulation of intragroup aggressiveness. Further elaboration of this theory and how it might be useful in understanding the evolution of sociality is discussed. Of course, the primacy of dominance and aggressive behavior in this theoretical proposal provides a target for critics who are dubious about the legitimacy of the dominance concept. McBride admittedly supports and makes use of a group-selection argument. Even though many will fault the premises and suggestions advanced by McBride, his is one of the few attempts to formulate a general theory of animal sociality.

REFERENCES

Conder, P. J. 1949. Individual distance. *Ibis* 91: 649–655.

Deegener, P. 1918. *Die Formen der Vergesellschaffung in Tierreiche. Ein systematisch-sozialogischer Versuch.* Leipsig: Veit, 420 pp.

Hediger, H. 1950. *Wild animals in captivity.* Butterworth & Co. London.

Rowell, T. E. 1974. The concept of social dominance. *Behav. Biol.* 11: 131–154.

Schein, M. W. 1975. *Social Hierarchy and Dominance.* Dowden, Hutchinson & Ross. Stroudsburg, Pa. 401 pp.

Wynne-Edwards, V. C. 1962. Animal Dispersion in Relation to Social Behaviour. Oliver and Boyd, Edinburgh and London, 653 pp.

1

Reprinted by permission of New York University Press from *Mutual Aid: A Factor of Evolution*, Peter Kropotkin, New York University Press, New York, 1972, pp. 27–33, 70–80

MUTUAL AID AMONG ANIMALS

Peter Kropotkin

THE conception of struggle for existence as a factor of evolution, introduced into science by Darwin and Wallace, has permitted us to embrace an immensely-wide range of phenomena in one single generalization, which soon became the very basis of our philosophical, biological, and sociological speculations. An immense variety of facts: – adaptations of function and structure of organic beings to their surroundings; physiological and anatomical evolution; intellectual progress, and moral development itself, which we formerly used to explain by so many different causes, were embodied by Darwin in one general conception. We understood them as continued endeavours – as a struggle against adverse circumstances – for such a development of individuals, races, species and societies, as would result in the greatest possible fullness, variety, and intensity of life. It may be that at the outset Darwin himself was not fully aware of the generality of the factor which he first invoked for explaining one series only of facts relative to the accumulation of individual variations in incipient species. But he foresaw that the term which he was introducing into science would lose its philosophical and its only true meaning if it were to be used in its narrow sense only – that of a struggle between separate individuals for the sheer means of existence. And at the very beginning of his memorable work he insisted upon the term being taken in its 'large and metaphorical sense including dependence of one being on another, and including (which is more important) not only the life of the individual, but success in leaving progeny'.[1]

1. *Origin of Species*, chapter 3.

While he himself was chiefly using the term in its narrow sense for his own special purpose, he warned his followers against committing the error (which he seems once to have committed himself) of overrating its narrow meaning. In *The Descent of Man* he gave some powerful pages to illustrate its proper, wide sense. He pointed out how, in numberless animal societies, the struggle between separate individuals for the means of existence disappears, how *struggle* is replaced by *cooperation*, and how that substitution results in the development of intellectual and moral faculties which secure to the species the best conditions for survival. He intimated that in such cases the fittest are not the physically strongest, nor the cunningest, but those who learn to combine so as mutually to support each other, strong and weak alike, for the welfare of the community. 'Those communities,' he wrote, 'which included the greatest number of the most sympathetic members would flourish best, and rear the greatest number of offspring' (2nd ed., p. 163). The term, which originated from the narrow Malthusian conception of competition between each and all, thus lost its narrowness in the mind of one who knew Nature.

Unhappily, these remarks, which might have become the basis of most fruitful researches, were overshadowed by the masses of facts gathered for the purpose of illustrating the consequences of a real competition for life. Besides, Darwin never attempted to submit to a closer investigation the relative importance of the two aspects under which the struggle for existence appears in the animal world, and he never wrote the work he proposed to write upon the natural checks to over-multiplication, although that work would have been the crucial test for appreciating the real purport of individual struggle. Nay, on the very pages just mentioned, amidst data disproving the narrow Malthusian conception of struggle, the old Malthusian leaven reappeared – namely, in Darwin's remarks as to the alleged inconveniences of maintaining the 'weak in mind and body' in our civilized societies (ch. 5). As if thousands of weak-bodied and infirm poets, scientists, inventors, and reformers, together with other thousands of so-called 'fools' and 'weak-minded enthusiasts', were not the most precious weapons used by humanity in its struggle for existence by intellectual and moral arms, which Darwin himself emphasized in those same chapters of *Descent of Man*.

17

It happened with Darwin's theory as it always happens with theories having any bearing upon human relations. Instead of widening it according to his own hints, his followers narrowed it still more. And while Herbert Spencer, starting on independent but closely-allied lines, attempted to widen the inquiry into that great question, 'Who are the fittest?' especially in the appendix to the third edition of the *Data of Ethics*, the numberless followers of Darwin reduced the notion of struggle for existence to its narrowest limits. They came to conceive the animal world as a world of perpetual struggle among half-starved individuals, thirsting for one another's blood. They made modern literature resound with the war-cry of *woe to the vanquished*, as if it were the last word of modern biology. They raised the 'pitiless' struggle for personal advantages to the height of a biological principle which man must submit to as well, under the menace of otherwise succumbing in a world based upon mutual extermination. Leaving aside the economists who know of natural science but a few words borrowed from second-hand vulgarizers, we must recognize that even the most authorized exponents of Darwin's views did their best to maintain those false ideas. In fact, if we take Huxley, who certainly is considered as one of the ablest exponents of the theory of evolution, were we not taught by him, in a paper on the 'Struggle for Existence and its Bearing upon Man', that,

from the point of view of the moralist, the animal world is on about the same level as a gladiators' show. The creatures are fairly well treated, and set to fight; whereby the strongest, the swiftest, and the cunningest live to fight another day. The spectator has no need to turn his thumb down, as no quarter is given.

Or, further down in the same article, did he not tell us that, as among animals, so among primitive men,

the weakest and stupidest went to the wall, while the toughest and shrewdest, those who were best fitted to cope with their circumstances, but not the best in another way, survived. Life was a continuous free fight, and beyond the limited and temporary relations of the family, the Hobbesian war of each against all was the normal state of existence.[2]

In how far this view of nature is supported by fact, will be seen

2. *Nineteenth Century*, February 1888, p. 165.

from the evidence which will be here submitted to the reader as regards the animal world, and as regards primitive man. But it may be remarked at once that Huxley's view of nature had as little claim to be taken as a scientific deduction as the opposite view of Rousseau, who saw in nature but love, peace, and harmony destroyed by the accession of man. In fact, the first walk in the forest, the first observation upon any animal society, or even the perusal of any serious work dealing with animal life (D'Orbigny's, Audubon's, Le Vaillant's, no matter which), cannot but set the naturalist thinking about the part taken by social life in the life of animals, and prevent him from seeing in Nature nothing but a field of slaughter, just as this would prevent him from seeing in Nature nothing but harmony and peace. Rousseau had committed the error of excluding the beak-and-claw fight from his thoughts; and Huxley committed the opposite error; but neither Rousseau's optimism nor Huxley's pessimism can be accepted as an impartial interpretation of nature.

As soon as we study animals – not in laboratories and museums only, but in the forest and the prairie, in the steppe and the mountains – we at once perceive that though there is an immense amount of warfare and extermination going on amidst various species, and especially amidst various classes of animals, there is, at the same time, as much, or perhaps even more, of mutual support, mutual aid, and mutual defence amidst animals belonging to the same species or, at least, to the same society. Sociability is as much a law of nature as mutual struggle. Of course it would be extremely difficult to estimate, however roughly, the relative numerical importance of both these series of facts. But if we resort to an indirect test, and ask Nature: 'Who are the fittest: those who are continually at war with each other, or those who support one another?' we at once see that those animals which acquire habits of mutual aid are undoubtedly the fittest. They have more chances to survive, and they attain, in their respective classes, the highest development of intelligence and bodily organization. If the numberless facts which can be brought forward to support this view are taken into account, we may safely say that mutual aid is as much a law of animal life as mutual struggle, but that, as a factor of evolution, it most probably has a far greater importance, inasmuch as it favours the development of such habits and characters as insure the maintenance and further development of the species, together with the greatest amount of welfare and enjoyment of life for the individual, with the least waste of energy.

Of the scientific followers of Darwin, the first, as far as I know, who understood the full purport of mutual aid *as a law of Nature and the chief factor of evolution*, was a well-known Russian zoologist, the late Dean of the St Petersburg University, Professor Kessler. He developed his ideas in an address which he delivered in January 1880, a few months before his death, at a Congress of Russian naturalists; but, like so many good things published in the Russian tongue only, that remarkable address remains almost entirely unknown.[3]

'As a zoologist of old standing', he felt bound to protest against the abuse of a term – the struggle for existence – borrowed from zoology, or, at least, against overrating its importance. Zoology, he said, and those sciences which deal with man, continually insist upon what they call the pitiless law of struggle for existence. But they forget the existence of another law which may be described as the law of mutual aid, which law, at least for the animals, is far more essential than the former. He pointed out how the need of leaving progeny necessarily brings animals together, and, 'the more the

3. Leaving aside the pre-Darwinian writers, like Toussenel, Fée, and many others, several works containing many striking instances of mutual aid – chiefly, however, illustrating animal intelligence – were issued previously to that date. I may mention those of Houzeau, *Les facultés mentales des animaux*, 2 vols., Brussels, 1872; L. Büchner's *Aus dem Geistesleben der Thiere*, 2nd ed., 1877; and Maximilian Perty's *Ueber das Seelenleben der Thiere*, Leipzig, 1876. Espinas published his most remarkable work, *Les Sociétés animales*, in 1877, and in that work he pointed out the importance of animal societies, and their bearing upon the preservation of species, and entered upon a most valuable discussion of the origin of societies. In fact, Espinas's book contains all that has been written since upon mutual aid, and many good things besides. If I nevertheless make a special mention of Kessler's address, it is because he raised mutual aid to the height of a law much more important in evolution than the law of mutual struggle. The same ideas were developed next year (in April 1881) by J. Lanessan in a lecture published in 1882 under this title: *La lutte pour l'existence et l'association pour la lutte*. G. Romanes's capital work, *Animal Intelligence*, was issued in 1882, and followed next year by the *Mental Evolution in Animals*. About the same time (1883), Büchner published another work, *Liebe und Liebes-Leben in der Thierwelt*, a second edition of which was issued in 1885. The idea, as seen, was in the air.

individuals keep together, the more they mutually support each other, and the more are the chances of the species for surviving, as well as for making further progress in its intellectual development'. 'All classes of animals,' he continued, 'and especially the higher ones, practise mutual aid', and he illustrated his idea by examples borrowed from the life of the burying beetles and the social life of birds and some mammalia. The examples were few, as might have been expected in a short opening address, but the chief points were clearly stated; and, after mentioning that in the evolution of mankind mutual aid played a still more prominent part, Professor Kessler concluded as follows:

I obviously do not deny the struggle for existence, but I maintain that the progressive development of the animal kingdom, and especially of mankind, is favoured much more by mutual support than by mutual struggle. ... All organic beings have two essential needs: that of nutrition, and that of propagating the species. The former brings them to a struggle and to mutual extermination, while the needs of maintaining the species bring them to approach one another and to support one another. But I am inclined to think that in the evolution of the organic world – in the progressive modification of organic beings – mutual support among individuals plays a much more important part than their mutual struggle.[4]

The correctness of the above views struck most of the Russian zoologists present, and Syevertsoff,* whose work is well known to ornithologists and geographers, supported them and illustrated them by a few more examples. He mentioned some of the species of falcons which have 'an almost ideal organization for robbery', and nevertheless are in decay, while other species of falcons, which practise mutual help, do thrive. 'Take, on the other side, a sociable bird, the duck,' he said; 'it is poorly organized on the whole, but it practises mutual support, and it almost invades the earth, as may be judged from its numberless varieties and species.'

The readiness of the Russian zoologists to accept Kessler's views seems quite natural, because nearly all of them have had opportunities of studying the animal world in the wide uninhabited regions

4. *Memoirs (Trudy) of the St Petersburg Society of Naturalists*, vol. xi, 1880.
* Nikolai Alekseevich Severtsov (Syevertsoff) (1827–85), Russian zoologist and geographer, noted for his exploration of Turkestan and the Pamirs. – *Ed.*

of Northern Asia and East Russia; and it is impossible to study like regions without being brought to the same ideas. I recollect myself the impression produced upon me by the animal world of Siberia when I explored the Vitim regions in the company of so accomplished a zoologist as my friend Polyakoff was.* We both were under the fresh impression of the *Origin of Species*, but we vainly looked for the keen competition between animals of the same species which the reading of Darwin's work had prepared us to expect, even after taking into account the remarks of the third chapter (p. 54). We saw plenty of adaptations for struggling, very often in common, against the adverse circumstances of climate, or against various enemies, and Polyakoff wrote many a good page upon the mutual dependency of carnivores, ruminants, and rodents in their geographical distribution; we witnessed numbers of facts of mutual support, especially during the migrations of birds and ruminants; but even in the Amur and Usuri regions, where animal life swarms in abundance, facts of real competition and struggle between higher animals of the same species came very seldom under my notice, though I eagerly searched for them. The same impression appears in the works of most Russian zoologists, and it probably explains why Kessler's ideas were so welcomed by the Russian Darwinists, whilst like ideas are not in vogue amidst the followers of Darwin in Western Europe.

[*Editor's Note:* Material has been omitted at this point.]

If the views developed on the preceding pages are correct, the question necessarily arises, in how far are they consistent with the theory of struggle for life as it has been developed by Darwin, Wallace, and their followers? and I will now briefly answer this important question. First of all, no naturalist will doubt that the idea of a struggle for life carried on through organic nature is the greatest generalization of our century. Life *is* struggle; and in that struggle the fittest survive. But the answers to the questions, 'By which arms is this struggle chiefly carried on?' and 'Who are the fittest in the struggle?' will widely differ according to the importance given to the two different aspects of the struggle: the direct one, for food and safety among separate individuals, and the struggle which Darwin described as 'metaphorical' – the struggle, very often collective, against adverse circumstances. No one will deny that there is, within each species, a certain amount of real competition for food – at least, at certain periods. But the question is, whether competition is carried on to the extent admitted by Darwin, or even by Wallace; and whether this competition has played, in the evolution of the animal kingdom, the part assigned to it.

The idea which permeates Darwin's work is certainly one of real competition going on within each animal group for food, safety, and possibility of leaving an offspring. He often speaks of regions being stocked with animal life to their full capacity, and from that overstocking he infers the necessity of competition. But when we look in his work for real proofs of that competition, we must confess that we do not find them sufficiently convincing. If we refer to the paragraph entitled 'Struggle for Life most severe between Individuals and Varieties of the same Species', we find in it none of that wealth of proofs and illustrations which we are accustomed to find in whatever Darwin wrote. The struggle between individuals of the same species is not illustrated under that heading by even one single instance: it is taken as granted; and the competition between closely-allied species is illustrated by but five examples, out of which one, at least (relating to the two species of thrushes), now proves to be doubtful.[34] But when we look for more details in order

34. One species of swallow is said to have caused the decrease of another swallow species in North America; the recent increase of the missel-thrush in Scotland has caused the decrease of the song-thrush; the brown rat has taken the place of the black rat in Europe; in Russia the small cockroach has everywhere driven before it its greater congener; and in Australia the imported hive-bee is rapidly exterminating the small stingless bee. Two other cases, but relative to domesticated animals, are mentioned in the preceding paragraph.

to ascertain how far the decrease of one species was really occasioned by the increase of the other species, Darwin, with his usual fairness, tells us:

We can dimly see why the competition should be most severe between allied forms which fill nearly the same place in nature; but probably in no case could we precisely say why one species has been victorious over another in the great battle of life.

As to Wallace, who quotes the same facts under a slightly-modified heading ('Struggle for Life between closely-allied Animals and Plants *often* most severe'), he makes the following remark (italics are mine), which gives quite another aspect to the facts above quoted. He says:

In *some* cases, no doubt, there is actual war between the two, the stronger killing the weaker; *but this is by no means necessary,* and there may be cases in which the weaker species, physically, may prevail by its power of more rapid multiplication, its better withstanding vicissitudes of climate, or its greater cunning in escaping the attacks of common enemies.

In such cases what is described as competition may be no competition at all. One species succumbs, not because it is exterminated or starved out by the other species, but because it does not well accommodate itself to new conditions, which the other does. The term 'struggle for life' is again used in its metaphorical sense, and may have no other. As to the real competition between individuals of the same species, which is illustrated in another place by the cattle of South America during a period of drought, its value is

---

While recalling these same facts, A. R. Wallace remarks in a footnote relative to the Scottish thrushes: 'Prof. A. Newton, however, informs me that these species do not interfere in the way here stated' (*Darwinism*, p. 34). As to the brown rat, it is known that, owing to its amphibian habits, it usually stays in the lower parts of human dwellings (low cellars, sewers, etc.), as also on the banks of canals and rivers; it also undertakes distant migrations in numberless bands. The black rat, on the contrary, prefers staying in our dwellings themselves, under the floor, as well as in our stables and barns. It thus is much more exposed to be exterminated by man; and we cannot maintain, with any approach to certainty, that the black rat is being either exterminated or starved out by the brown rat and not by man.

impaired by its being taken from among domesticated animals. Bisons emigrate in like circumstances in order to avoid competition. However severe the struggle between plants – and this is amply proved – we cannot but repeat Wallace's remark to the effect that 'plants live where they can', while animals have, to a great extent, the power of choice of their abode. So that we again are asking ourselves, To what extent does competition really exist within each animal species? Upon what is the assumption based?

The same remark must be made concerning the indirect argument in favour of a severe competition and struggle for life within each species, which may be derived from the 'extermination of transitional varieties', so often mentioned by Darwin. It is known that for a long time Darwin was worried by the difficulty which he saw in the absence of a long chain of intermediate forms between closely-allied species, and that he found the solution of this difficulty in the supposed extermination of the intermediate forms.[35] However, an attentive reading of the different chapters in which Darwin and Wallace speak of this subject soon brings one to the conclusion that the word 'extermination' does not mean real extermination; the same remark which Darwin made concerning his expression: 'struggle for existence', evidently applies to the word 'extermination' as well. It can by no means be understood in its direct sense, but must be taken 'in its metaphoric sense'.

If we start from the supposition that a given area is stocked with animals to its fullest capacity, and that a keen competition for the sheer means of existence is consequently going on between all the inhabitants – each animal being compelled to fight against all its congeners in order to get its daily food – then the appearance of a new and successful variety would certainly mean in many cases (though not always) the appearance of individuals which are enabled to seize more than their fair share of the means of existence;

35. 'But it may be urged that when several closely-allied species inhabit the same territory, we surely ought to find at the present time many transitional forms. . . . By my theory these allied species are descended from a common parent; and during the process of modification, each has become adapted to the conditions of life of its own region, and has supplanted and exterminated its original parent-form and all the transitional varieties between its past and present states' (*Origin of Species*, 6th ed., p. 134); also p. 137, 296 (all paragraph 'On Extinction').

and the result would be that those individuals would starve both the parental form which does not possess the new variation and the intermediate forms which do not possess it in the same degree. It may be that at the outset, Darwin understood the appearance of new varieties under this aspect; at least, the frequent use of the word 'extermination' conveys such an impression. But both he and Wallace knew Nature too well not to perceive that this is by no means the only possible and necessary course of affairs.

If the physical and the biological conditions of a given area, the extension of the area occupied by a given species, and the habits of all the members of the latter remained unchanged – then the sudden appearance of a new variety might mean the starving out and the extermination of all the individuals which were not endowed in a sufficient degree with the new feature by which the new variety is characterized. But such a combination of conditions is precisely what we do not see in Nature. Each species is continually tending to enlarge its abode; migration to new abodes is the rule with the slow snail, as with the swift bird; physical changes are continually going on in every given area; and new varieties among animals consist in an immense number of cases – perhaps in the majority – not in the growth of new weapons for snatching the food from the mouth of its congeners – food is only one out of a hundred of various conditions of existence – but, as Wallace himself shows in a charming paragraph on the 'divergence of characters' (*Darwinism*, p. 107), in forming new habits, moving to new abodes, and taking to new sorts of food. In all such cases there will be no extermination, even no competition – the new adaptation being *a relief from competition, if it ever existed*; and yet there will be, after a time, an absence of intermediate links, in consequence of a mere survival of those which are best fitted for the new conditions – as surely as under the hypothesis of extermination of the parental form. It hardly need be added that if we admit, with Spencer, all the Lamarckians and Darwin himself, the modifying influence of the surroundings upon the species, there remains still less necessity for the extermination of the intermediate forms.

The importance of migration and of the consequent isolation of groups of animals, for the origin of new varieties and ultimately of new species, which was indicated by Moritz Wagner, was fully.

recognized by Darwin himself. Consequent researches have only accentuated the importance of this factor, and they have shown how the largeness of the area occupied by a given species – which Darwin considered with full reason so important for the appearance of new varieties – can be combined with the isolation of parts of the species, in consequence of local geological changes, or of local barriers. It would be impossible to enter here into the discussion of this wide question, but a few remarks will do to illustrate the combined action of these agencies. It is known that portions of a given species will often take to a new sort of food. The squirrels, for instance, when there is a scarcity of cones in the larch forests, remove to the fir-tree forests, and this change of food has certain well-known physiological effects on the squirrels. If this change of habits does not last – if next year the cones are again plentiful in the dark larch woods – no new variety of squirrels will evidently arise from this cause. But if part of the wide area occupied by the squirrels begins to have its physical characters altered – in consequence of, let us say, a milder climate or desiccation, which both bring about an increase of the pine forests in proportion to the larch woods – and if some other conditions concur to induce the squirrels to dwell on the outskirts of the desiccating region – we shall have then a new variety, i.e. an incipient new species of squirrels, without there having been anything that would deserve the name of extermination among the squirrels. A larger proportion of squirrels of the new, better-adapted variety would survive every year, and the intermediate links would die *in the course of time*, without having been starved out by Malthusian competitors. This is exactly what we see going on during the great physical changes which are accomplished over large areas in Central Asia, owing to the desiccation which is going on there since the glacial period.

To take another example, it has been proved by geologists that the present wild horse (*Equus Przewalskii*) has slowly been evolved during the later parts of the Tertiary and the Quaternary period, but that during this succession of ages its ancestors were *not* confined to some given, limited area of the globe. They wandered over both the Old and New World, returning, in all probability, after a time to the pastures which they had, in the course of their migrations, formerly left.[36] Consequently, if we do not find now, in Asia, all the intermediate links between the present wild horse and its Asiatic Post-Tertiary ancestors, this does not mean at all that the intermediate links have been exterminated. No such extermination has ever taken place. No exceptional mortality may even have occurred among the ancestral species: the individuals which belonged to intermediate varieties and species have died in the usual course of events – often amidst plentiful food, and their remains were buried all over the globe.

In short, if we carefully consider this matter, and carefully re-read what Darwin himself wrote upon this subject, we see that if the word 'extermination' be used at all in connection with transitional varieties, it must be used in its metaphoric sense. As to 'competition', this expression, too, is continually used by Darwin (see, for instance, the paragraph 'On Extinction') as an image, or as a way of speaking, rather than with the intention of conveying the idea of a real competition between two portions of the same species for the means of existence. At any rate, the absence of intermediate forms is no argument in favour of it.

In reality, the chief argument in favour of a keen competition for the means of existence continually going on within every animal species is – to use Professor Geddes' expression – the 'arithmetical argument' borrowed from Malthus.

But this argument does not prove it at all. We might as well take a number of villages in South-East Russia, the inhabitants of which enjoy plenty of food, but have no sanitary accommodation of any kind; and seeing that for the last eighty years the birth-rate was sixty in the thousand, while the population is now what it was eighty years ago, we might conclude that there has been a terrible competition between the inhabitants. But the truth is that from year to year the population remained stationary, for the simple reason that one third of the new-born died before reaching their sixth month of life; one half died within the next four years, and out of

36. According to Madame Marie Pavloff, who has made a special study of this subject, they migrated from Asia to Africa, stayed there some time, and returned next to Asia. Whether this double migration be confirmed or not, the fact of a former extension of the ancestor of our horse over Asia, Africa and America is settled beyond doubt.

each hundred born, only seventeen or so reached the age of twenty. The newcomers went away before having grown to be competitors. It is evident that if such is the case with men, it is still more the case with animals. In the feathered world the destruction of the eggs goes on on such a tremendous scale that eggs are the chief food of several species in the early summer; not to say a word of the storms, the inundations which destroy nests by the million in America, and the sudden changes of weather which are fatal to the young mammals. Each storm, each inundation, each visit of a rat to a bird's nest, each sudden change of temperature, take away those competitors which appear so terrible in theory.

As to the facts of an extremely rapid increase of horses and cattle in America, of pigs and rabbits in New Zealand, and even of wild animals imported from Europe (where their numbers are kept down by man, not by competition), they rather seem opposed to the theory of over-population. If horses and cattle could so rapidly multiply in America, it simply proved that, however numberless the buffaloes and other ruminants were at that time in the New World, its grass-eating population was far below what the prairies could maintain. If millions of intruders have found plenty of food without starving out the former population of the prairies, we must rather conclude that the Europeans found a *want* of grass-eaters in America, not an excess. And we have good reasons to believe that want of animal population is the natural state of things all over the world, with but a few temporary exceptions to the rule. The actual numbers of animals in a given region are determined, not by the highest feeding capacity of the region, but by what it is every year under the most unfavourable conditions. So that, for that reason alone, competition hardly can be a normal condition; but other causes intervene as well to cut down the animal population below even that low standard. If we take the horses and cattle which are grazing all the winter through in the Steppes of Transbaikalia, we find them very lean and exhausted at the end of the winter. But they grow exhausted not because there is not enough food for all of them – the grass buried under a thin sheet of snow is everywhere in abundance – but because of the difficulty of getting it from beneath the snow, and this difficulty is the same for all horses alike. Besides, days of glazed frost are common in early spring, and if several such days come in succession the horses grow still more exhausted. But then comes a snow-storm, which compels the already weakened animals to remain without any food for several days, and very great numbers of them die. The losses during the spring are so severe that if the season has been more inclement than usual they are even not repaired by the new breeds – the more so as *all* horses are exhausted, and the young foals are born in a weaker condition. The numbers of horses and cattle thus always remain beneath what they otherwise might be; all the year round there is food for five or ten times as many animals, and yet their population increases extremely slowly. But as soon as the Buriate owner makes ever so small a provision of hay in the steppe, and throws it open during days of glazed frost, or heavier snow-fall, he immediately sees the increase of his herd. Almost all free grass-eating animals and many rodents in Asia and America being in very much the same conditions, we can safely say that their numbers are *not* kept down by competition; that at no time of the year they can struggle for food, and that if they never reach anything approaching to over-population, the cause is in the climate, not in competition.

The importance of natural checks to over-multiplication, and especially their bearing upon the competition hypothesis, seems never to have been taken into due account. The checks, or rather some of them, are mentioned, but their action is seldom studied in detail. However, if we compare the action of the natural checks with that of competition, we must recognize at once that the latter sustains no comparison whatever with the other checks. Thus, Mr Bates mentions the really astounding numbers of winged ants which are destroyed during their exodus. The dead or half-dead bodies of the formica de fuego (*Myrmica saevissima*) which had been blown into the river during a gale 'were heaped in a line an inch or two in height and breadth, the line continuing without interruption for miles at the edge of the water'.[37] Myriads of ants are thus destroyed amidst a nature which might support a hundred times as many ants as are actually living. Dr Altum, a German forester, who wrote a very interesting book about animals injurious to our forests, also gives many facts showing the immense importance of natural checks. He says that a succession of gales or cold and damp weather

37. *The Naturalist on the River Amazon*, ii, 85, 95.

during the exodus of the pine-moth (*Bombyx pini*) destroy it to incredible amounts, and during the spring of 1871 all these moths disappeared at once, probably killed by a succession of cold nights.[38] Many like examples relative to various insects could be quoted from various parts of Europe. Dr Altum also mentions the bird-enemies of the pine-moth, and the immense amount of its eggs destroyed by foxes; but he adds that the parasitic fungi which periodically infest it are a far more terrible enemy than any bird, because they destroy the moth over very large areas at once. As to various species of mice (*Mus sylvaticus*, *Arvicola arvalis*, and *A. agrestis*), the same author gives a long list of their enemies, but he remarks: 'However, the most terrible enemies of mice are not other animals, but such sudden changes of weather as occur almost every year.' Alternations of frost and warm weather destroy them in numberless quantities; 'one single sudden change can reduce thousands of mice to the number of a few individuals'. On the other side, a warm winter, or a winter which gradually steps in, makes them multiply in menacing proportions, notwithstanding every enemy; such was the case in 1876 and 1877.[39] Competition, in the case of mice, thus appears a quite trifling factor when compared with weather. Other facts to the same effect are also given as regards squirrels.

As to birds, it is well known how they suffer from sudden changes of weather. Late snow-storms are as destructive of bird-life on the English moors, as they are in Siberia; and C. Dixon saw the red grouse so pressed during some exceptionally severe winters, that they quitted the moors in numbers, 'and we have then known them actually to be taken in the streets of Sheffield. Persistent wet', he adds, 'is almost as fatal to them.'

On the other side, the contagious diseases which continually visit most animal species destroy them in such numbers that the losses often cannot be repaired for many years, even with the most rapidly multiplying animals. Thus, some sixty years ago, the *sousliks* suddenly disappeared in the neighbourhood of Sarepta, in South-Eastern Russia, in consequence of some epidemics; and for years no *sousliks* were seen in that neighbourhood. It took many

38. Dr B. Altum, *Waldbeschädigungen durch Thiere und Gegenmittel* (Berlin, 1889), pp. 207 ff.
39. Dr B. Altum, op. cit., pp. 13 and 187.

years before they became as numerous as they formerly were.[40]

Like facts, all tending to reduce the importance given to competition, could be produced in numbers.[41] Of course, it might be replied, in Darwin's words, that nevertheless each organic being 'at some period of its life, during some season of the year, during each generation or at intervals, has to struggle for life and to suffer great destruction', and that the fittest survive during such periods of hard struggle for life. But if the evolution of the animal world were based exclusively, or even chiefly, upon the survival of the fittest during periods of calamities; if natural selection were limited in its action to periods of exceptional drought, or sudden changes of temperature, or inundations, retrogression would be the rule in the animal world. Those who survive a famine, or a severe epidemic of cholera, or smallpox, or diphtheria, such as we see them in uncivilized countries, are neither the strongest, nor the healthiest, nor the most intelligent. No progress could be based on those survivals – the less so as all survivors usually come out of the ordeal with an impaired health, like the Transbaikalian horses just mentioned, or the Arctic crews, or the garrison of a fortress which has been compelled to live for a few months on half rations, and comes out of its experience with a broken health, and subsequently shows a quite abnormal mortality. All that natural selection can do in times of calamities is to spare the individuals endowed with the greatest endurance for privations of all kinds. So it does among the Siberian horses and cattle. They *are* enduring; they can feed upon the Polar birch in case of need; they resist cold and hunger. But no Siberian horse is capable of carrying half the weight which a European horse carries with ease; no Siberian cow gives half the amount of milk given by a Jersey cow, and no natives of uncivilized countries can bear a comparison with Europeans. They may better endure hunger and cold, but their physical force is very far below that of a well-fed European, and their intellectual progress is despairingly slow. 'Evil cannot be productive of good', as Tchernyshevsky wrote in a remarkable essay upon Darwinism.[42]

40. A. Becker in the *Bulletin de la Société des Naturalistes de Moscou*, 1889, p. 625.   41. See Appendix 5.

42. *Russkaya Mysl*, Sept. 1888: 'The Theory of Beneficency of Struggle for Life, being a Preface to various Treatises on Botanics, Zoology, and Human Life', by an Old Transformist.

Happily enough, competition is not the rule either in the animal world or in mankind. It is limited among animals to exceptional periods, and natural selection finds better fields for its activity. Better conditions are created by the *elimination of competition* by means of mutual aid and mutual support.[43] In the great struggle for life – for the greatest possible fullness and intensity of life with the least waste of energy – natural selection continually seeks out the ways precisely for avoiding competition as much as possible. The ants combine in nests and nations; they pile up their stores, they rear their cattle – and thus avoid competition; and natural selection picks out of the ants' family the species which know best how to avoid competition, with its unavoidably deleterious consequences. Most of our birds slowly move southwards as the winter comes, or gather in numberless societies and undertake long journeys – and thus avoid competition. Many rodents fall asleep when the time comes that competition should set in; while other rodents store food for the winter, and gather in large villages for obtaining the necessary protection when at work. The reindeer, when the lichens are dry in the interior of the continent, migrate towards the sea. Buffaloes cross an immense continent in order to find plenty of food. And the beavers, when they grow numerous on a river, divide into two parties, and go, the old ones down the river, and the young ones up the river – and avoid competition. And when animals can neither fall asleep, nor migrate, nor lay in stores, nor themselves grow their food like the ants, they do what the titmouse does, and what Wallace (*Darwinism*, chapter 5) has so charmingly described: they resort to new kinds of food – and thus, again, avoid competition.[44]

'Don't compete! – competition is always injurious to the species, and you have plenty of resources to avoid it!' That is the *tendency* of nature, not always realized in full, but always present. That is the watchword which comes to us from the bush, the forest, the river, the ocean. 'Therefore combine – practise mutual aid! That is the

43. 'One of the most frequent modes in which Natural Selection acts is, by adapting some individuals of a species to a somewhat different mode of life, whereby they are able to seize unappropriated places in Nature' (*Origin of Species*, p. 145) – in other words, to avoid competition.

44. See Appendix 6.

surest means for giving to each and to all the greatest safety, the best guarantee of existence and progress, bodily, intellectual and moral.' That is what Nature teaches us; and that is what all those animals which have attained the highest position in their respective classes have done. That is also what man – the most primitive man – has been doing; and that is why man has reached the position upon which we stand now, as we shall see in the subsequent chapters devoted to mutual aid in human societies.

2

Reprinted from *Human Biology and Racial Welfare*, E. V. Cowdry, ed.,
Paul B. Hoeber, Inc., New York, 1930, pp. 139–155

SOCIETAL EVOLUTION

W. M. WHEELER

WHEN as children we first escape from the "big, buzzing, booming confusion," which to our infantile consciousness represents the surrounding world, we distinguish an indefinite variety of different things. Somewhat later we notice that our world also contains a vast number of very similar objects. All this most of us take for granted and never give it second thought during the remainder of our lives. But if we happen to become philosophers or scientists, this composition of reality strikes us as worthy of closer study, though we may entertain little hope of learning why our world should be made up of such an extraordinary number of similars and dissimilars. As our knowledge increases, we observe a pronounced tendency in the numerous like objects to form cohering aggregates, and this tendency seems to be universal in its range from the electrons that make the atoms, the atoms that make the molecules, the molecules that make the masses, from sand dunes and oceans to planets and suns, and their aggregates, the constellations and nebulae. When we turn to living things we find the tendency even more pronounced so that the like entities cohere to form peculiar integrated systems known as organisms which, on analysis, reveal themselves as hierarchies of living entities. We find living molecules, which are themselves systems of inorganic molecules, atoms and electrons, organizing themselves to form cells, cells to form persons, persons to form societies consisting of single families and finally multi-familial or group societies like the one into which we are born and in which we are constrained to live till the end of our days.

Yet closer observation has revealed the startling fact, emphasized only within recent years, that the similar entities when integrated or organized as wholes, i.e. as systems or organisms, exhibit new and unpredictable behavior (qualities) as compared with the behavior of their components. Thus when sodium and chlorine combine chemically to form common salt, we observe that it behaves in a manner very different from either of its constituent substances in isolation. Similarly, a personal organism behaves very differently from its individual cells. A new phenomenal "level" has been created, so to speak, which is not a mere sum or resultant of the component units but a novelty, or "emergent." This term, like the noun "emergence," has in this connection the meaning of "emergency" and is not to be understood in the ordinary sense which implies simply a manifestation or revelation of behavior, or properties previously existing in the components of the system or organism. It should also be noted that in order to bring this consideration into harmony with present physical theory, we must not regard the various components and emergent wholes (systems and organisms) as static things, or as so many lumps of inert matter, but as activities or movements, albeit of very various velocities. Such an attitude enables the scientist to avoid the embarrassing contradictions and inconsistencies with which our thinking has been seriously infected by age-long indulgence in dualistic (materialistic and spiritualistic) notions of reality.

Leaving the physicists, chemists and astronomers to deal with the inorganic aggregations and systems, we may turn to their counterparts, the associations and societies among living things. Here the cohesion and organization of like elements, or components, is indeed astonishingly diverse and complicated. Some of the wholes which they constitute are very loose and temporary and may be called aggregations, like the swarms of dancing midges or the collections of hibernating lady-bird bettles in the mountains of the Pacific States. Others are very persistent and consist of very interdependent, and therefore very intricately organized, parts, like the multicellular bodies of most plants (Metaphyta) and animals (Metazoa). Less highly organized are the wholes represented by the colonies of the social insects, the flocks and herds of birds and mammals, and the societies of man. Table i enumerates the various categories of associations and societies.

36

TABLE I

TYPES OF ASSOCIATIONS AND SOCIETIES

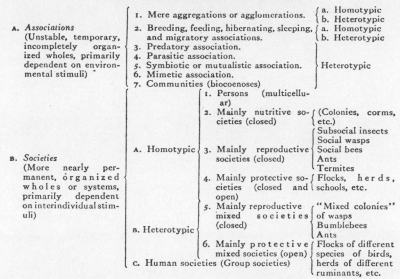

A. Associations (Unstable, temporary, incompletely organized wholes, primarily dependent on environmental stimuli)	1. Mere aggregations or agglomerations.		a. Homotypic / b. Heterotypic
	2. Breeding, feeding, hibernating, sleeping, and migratory associations.		a. Homotypic / b. Heterotypic
	3. Predatory association.		
	4. Parasitic association.		
	5. Symbiotic or mutualistic association.		Heterotypic
	6. Mimetic association.		
	7. Communities (biocoenoses)		
B. Societies (More nearly permanent, organized wholes or systems, primarily dependent on interindividual stimuli)	A. Homotypic	1. Persons (multicellular)	
		2. Mainly nutritive societies (closed)	(Colonies, corms, etc.)
		3. Mainly reproductive societies (closed)	Subsocial insects / Social wasps / Social bees / Ants / Termites
		4. Mainly protective societies (closed and open)	Flocks, herds, schools, etc.
	B. Heterotypic	5. Mainly reproductive mixed societies (closed)	"Mixed colonies" of wasps / Bumblebees / Ants
		6. Mainly protective mixed societies (open)	Flocks of different species of birds, herds of different ruminants, etc.
	C. Human societies (Group societies)		

The associations, of course, vary greatly in the number of their component individuals, from many millions as in the migrating swarms of locusts, to as low as two, in most host and parasite associations among insects. Many species often assemble to form aggregations on the same tree or flower, or under the same stone, and these aggregations may be either homotypic; i.e. consisting of members of the same species, or heterotypic, when individuals of more than one species assemble. These and other aggregations may also result from very important or urgent activities of the individuals, such as feeding, breeding, hibernating, sleeping or migrating. Comprehensive reviews of such cases, with citation of the pertinent literature, have been recently published by Deegener (1918), Allee (1927) and Brues (1926). Special instances of the very small associations are also seen to center about nutrition in the cases of predators, i.e. carnivorous animals and their prey, between parasites and their hosts, insects and the plants they pollinate, mimetic organisms, mainly insects, and their models and in what are known as the communities, or biocoenoses, which are great heterotypic associations of numerous animals and plants occupying the same type of environment and entering into the most diverse and intricate relations with one another, e.g. the biota (fauna and flora) of a tropical rainforest regarded as a whole, that of a bog, sand dune, desert, lake, etc. In all these cases the associations are more or less unstable and temporary, because not very highly integrated. Their integration, in fact, seems to be largely determined by general, extrinsic or environmental stimuli.

The societies, as distinguished from the associations, are more permanent, organized wholes which depend primarily on the behavior of the component individuals towards one another. In order to be with its fellows the social individual will not infrequently seek to adjust itself even to a harmful or fatal environment or situation. Hence societies, as a rule, can be established only between individuals of the same species, i.e. of the same genetic origin, but there are exceptions in which individuals of two or more species may form single societies (ants, bees, wasps, compound flocks and herds of birds and mammals). We may therefore distinguish homotypic and heterotypic societies. Human societies are in many ways so peculiar that they may be assigned to a third category by themselves.

Now it is obvious that all associations and societies are merely peculiar expressions of the most general and fundamental activities of living things, namely adaptation, and it is also apparent that the associative and social adaptations are referable to the basic physiological responses of the individual organisms to stimuli emanating from their fellows or their general environment. We may roughly divide these responses into three general categories, those which satisfy the nutritive, reproductive and protective (defensive and offensive) needs of the individual organism, respectively. The different societies may therefore be classified according to the preponderance of these several needs, in their behavior. Thus such societies as the human person, which consists of some 60,000,000,000,000 cells and such compound organisms as the Portuguese man-of-war, tapeworm, etc. are of the predominant nutritive type. All their individuals are in contact or interconnected in such a manner that certain ones, specialized for the purpose, secure and distribute nutriment to the whole. The *raison d'être* of the society seems to be primarily the facilitation of this function. In other societies, however, like those of the social insects (wasps, bees, ants, termites) nutrition seems to be subordinated to producing and rearing as many young as possible, so that reproduction and all that it implies would appear to be the principal adaptive peculiarity of such societies. Among the flocks and herds of birds and mammals, nutrition and reproduction are less conspicuous than the forms of social behavior connected with protection. For discussions of these societies the reader may be referred to the works of Espinas (1924) and Petrucci (1906) and the recent volume of Alverdes (1927). As would be expected, primitive human societies have their closest analogues among certain gregarious mammals, and notably among the anthropoid apes.

The problem of greatest interest to the student of animal associations and societies is concerned with the precise nature of the communal bonds, or social cohesion which causes the individuals to assemble and remain together for a longer or shorter period. The aggregations in some cases are obviously the result of mere accidental propinquity due to the individuals hatching simultaneously from a batch of eggs deposited by the mother organism directly on food suitable for the young. Thus the larvae of such insects as the gypsy moth and potato bettle are too feeble to stray far from the egg cluster from which they hatch and really need not stray far from one another because they are surrounded by an abundant supply of nutriment. The same is true of plant lice which are born alive by their feeble wingless mothers and the sluggish, legless larvae of Drosophila and blow flies which hatch from numbers of eggs laid almost simultaneously in fermenting fruit or decomposing flesh. We need not assume, therefore, that such aggregations of larvae are due to fondness for one another's company or are kept together by any other bond than a simple chemotropic response to their common nutritive environment. But some aggregations and associations undoubtedly depend on stimuli emanating from the different individuals. All animals give off heat, moisture, carbon dioxide, secretions and excretions and make movements. Simple tropistic or reflex responses to these, such as those designated by the terms thermotropism, hygrotropism, chemotropism and stereotropism, are probably sufficient to account for many aggregations and associations. Migratory crickets have been observed to huddle together for mutual warmth when the cool of evening comes on; slaters (Oniscus) are induced to assemble by the moisture which they give off, and resting locusts may be stimulated to flight by the movements of their fellows. Rhythmic emission of light in fireflies or of chirping in crickets may excite rhythmic, and according to some authors synchronous, responses of the same kind in other individuals in the immediate neighborhood ("physiological sympathy" of Ribot). Some simple aggregations are evidently the result of a number of tropistic responses. One example will suffice.

The larvae of the common blackfly (Simulium) often congregate in dense masses on stones in the more torrential parts of our streams, stand erect on their posterior ends and capture with their out-spread, rake-like mouthparts diatoms and other microorganisms as they float past. In this case we may distinguish stereotropic responses of the larvæ to solid bodies (the stones), rheotropic responses to the current and probably also chemotropic responses to the higher oxygen content of the more rapidly moving water. It will be noticed that this combination of tropistic responses constitutes an exquisite adaptation because it places the stationary larvae in the optimal enviroment for securing their food, since much more of it passes within their reach in a given time in the torrential than in the sluggish portions of the stream. Although aggregation here actually brings about a competition for the food among the individuals, this disadvantage is more than compensated by the increased supply due to the swiftness of the stream. There is an extensive and interesting literature, much of which has been reviewed by Allee, dealing with the effects of aggregations on their component individuals, but the subject cannot be further elaborated in this article.

Some authors have endeavored to derive the societies from the associations, but it is difficult to find any cogent proof of their contentions. The societies really represent very different emergent levels from the associations and have arisen in a different way, though, of course, ancient aggregative or associative proclivities may have been retained by many species and may serve to reinforce their specifically social behavior. The members of societies, as distinguished from the associations, are primarily concerned with their adaptations to one another, i.e. with neutralizing their individual antagonisms, and with their mutual adjustment and cooperation. The mass of stimuli which elicit these adaptive responses may be called the social medium. It constitutes a very complex and unstable environment for the individuals, and successful and enduring adjustment to it presupposes a high sensitivity and considerable behavioristic plasticity on the part of the consociated organisms, and this in turn presupposes a highly organized neuromuscular apparatus. It is clear therefore that societies can be constituted only by species in which the sense-organs, brain and muscular system have attained a high degree of specialization, and not by animals that have never succeeded in transcending the merely tropistic and reflex level. Social life demands at least a rudimentary memory and intelligence, if we understand by the latter the ability to respond adaptively to new situations on the basis of previous experience, or in other words, some ability to learn. It is obvious, moreover, that to such organisms social life furnishes the only adequate opportunity for much further perfecting of the intelligent activities.

Though a rather highly developed neuromuscular system is a *sine qua non* of social life, it is far from true that all animals thus equipped must become social. Tigers, hawks, spiders and tiger-beetles are richly endowed organisms, but they do not live in societies. Moreover, when we study the positions of social species in the animal hierarchy we find that they are confined to certain sporadic groups of species and that they often differ externally in no respect from the most closely related, highly specialized, nonsocial forms. A worker honey-bee or hornet is quite unable to live except in a society, and yet no one could infer this fact from their structure, which differs in no essential character from that of their solitary congeners. It would seem, therefore, that some other peculiar condition in addition to the high development of the neuromuscular system is essential to the formation of true societies. This I believe to be the development of the family. So long as its members remain together, a family is, of course, a rudimental society, with reproductive, nutritive and protective functions and an unmistakable differentiation, or division of labor in its components. All the societies of insects are merely single families in origin though they may become very populous and acquire an extraordinary differentiation of their members. The family origin of the flocks and herds of birds and mammals and hordes and tribes of primitive man is also apparent; though in these societies the family is open and not closed as in insects and there is a retention in the flocks, herds and hordes of primitive aggregative or associative tendencies which seem to hark back to the ancestral fish and tadpole stages. This retention is apparent in important further developments to be briefly considered in a later paragraph.

The family implies the vital affiliation of the offspring with the parents and this can only be accomplished on the condition that the adult life of the parents is sufficiently prolonged to admit of rearing the offspring to maturity. This increase in parental longevity also permits a corresponding extension of the care of the offspring and gives the latter time for a more complicated development and greater opportunities for learning and therefore of preparation for adult life. The latter consideration has been often discussed by sociologists, psychologists and educators, but the increase of the adult life of the parents as a prerequisite to that of the young has been overlooked. It is just this latter condition which enables us to account for the beginnings and further development of the families which become the elaborate closed societies of the ants, wasps, bees and termites. Most mother insects die soon after oviposition and the young are left to shift for themselves, but in certain groups, owing to peculiarities of food or environment, the adult life of the mother or of both parents is considerably prolonged so that it overlaps the larval period or even a part or the whole of the adult life of the offspring and thus furnishes an opportunity for close relations between the members of two successive generations. I find these conditions realized in at least thirty insect groups, which are often so remotely related to one another or related in such a manner that the family must be supposed to have arisen independently (polyphyletically) on at least as many separate occasions during the long racial history of the Hexapoda, which extends over some 300,000,000 years from the Upper Carboniferous to the present time. These families vary greatly in complexity and stability. In most cases the parents are deserted by the progeny while the latter are still young, and the rudimentary society dissolves, a condition observed in what I have called the subsocial insects (certain beetles, wasps, bees, Embiids, earwigs, etc.). In the termites, ants, higher wasps and bees, however, the affiliation of the progeny with the mother (Hymenoptera) or with both parents (termites) becomes much more intimate and prolonged, so that at least the worker caste, which constitutes the great majority of the personnel of the society, never dissolves its consociation with the parents. The single family is thus enabled to remain a society, though capable in some cases of growth to a population of hundreds of thousands of individuals. This is accomplished by partial starvation of most of the offspring so that they fail as larvae to develop their reproductive organs (alimentary castration) and even as adults, in their capacity as nurses, inhibit the further development of their gonads by starving themselves as a result of feeding the successive broods of larvae, the queen and the other adult members of the colony (nutricial castration).

In the foregoing account of insect societies nothing is said about the nature of the bonds which unite the parents and offspring and thus initiate the family or about the nature of the social medium which regulates the social behavior. I believe that we may detect these recondite factors in what I have called "trophallaxis," or exchange of food.

The larvae of many social insects (ants, wasps and termites) are not only fed by the adult members (parents and workers) of the colony, but may in turn feed their nurses with salivary or other secretions. The young are therefore a source of food for the adults and *vice versa*, and the "fondness" of the social insects for their young proves to be not some altruistic "instinct," such as love or affection, but the hunger of the individual and therefore an egoistic appetite. There is also a trophallactic relation between the adult members of the colony, which are constantly feeding one another with regurgitated liquid foods (ants) or with regurgitated semisolid foods or feces or substances (exudates) secreted from the surfaces of their own bodies (termites). So powerful is this habit that it is extended even to many of the heterogeneous insects which have managed to live in the nests of the social species, i.e. to the true guests among the myrmecophiles, which live with ants, and the highly specialized guests of termites (termitophiles). Furthermore, since the senses of taste and smell are not differentiated in insects as they are in the higher vertebrates, and since, in the former, we may therefore more properly speak of a single chemical sense, we are justified in including under trophallaxis also an exchange of odors as one of the important cohesive bonds in insect societies. There are, in fact, individual, colonial and nest odors, which the social insects are able to recognize and distinguish and therefore serve to determine many of their reactions and much of their behavior. Both the food and the odors thus constitute a regulative, circulating social medium which not only functions as a social cohesive, but in the case of the food, furthers the growth and maintenance of the society in a manner analogous to the circulating blood stream in the body of a higher animal, which is also a society of cells. Of course, the word "food" is here used in a general physiological sense, both as nutriment and as a stimulant, or excitant, because the amounts of the substances exchanged may be extremely small though of such a chemical composition as to produce pronounced reactions, just as a very small amount of alcohol may produce much more violent reactions in some people than large quantities of rice or potatoes.

The flocks, packs and herds of the higher vertebrates constitute peculiar societies, quite unlike those of insects, and might more properly be called peoples, populations or "peuplades," to use Espinas' term. They may consist of a number of associated families, of individuals of one or both sexes or of the young only. They may be loosely or intensively organized, temporary or more permanent, and either closed or open, to use the classification of Alverdes, i.e. they may either repel outsiders or admit them, even when they belong to alien species, and permit them to become members of the society in good standing. Our knowledge of these peculiar organizations is far from complete, but some of them have been recently studied with results that seem to have important bearings on human societies, which are really of the same fundamental structure. The peuplades are held together by what has been usually called the gregarious, or herd instincts, but the investigations to which I have just referred seem to show that the cohesion may be due to more concrete and observable behavioristic factors.

In their studies of birds, Schjelderup-Ebbe, Katz and Fischel have shown that the flocks are organized according to a peculiar "pecking order," in which each individual has its own status depending on whether it may peck other individuals or submit to being pecked by them. To quote Alverdes, "Schjelderup-Ebbe has shown how an order of precedence comes into existence within societies. A flock of fowls in a fowl run is not exclusive in the sense that its members make common cause against a new arrival, leaving the latter isolated. The new comer may safely attach itself to the flock, but the position it is to hold therein must first be won by fighting. For no two hens ever live side by side in a flock without having previously settled, either for the time being, or for good, which is the superior and which the inferior; the "pecking order" thus established decides which of the birds may peck the other without fear of being pecked in return. Similar pecking codes exist, according to Schjelderup-Ebbe, among sparrows, wild ducks, and possibly among many other kinds of animals. Pecking among cocks is governed by the same rules as among hens, except that the cocks exhibit greater ferocity. Such "pecking orders"

give the society concerned a certain degree of organization."
This "pecking order" which would seem to be the cohesive
among the peuplades, corresponding to the very different
cohesive, trophallaxis, among the societies of insects, leads
naturally to a complex, organized hierarchy of individuals,
depending on their age, sex, vigor, and bluffing capacity.
It is so suggestive of human communities, in which we have
a similar hierarchy of social status based on the bivalent
self-assertion and self-abasement, or sadistic and masochistic
motives of the individual, that the results of further investigations on the peuplades of other Vertebrates will be awaited
with interest. Perhaps what we call government in human
societies is really only a glorified "pecking order!"[1]

When we turn to the societies of man we are confronted
with an emergent level so much higher and so much more
complicated than that of any of the other social animals
that it seems to transcend analysis. Biologically it is obvious
that it consists of a great number of genetically related
families, and though there is among the individuals of each
of these a division of labor essentially like that of the animal
family, there is superadded a more elaborate division of
labor which traverses the families and is quite unlike that
of the unifamilial society of insects, since it is a product of
learning and custom and has not become hereditary. Furthermore, his much more highly developed neuromuscular
system, intelligence, memory, and language have enabled
man to create and transmit from generation to generation
vast accumulations in the form of stores of knowledge,
elaborate institutions, constructions, mores, arts, sciences,
etc., which the animals, restricted to their limited hereditary
endowments and feeble individual plasticity of response to
their inorganic and living environment, could neither develop
nor transmit. This tradition, or social memory, has therefore
been regarded as the leading peculiarity of human societies,
but it must be admitted that there are some very rudimental
indications of it even in the social insects.

[1] In this connection it is interesting to note that the domestication of
animals depends on a similar order. Nearly all our domestic animals belong
to social species and their successful subjugation implies, so to speak, a realization on their part of their defenseless inferiority in the presence of man.

Naturally the question as to what brings about the cohesion of the individuals is far from being as easily answered in human as in animal societies. To this question, which also involves the causes of the maintenance or continuation as well as the origin of human societies, the philosophers and sociologists of the past have given a number of different answers. These are all hypothetical, and none of them is altogether satisfactory. Several of them, in fact, are quite inadequate and at present obsolete, but it may be of interest to consider them *seriatim*.

1. The earliest hypothesis, if it deserves so dignified a name, is, of course, that of Genesis, according to which the male of our species was made out of clay by divine fiat on the sixth day and the female from his rib by the same process somewhat later. This statement has some extraordinary implications, only two of which need be mentioned. First, man having been created complete, he was necessarily a social being from the beginning, and inquiry into the causes of society must be useless. Secondly, owing to his special creation, man is definitively set over against the other animals and Nature in general. This is the view still taken for granted by many theologians so that for them all inquiry into social evolution and cohesion in a biological sense must be meaningless.

2. Some of the Greek philosophers, including Plato, entertained a similar supernatural view of the origin of man and his society, mainly on ethical grounds. But Plato seems also to have been inclined to regard society as having had a natural origin and growth and this view was definitively developed by Aristotle and much later by other philosophers, including de Maistre and Kant.

3. In the seventeenth and eighteenth centuries an hypothesis of the origin of human society was expounded by Hobbes, Locke and Rousseau, which seems very strange to us. These philosophers imagined that men had formed and continued to maintain their society by mutual agreement, or compact. That anything like society could owe its beginning and cohesion to such frail intellectual motives was quite in harmony with the thinking of those centuries. Of course, the corollary that men might dissolve by intellec-

tual agreement what they had built up by the same means was one of the arguments in favor of the French Revolution. At the present time the doctrine has no sociological importance, except possibly to a certain type of reformer.

4. Another intellectualistic hypothesis of societal origins was excogitated by Starcke who believed that early man took up social life because he had observed its advantages among the social insects in his environment. Similar fanciful notions have been occasionally advanced to account for more modest human inventions, e.g. the supposition that the Australian aborigines derived the boomerang from observing the circular paths described by the falling sickle-shaped leaves of certain species of Eucalyptus. While this latter hypothesis cannot be dismissed as altogether improbable, Starcke's notion becomes absurd when we consider that man must have been a social animal long before he had sufficient intelligence to observe and imitate the social insects and that, even had he conceived society as the result of such casual observations, that fact would not have enabled him to maintain it throughout his whole subsequent racial history in all parts of the world.

5. Darwin's hypothesis of evolution through natural selection is, of course, quite a different matter. According to it, social origins are really accidental, but when men had once established social relations with one another, the advantages accruing would lead to the survival of the individuals that adhered to the social habit, while those who reverted to a solitary mode of life would be eliminated. Naturally, from a human point of view, the advantages of society are enormous, and existing man has never experienced any other mode of life, but the selection hypothesis, though logical, does not go to the root of the matter. The merely evolutionary, or transformist core of the hypothesis, however, is immensely important. We should not be wrong in stating that evolution may be more easily demonstrated in ethnology, archeology, and history than in the study of living and extinct organisms.

6. If we revert to the principle of emergence, briefly considered in the introduction to this chapter, we might say that human society arose rather suddenly and discon-

tinuously when the primitive family expanded by some natural process of growth, affiliation and differentiation of individuals into the clan or tribe, but even this leaves us in the dark in regard to the actual factors which brought about the expansion and still maintain the solidarity of the individuals in the great societies of the present time.

7. Psychologists, psychopathologists and sociologists are now unanimous in maintaining that social cohesion, or what some have called the "social mind," must be constituted by the wealth of non-rational behavior which has been variously designated as the appetites, cravings, instincts, interests and emotions of the individual. Some have postulated a special "herd instinct" (Trotter) or "gregarious instinct" (Drever), while others have based human solidarity on "consensus" (Comte), synergy, or cooperation (Spencer), on altruism, sympathy, affection or even egoism (Le Dantec). It will be seen that all of these bonds are of a physiological or primitively psychological character and therefore quite different from those which we call intellectual, or rational. They are, no doubt, fundamentally the same as the primitive associative tendencies which we observe in the uniform members of the flocks and herds of birds and mammals, and may therefore be traceable to the instinctive bonds which unite the members of the family.

8. Durkheim, while accepting these tendencies as the basis of social cohesion in the more primitive human societies, has pointed out that as society develops, the strongest bonds are those produced by the continued action and intensification of the social division of labor. The associated individuals necessarily become more and more heterogeneous psychically, and therefore more and more interdependent. This increase in interdependence brings about both the cooperation and the constraint which are such conspicuous features in highly developed societies. Cooperation is not, therefore, a primitive condition, but supervenes after a certain differentiation has been developed by division of labor, or specialization among the individuals; and the constraint, restraint, inhibitions and repressions which the social unit is bound to exercise and endure have had much to do with the creation of the traditions (social heredity),

mores, laws, religious institutions, etc. which in turn constrain their creators. Durkheim's view has the advantage of referring the integration, or solidarity of society, to a principle which is universal, not only in all animal societies, but also in all multicellular organisms. This principle, the division of labor, was first recognized and named by the economist Adam Smith and only later introduced into biology by Milne-Edwards.

The very significant rôle of the primitively psychological and the relative insignificance, even in our present civilization, of the specifically intellectual processes have been most impressively set forth by Pareto in his "Traité de Sociologie" and by Sumner and Keller in their "Science of Society." A study of these works might be said to constitute a liberal education. Pareto designates the irrational foundations of social behavior as the "residues," the rationalizations of them in which we are constantly indulging, as the "derivations." Sumner and Keller's remarkable picture of the mores and of their fundamental significance, stability, and tenacity, based on exhaustive ethnological studies, forms an admirable background for Pareto's contentions, which he illustrates mainly with materials drawn from the ancient and contemporary history of European peoples. Both works are important also because they lift sociology entirely out of the valuative and moralizing slough, in which it has long floundered, onto the scientific plane. The strange light which these and many other similar studies of human society cast on our zealous social reformers and propagandists enables us to appreciate, on the one hand, the impulsive, irrational, wishful thinking which is the true drive of their own activities and, on the other hand, the extraordinary magnitude and inertia of the behavior they are trying to control and reform.

REFERENCES

ALLEE, W. C. 1927. Animal aggregations. *Quart. Rev. Biol.*, 2: 267–398.
ALVERDES, F. 1927. Social Life in the Animal World. N. Y., Harcourt Brace.
BRUES, C. T. 1926. Remarkable abundance of a cistelid beetle, with observations on other aggregations of insects. *Amer. Natural.*, 60: 526–545.
DEEGENER, P. 1918. Die Formen der Vergesellschaftung im Tierreiche. Leipzig, Veit.

DURKHEIM, E. 1922. De la Division du Travail Social. Paris, Alcan.

ESPINAS, A. 1924. Des Sociétés Animales. Ed. 3. Paris, Alcan. (Ed. 1, 1877.)

FERRIÈRE, A. 1915. La Loi du Progrès en Biologie et en Sociologie et la Question de l'Organisme Social. Paris, Giard & Briéve.

FISCHEL, W. 1927. Beiträge zur Sociologie des Haushuhns. *Biol. Zentralbl.* 47: 678–696.

KATZ, D. 1922. Tierpsychologie und Soziologie des Menschen. *Zeitschr. f. Psychol.* 88.

——— 1926. Socialpsychologie der Vögel. *Ergebn. der Biol.*, 1: 447–477.

LE DANTEC, F. 1911. L'Egoisme seule base de toute Société. Paris, Flammarion.

PARETO, V. 1917. Traité de Sociologie Générale. French ed. by P. Boven. 2 vols. Paris, Payot & Co.

PETRUCCI, R. 1906. Origine Polyphylétique, Homotypie et Noncomparabilité des Sociétés Animales. Fasc. 7. Notes et Mém. Inst. Solvay.

SCHJELDERUP-EBBE, T. 1922. Beiträge zur Sozialpsychologie des Haushuhns. *Ztschr. f. Psychol.*, 88.

——— Das Leben der Wildente in der Zeit der Paarung. *Psychol. Forsch.*, 3.

SUMNER, W. C. and KELLER, A. G. 1927. The Science of Society. 4 vols. New Haven, Yale Univ. Press.

WHEELER, W. M. 1923 Social Life Among the Insects. N. Y., Harcourt Brace.

——— 1928. Emergent Evolution and the Development of Societies. N. Y., Norton.

——— 1928. Insect Societies, their Origin and Evolution. London, Keegan Paul.

3

Copyright © 1940 by Scientia

Reprinted from *Scientia* (Milan), 34:154–160 (Apr. 1940)

CONCERNING THE ORIGIN
OF SOCIALITY IN ANIMALS

W. C. Allee

University of Chicago

The phases of biology can be divided roughly into two categories: (1) those which deal with the individual organism, or with some part of the organism, and (2) those which are concerned with groups of organism as a fundamental unit. These latter include such groupings as species, aggregations, populations or communities, which have at least some degree of supra-individualistic unity.

Without putting forward here a precise definition of individuality, we may say that the concept is fairly definite and can usually be applied within tolerable limits of accuracy; in spite of the fact that there exist a considerable series of colonial forms, of which sponges supply an example, in which individuality can be recognized with difficulty if at all.

Similar precision of delimitation of supra-individualistic units is frequently impossible in concrete situations. Species have been well studied and the species concept has validity, but its concrete application seems frequently to the non-specialist, at least, to be an art rather than a science. As with the concept of species, general ideas concerning what constitutes sociality show a high degree of agreement. But there is less meeting of the minds concerning the criteria by which sub-social animals may be distinguished from truly social animals.

Three general types of definition of sociality have been developed. Sociologists have worked out the most prominent of these. Focusing on the human aspect, they state that true sociality occurs only in the presence of abstract values of which members of the group are more or less conscious. This definition restricts sociality to an intellectual level and, as far as we know, limits true social life to man.

A second type of definition is based on the assumption of a social instinct. This idea can be re-stated by those wo rightly distrust the word « instinct » to make sociality co-extensive with the existence of an innate inherited pattern of a certain specialized appetite. It is the definition preferred by those who find much in common between the social tendencies of insects and of many vertebrates, but who would still restrict sociality to relatively high levels in these two classes of animals.

A third and still more inclusive definition of sociality says that all integrations of two or more organisms into a supra-individualistic unity on which natural selection can act marks at least a beginning of social life. Such low or feeble social units may be poorly integrated, but still possess demonstrable survival values, or they may be well integrated pairs, or small or large groups, which may or may not show an obvious survival value.

Advocates of this inclusive definition usually maintain that while various types of organization are readily distinguished, as well as different levels of group integration, no valid criteria exist to justify the restriction of the concept of sociality to the higher levels only. In other words, while some social groups seem to be integrated by ideas or emotions, and others by inherited social patterns, and while both of these may have common social aspects, such as division of labor, which partially set them off from less closely knit animal aggregations, still all groupings of individuals which are sufficiently integrated so that natural selection can act on them as units show at least the beginnings of sociality.[1]

It has been recognized for years[2],[3],[4] that this third attitude towards social essentials constitutes a promising basis for the general study of the social life of animals. It is from this point of view that recent experimental attacks have been made on fundamental attributes of sociality [5],[6],[7], by testing some of the mutual interactions of numbers of organisms while living together in a more or less limited microcosm. The groupings may be natural or the organisms may be artificially placed together in experimental populations.

Many of the physiological processes associated with population density are adequately summarized by the curves shown in Figure 1. The monophasic curve A summarizes results of biological process(es) in which the highest recorded value is given by the fewest organisms that can exhibit the process in question. With sexual reproduction, a single pair is the smallest unit possible: with asexual reproduction and many other processes, the reactions of single individuals can be followed. The diphasic curve B, on the other hand, is a summary of responses when the highest value is given by some population density above the minimum. Inso-

---

[1] I believe I am indebted to my colleague, Professor A. E. Emerson for this compact summarizing formula. In this connection see A. E. EMERSON, 1939. *Social co-ordination and the superorganism.* « The Am. Midland Naturalist », 21; 181-209. I am also indebted to Drs. Emerson, Ralph Buchsbaum and Thomas Park for reading this manuscript critically.

[2] A. V. ESPINAS, 1878. *Des sociétés animales.* Paris, 588 pp.

[3] W. M. WHEELER, 1923. *Social life among the insects.* New York. 375 pp. 1939. *Essays in philosophical biology.* Harvard, 261 pp.

[4] W. C. ALLEE, 1931. *Animal aggregations. A study in general sociology.* Chicago. 431 pp.

[5] W. C. ALLEE, 1931. *Animal aggregations. A study in general sociology.* Chicago. 431 pp.

[6] W. C. ALLEE, 1934. *Recent advance in mass physiology.* « Biol. Rev. », Vol. 9, pp. 1-48.

[7] W. C. ALLEE, 1938. *The social life of animals.* New York, 293 pp.

far as quantitative data can be interpreted in terms of optimal population size, curve A results when the optimum response is given by the smallest population which can give that particular reaction while for curve B, the optimal size is somewhat greater then the minimum.

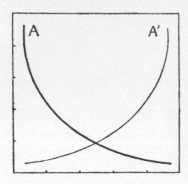 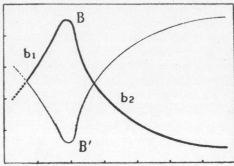

Fig. 1.

Summarizing curves showing some of the relations between the measurement of some biological process or processes (ordinates) and the numbers 'of associated organisms (abscissae). The lower left hand corner represents zero for both axes.

Résumé des courbes qui montrent quelques-unes des relations entre la mesure de quelque(s) processus biologique(s) (ordonnées) et le nombre d'organismes associés (abscisses). Le coin inférieur gauche représente le zéro pour les deux axes.

Although it is too early to make a definitive statement concerning which processes may be expected to be summarized by curve A and which by curve B, we do know that the former frequently expresses fairly accurately many of the relations between population size and fecundity among bisexual animals: that is, the more animals there are present in a limited universe, the fewer offspring there will be per female per day. [1] Curve B shows general experience with survival of many types of organisms in the presence of a somewhat adverse environment. The longevity of many different animals, for example, shows a curve of this type when they are exposed to a dilute concentration of some toxic material such as colloidal silver. [2] Other processes which give a similar response include the speed of cleavage in certain sea-urchin eggs, the rate of asexual division in cultures of some protozoans (this is not given under all conditions), the relation between litter size and survival in guinea pigs, [3] and the expected rate of evolutionary change in natural populations. [4, 5]

If some opposite aspect of the process in question is considered, death-rate rather than longevity for example, curve B becomes inverted or curve

[1] W. C. ALLEE, 1931. *Animal aggregations. A study in general sociology.* Chicago. 431 pp.

[2] W. C. ALLEE, 1938, *The sociol life of animals.* New York, 293 pp.

[3] SEWALL WRIGHT, 1922. *The effects of inbreeding and cross breeding on guinea pigs.* « U. S. Dept. Agric. Bull. », 1090, 63 pp. S. WRIGHT and O. N. EATON, 1929. *The persistence of differentiation among inbred families of guinea pigs.* « U. S. Dept. Agric. Tech. Bull. », 103, 45 pp.

W. C. ALLEE, 1938. *The social life of animals.* New York, 293 pp.

[5] T. G. DOBZHANSKY, 1937. *Genetics and the origin of species.* 364 pp. New York.

A becomes curve A'. Since these changes in representation do not introduce anything that is essentially different, they need only this passing
mention.

In further discussion of diphasic curve B it will be convenient to speak
of the left hand limb as curve b_1 and the right hand limb as curve b_2. Both
limbs represent effects of increased crowding. Curve b_2 is obviously associated with over-crowding and curve b_1 with what we may call under-
crowding. Since curve A and curve b_2 may summarize fundamentally
the same types of relationship, much of the continuing discussion can
be simplified, without loss of accuracy, by contrasting and attempting an
interpretation of the two limbs of curve B.

As has been stated previously, these curves are fairly accurate summa-
ries of quantitative observations of some of the simpler interrelations between various biological processes and population size. Cautious conservative biologists apparently are content to use such curves for descriptive
or comparative purposes only or, at all events, to proceed no further than
to find the mathematical formula of which the curve in question is the
concrete expression. Curve are helpful for such comparisons and it may
well be that such use is all that can be made with entire safety either of
data obtained from quantitative studies of populations or with any other
quantitative biological information.

If we are to make progress with scientific interpretation, it is necessary to inquire further into the general significance of such basic data.
Such an inquiry is admittedly hazardous in the situation which is under
examination. It is no more so, however, than is the attempt of find appropriate hypotheses and theories from scientific data in general and the
construction of such hypotheses is, of course, a standard and commonplace
process in the attempt to get a general synthesis from reputable scientific
data. It is understood that these interpretations are not of the same
order of certainty or of objectivity as are the data on which they are based.
Such considerations are fundamental for scientific hypotheses and are so
generally understood that they are usually safely taken for granted.

With this introduction, let us attempt an interpretation of the basic
interrelation between members of a population, experimental or natural.
Curve A, or curve b_2, is a somewhat close approximation of the generalization that is commonly known as Farr's law. [1] This states in substance that with human populations in England, the death rate is highest
the more dense the population. This law has been generally accepted as
an illustration of the doctrine of Malthus, hence curve b_2 becomes associated with this doctrine. This is an association with interesting implications since the Malthusian doctrine greatly influenced Charles Darwin in
the thinking which led to his theory of evolution by natural selection. It
was especially potent as regards his conclusions concerning the relation
between over-population and the resulting struggle for existence. Phenomena which can be summarized by curve b_2 have been interpreted for

[1] W. C. ALLEE, 1931. *Animal aggregations. A study in general sociology.* Chicago.
431 pp.

years in terms of the disoperative [1] phase of competition which Darwin thought to be extremely important in his scheme of natural selection. This part of the interpretation of the phenomena that underlie these curves has already been made for us by inference at any rate even if not specifically and, so far as I am aware, has not been seriously questioned.

If curves of the general type of b_2 have these implications, then those of the type of b_1 must have other meanings, since with b_2 the biological reaction(s) being measured shows decreases with increasing numbers while with b_1, the same or similar reaction(s) has the opposite trend as the nembers of organisms become larger. If we designate as X that which curve b_2 implies, then curve b_1 implies processes which are the opposite of X. If X is interpreted in terms of disoperative aspects of Darwinian competition, then curve b_1 necessarily suggests some process or processes which run counter to such competition. Historically, Kesslerian co-operation has been interpreted as opposing Darwinian competition at least in large part. Espinas, [2] long ago, and Wheeler [3] later recognized the validity of the concept of natural unlearned co-operation largely on the basis of observational data and philosophical considerations. More recently [4],[5] with some hesitation on the grounds of semantics only, I have followed them and have written of *unconscious* co-operation or *automatic* mutualism. What I have attempted to imply in these terms, and all I have attempted to imply, is that there are fundamental tendencies in the biological universe which are radically different from the disoperative competition based primarily on an individualistic struggle for existence such as Darwin thought resulted from over-population.

If curve b_2 represents effects of over-population, then curve b_1 shows results from under-population and in the phenomena associated with under-population we have biological evidence of simple supra-individualistic groupings on which evolution can act to produce more complicated and better knit social systems. This means that between the individual members which compose such groups, there are unconscious interactions which, under certain conditions, tend to preserve the lives of the members and to do all the other things which have been experimentally demonstrated [6],[7],[8] and which are summarized in general terms by curve b_1. To this extent we have supplied evidence of sociality, in the sense of the third definition, which is practically co-extensive with animal life and which has been found, at least in simplest terms, among many plants as well.

[1] In an illuminating discussion, CLEMENTS and SHELFORD (*Bioecology*, New York, 425 pp., 1939) use the terms disoperation and competition with a somewhat different connotation.

[2] A. W. ESPINAS, 1878. *Des sociétés animales*. Paris, 588, pp.

[3] W. M. WHEELER, 1923. *Social life among the insects*. New York, 375 pp. 1939. *Essays in philosophical biology*. Harvard, 261 pp.

[4] W. C. ALLEE. 1931. *Animal aggregations. A study in general sociology*. Chicago. 431 pp..

[5] W. C. ALLEE, 1938. *The social life of animals*. New York, 293 pp.

[6] W. C. ALLEE, 1931. *Animal aggregations. A study in general sociology*. Chicago. 431 pp.

[7] W. C. ALLEE, 1934. *Recent advance in mass physiology*. « Biol. Rev. ». Vol. 9. pp. 1-48.

[8] W. C. ALLEE, 1938. *The social life of animals*. New York, 293 pp.

The conclusion seems unavoidable that the two limbs of curve B have decidedly different biological implications. If we are to interpret them at all, it is imperative to interpret some of the phenomena associated with under-crowding as opposed in many major aspects to those associated with over-crowding. It is obviously unfair to attempt or to accept a general interpretation that is based on one limb only of the whole curve (b_2 for example) without making some effort to understand the implications of the other limb. Since phenomena associated with under-crowding have been demonstrated in all major groups of living animals and since they imply the existence of supra-individualistic groupings on which natural selection can act, we have here evidence of a widespread phenomenon and one from which the best integrated phases of social living can evolve.

There does not seem to be anything inherently radical in this analysis. Essentially all that has been done is to recognize, as many geneticists are now doing, [1] that populations can be selected as well as individuals. The radical aspect of the interpretation comes in its implications. If the present suggestion is sound, then sociality is seen to be a phenomenon whose potentialities are as inherent in living protoplasm as are the potentialities of destructive competition. This means that social living is not an isolated incident which appears as a sudden emergent in certain highly evolved animals. It is rather, as Espinas and Wheeler declared, a normal and basically widespread phenomenon. And it should be remembered that Espinas and Wheeler reached their conclusions without the benefit of the recently accumulated mass of experimental evidence.

If the fundamental thesis of the present paper turns out eventually to be correct, that the implications of curve b_1, as opposed to curve b_2 (or A) justify us in seeing some innate tendency toward sociality as being widely distributed among living things, such a conclusion will carry other far-reaching suggestions. One of the most interesting of these is the idea that the currently popular pose of biologists in embracing human defeatism and general pessimism, is not the only interpretation which can be put on our painfully acquired store of biological knowledge. Nature red in tooth and claw, while real enough, is not the only reality in nature. While there are ample grounds for being more pessimistic than optimistic over human nature and the biologist's world in general, the facts, or at least reasonable interpretations of apparent facts, do not of necessity push us in that direction.

Having started with a discussion of the extent of the phenomena of sociality in animals, we arrive at suggestions which permit, and even seem to encourage, a certain amount of naturalistic optimism. In any event, and by supposedly objective processes of analysis and synthesis, this is my present interpretation of the available evidence.

Lest the relationship between co-operation, whether learned or unlearned, and other phases of the interactions of organisms be misunderstood, it should be stated in conclusion that not all phases of competition are destructive and competition may in fact be beneficial under many conditions. This is not the place and, in fact, there is not at present sufficient critical

[1] T. G. DOBZHANSKY, 1937. *Genetics and the origin of species.* 364 pp. New York.

evidence upon which to base an accurate comparison of the relative importance of cooperation and competition. Given an underlying stratum of
co-operation, competition, by establishing degrees of dominance and subordination, whether in the ecological [1] or human sense, may furnish
a basis for more effective community co-operation. Such organized communities may then compete with each other and the less fit co-operating
systems may be weeded out to the general advantage of the survivors
at least. All this may occur without affecting in the slightest the general
conclusion concerning the essential antithesis between co-operation and
disoperative aspects of competition or the fundamental importance of
innate co-operation in nature.

[1] See discussion by CLEMENTS and SHELFORD; citation in footnote 1, pag. 158.

4

Reprinted by permission of the American Psychological Association from
J. Abnorm. Social Psychol., 41:385–402 (1946)

PROBLEMS IN THE BIOPSYCHOLOGY OF SOCIAL ORGANIZATION

BY T. C. SCHNEIRLA

Fellow of the John Simon Guggenheim Memorial Foundation

THERE is an ancient and understandable tendency to draw moral conclusions from apparent similarities between man and lower animals, for which the social insects have served as convenient material. Solomon's advice to the sluggard, "Go to the ant . . . ," comes readily to mind as an outstanding instance of seemingly infallible repute. Unfortunately, in the light of present knowledge about individual differences among insects in social participation (Combes, 24; Chen, 21), this moral cuts two ways, for the sluggard might well find all his time taken up in contemplating the leisurely ways of those relatively sessile and less productive members which any ant-hill is almost certain to contain. On the other hand, we who are not sluggards may learn much from a more careful comparison of social activities in lower animals and in man, very possibly to the great advantage of better insight into man's social capacities and potentialities.

Scientists exhibit a growing tendency to study comparatively the makeup of what are considered different levels of organization in the inorganic and organic worlds (Redfield, 61). Simple and complex levels are recognized among inorganic phenomena, and in an evolutionary sense certain of the inorganic levels are recognized as prerequisite to the occurrence of organic wholes such as viruses which are regarded as primary. Among biologists and students of behavior there is an increasing alertness for what can be learned from the investigation of one level or type of organization that will assist in the improved understanding of others. An interest in comparative study leads to a closer examination of the various instances of organization describable as levels, for example the individual organism, the animal aggregation, social group and society (Allee, 2). We shall want to inquire how far such studies have advanced beyond the stage of description and naming, exemplified by Alverdes' (5) general survey, and are searching out the essential qualities of different animal organizations. And since there is implicit in the contemporary study of both individual organisms and social groups a conviction that such phenomena must be regarded as unitary wholes in some important sense, we should be concerned about the meaning and validity of this doctrine in the study of different social organizations. In what sense is an individual ant a unitary whole, or an ant colony, a species, a human city, nation, or a "United Nations"?

SUPERORGANISM CONCEPT AND ANALOGY

The first stage of comparative study generally involves a focusing upon similarities between observed phenomena. It is thus convenient to open our consideration of social-group comparisons with the concept of "superorganism" as it has been developed by certain students of insect behavior (Wheeler, 81, 84; Emerson, 31). This concept derives from an old notion entertained by Plato and Aquinas, that a society exhibits the principal attributes of an individual and may be considered a superior type of individual. In the writings of Wheeler and Emerson we find the idea developed to a high degree of elaborateness and in great detail as a technique for analogical study of organic systems. How interesting the study of insect social attributes becomes when approached from this point of view, may be represented by a brief sketch of Emerson's description of the insect "superorganism." His scheme of the insect colony as "superorganism" in relation to organisms and other "superorganisms" on different levels rests upon the premise that scientific methodology is fundamentally the same whether it applies to the mechanisms of the individual organism, the insect society, or human society. A careful comparison and correlation of these levels of organization leads, he believes, to the formulation of important principles; thus a detailed examination of the insect society leads to the discovery of many parallels to the properties of organisms. Some of these parallels we may outline as follows, condensed from his detailed survey (31):

1. Division of labor: social insects commonly have castes, such as the differentiation of fertile individuals from infertile workers, analogized to the differentiation of gametes and somatic cells in the individual organisms; or the special function of infertile workers may be analogized to the function of cells in the individual's gastrovascular tract.

2. Intersectional communication: insect communities are aroused by an interindividual transmission of impulses (*e.g.,* excited termite soldiers rap their heads against wooden gallery walls, thereby exciting others; ants transmit excitation to nestmates by touching antennae), and such effects are compared with the intra-organismic transmission of nervous impulses, or to a chemical transmission among tissues through the blood stream.

3. Rhythmic periodicities: social insects characteristically produce their sexual forms seasonally, a process compared with the rhythmic production of gametes in an individual organism.

4. Life cycles: birth, growth, and senescence in the individual organism are paralleled by colony foundation, expansion, and degeneration in social insects.

5. Organismic and superorganismic phylogeny: Emerson believes that the above similarities and others are not accidental, but that in the two cases natural selection operates in comparable ways, eliminating given mechanisms (*i.e.,* of individual structure or those underlying group functions) according to their relative adaptive value.

Thus it is Emerson's conviction that the method of analogies is a valid way of studying organization on different levels, for in his view essentially the same biological processes have accounted for the evolution of organism and superorganism alike. He points to the comparable operation of selection as a factor in the evolution of individual and group at various integrative levels as a working assumption used by many biologists (*e.g.,* Maidl, 49; Sturtevant, 73), and by Fisher (35) as a biometrician. However, the manner in which natural selection operates at different levels is very incompletely known at present. Simpson (71), for instance, has insisted that the superorganismic-organismic analogical procedure cannot be justified beyond the field of observation, for individual and social group are entities in very different senses of the word, differently subject to selection in evolution. It is questionable whether the deductive leaps of the analogy method provide a legitimate means of studying either the past history of animal societies or the nature of their present integrations.

CRITICAL DIFFERENCES BETWEEN LEVELS

Although it may be granted that the superorganism concept is a serviceable device for teaching the observed characteristics of organism and of group levels, beyond the unsettled question of whether similar evolutionary processes actually have been involved there is the further question, upon which we wish to concentrate in this discussion, whether the organization of different levels as they exist really involves similar factors (biological and psychological) of equivalent significance in the respective cases. In a recent critique of the superorganismic concept, Novikoff (58, 59) has concluded that it is inadequate both for discovering the actual nature of qualitative differences among levels of organization and for revealing the nature of the part-whole relationship in each given level. Although many students believe, as does Collias (23), that "integrated systems at one level may themselves be units in a more inclusive grouping," we may ask whether lower-level processes *as such* are really incorporated into higher levels of organization. To what extent are lower-level principles adequate for the understanding of higher levels? In what sense is the insect colony a *lower*-level phenomenon and human society a *higher*-level phenomenon, and to what extent does a study of the former contribute to understanding the latter?

While it is true as Gerard and Emerson (39) maintain that the method of analogy has an important place in scientific theory, its usefulness must be considered introductory to a comparative study in which differences may well be discovered which require a reinterpretation of the similarities first noted. As Emerson (32) has admitted, striking differences are found between insect and human society which may be too great to make the analogy significant. Perhaps the most significant differences come to light in a closer examination of some of the analogies described above. Outstanding contrasts are found especially in the nature of communication, in the occurrence of castes, and in "tradition."

The essential characteristics of communication in human society are (1) its symbolic and conventionalized character, (2) its directive function (*i.e.,* using symbols to influence others), (3) its intentional use in social situations, and (4) its capacity for arrangement or rearrangement according to the requirements of meaning (cf. Bierens de Haan, 9). These characteristics are mastered in the socialization of the child, and are apparent both in linguistic and written communication, and in necessary modifications such as the finger codes of speechless individuals. In the social insects not one of these characteristics has been demonstrated in careful studies such as those of von Frisch (37) on bees, Eidmann (29) on ants, and Emerson (30) on termites. Instead, there is a direct transmission of excitement from individual to individual through antennal (and sometimes also front-leg) stroking as in most ants, special air-transmitted stridulatory vibrations in ants such as the tropical American *Paraponera clavata,* and through body vibrations transmitted via the substratum as in termites. These effects are not codified in any real sense, do not convey "information," and lack any directive effect in themselves. If the recipient is influenced in any describable way beyond being aroused to some activity, it is because of incidental stimuli such as the chemical effects described by von Frisch among flower gatherers, the existence of a chemical trail as Eidmann demonstrated, or because of previous individual learning (Maier and Schneirla, 50). Although the acquisition of the transmissive function by an individual appears to require a simple initial process of learning, which may occur through a modification of the first feeding reactions as Heyde (42) found in ants (Maier and Schneirla, 50), it cannot be regarded as psychologically comparable to human language acquisition (Révész, 62).

The occurrence of castes is a common characteristic of insect societies (Wheeler, 84). In addition to the sexual dichotomy of male and fertile females, there may be structurally different sub-types in the caste or castes of completely infertile or only partially fertile individuals. In many ants there occur polymorphic differences among the workers, sometimes in a continuous series from workers major on one extreme to workers minor on the other as in the leaf-cutter ant *Atta cephalotes,* sometimes in a discontinuous series with major and minor workers only. Because of their structural differences the various polymorphic castes differ in function: males and fertile females (queens) in sexual function, the workers serving as colony defenders in the case of the majors, as brood-tenders within the nest in the case of the minors, and the intermediates typically as foragers in the surrounding area. In the ants the basis of castes is typically biologically established, as follows,— queens appearing from overfed fertilized eggs and males from unfertilized eggs, the various polymorphic worker castes from fertilized eggs which have

been underfed to different extents. The differences in feeding which determine whether a queen or a worker type will develop from a fertilized egg are effective during the larval stage, and when the individual emerges as a young adult, fully grown at birth so to speak, its functional capacities may be predicted on the basis of its organic makeup. Thus through very long series of generations in a given species, the caste functions of individuals are stereotyped, and essentially changeless unless new developments occur through genetic evolution.

On the human level, in contrast to insects, castes in the sense of social or functional classes of different social ranks exist essentially on a psychological basis. The one clear biological foundation for a human functional "caste" differentiation is a sexual dichotomy, in which the organic differentiation of sexes imposes a qualitative differentiation of general reproductive function. It does not, however, impose any inevitable behavioral or psychological differentiation of social function; these are matters which anthropologists find attributable to social heritage and cultural pattern.[1] The social emancipation of women in relatively recent times, and their successful entrance into a variety of social and professional functions formerly considered the nature-given right of man alone, offer countless examples of the relatively limited extent to which human sex biology *in itself* channelizes social function. In insects, biological factors alone determine the channelizing (Schneirla, 66). Male ants or drone

[1] That is to say, genetico-physiological differences such as muscular strength existing between the sexes would facilitate certain differences in social function, but such differences are quantitative or relative rather than absolute, and are subject to alteration under appropriate social conditions.

bees lack any organic basis for brood-tending or foraging, which are set functions of the hereditarily and trophically differentiated workers; whereas human males under appropriate conditions may even excel females in baby-tending, in cookery, and in almost the entire range of domestic functions except giving birth.

In man, castes are not biologically differentiated in fixed ways as in insects, but exist essentially on an ideological and traditional basis. The fact that, whereas in the social insects germination has become highly centralized in a few individuals, in man it is a general property of individuals, Crowell (27) believes, is a biological characteristic contributing in man to a generalization of individual function in the family and hence in the social sphere. Thus the generalization of human reproductive functions throughout the population, other things being equal, would militate against a generation-to-generation stereotypy in human social function, facilitating the influence of the major biological factor basic to plastic adjustments—cerebral cortex.

It is often asserted that, while evidence is lacking for an anatomical specialization of human castes comparable to that of insects, an equivalent differentiating factor exists in the organic basis of intelligent behavior. For instance, R. B. Cattell (20) has reported intelligence-test results which differ according to the social level of the subjects, with the implications that biological factors necessarily play a maximal role and social factors a minimal role in accounting for the differences. Without adequate control (e.g., special tests) of the factors pertaining to social and economic background which are known to influence intelligence-test performance, such conclusions are widely open to question. As long as a satisfactory technique for such controls is lacking, scores on intelligence tests can have no clear meaning as to what native differences may exist among cultural, professional, or class groups (Neff, 55; Loevinger, 48; Mann, 53). In contrast to the stereotyped organic basis of insect social castes, the ·differentiation of human social castes appears to be dominantly influenced by essentially non-genetic factors.[2] However, by introducing vaguely hypothecated native factors, theories of "social instincts" such as that offered by McDougall tend to obscure the *relatively* homogeneous biology of the human population, and its great psychological plasticity as evidenced by the shifting of individuals across class lines. Such theories, as Brown (13) makes clear, stem from the ideology of a particular caste rather than from a scientific study of social organization. It is definitely not established that native (*i.e.,* biological) factors play any major role in restricting human societal differentiations as found in the castes of India and the class hierarchies of Western countries.

A third principal difference between insect and human societies arises through man's extensive human capacity for learning and reasoning based especially upon the elaboration of cerebral cortex in evolution. Dependent upon the use of these capacities through many generations but within a relatively short space of time as compared with insect social evolution, mankind has worked out many highly diversified societies, with very different

[2] This statement is of course consistent with the view that genetically based individual differences in intelligence also play a selective role, according to the relative strength of other factors. The role of the diversified procedures which have been in social practise from ancient times to the present to limit or block upward shifting across the traditional class lines of a given society has been discussed by Veblen (78) and numerous other writers.

forms of organization, institutions, and traditions (Benedict, 8). In each cultural setting man exhibits somewhat differently his capacity for transmitting the conceptualized traditions of previous generations to his descendants; yet all are alike in possessing some form of "social heritage."

Moreover, the members of each generation may increase, change, or even displace given aspects of the social heritage, rather than merely transmit the given system passively. Men are capable of a degree of plastic learning and reasoning which inevitably dynamizes and revises this process for better or worse. Furthermore, on the human level, not only do changes occur in the heritage of a given society on an internal basis, but also through interactions among societies (Malinowski, 51). Insect societies, on the other hand, appear to be limited essentially to passing on given social patterns on a biological basis, with changes appearing only through genetic evolution. Thus Wheeler (84) suggests the probability that the principal patterns of insect social behavior now extant were in substantially their present condition in remote Tertiary times. This is because insect social transmission is gametic transmission, and changes can be effected only through that process. A learning capacity is present, but ineffective in such matters. In certain genera of ants such as Formica the workers are capable of mastering fairly complex maze patterns (Schneirla, 64), and similarly they learn individually different routes in foraging outside the nest under natural conditions. Yet the learning process is stereotyped and rote in character, and as a process is limited to the individual and to the given situation (Schneirla, 68). Consequently new advances by the individual in learning a route contribute substantially

nothing to the colony except additional increments of food, and the special learning of each individual dies with it. The activity of an insect colony, beyond furnishing food which maintains the population, offers little to further generations except an existing nest structure and the effect of certain simple interindividual behavior patterns, contributions of minimal plasticity rather directly depending upon the germ-plasm pattern of the social species. To know the contribution of a given ant colony to its brood, one need only know the taxonomic group and the given environment, and not the century in time beyond the Tertiary period; to know the contribution of given human parents, it is necessary to know very specifically not only the year and locality of residence, but also innumerable data concerning the given family, cultural influences, social affiliations, and the experience background of various individuals involved.

These differences are far-reaching ones, and in view of them we may say that, whereas social insects are biosocial, man is psycho-social. The differences appear to be far more significant than the similarities for understanding the characteristic nature of the respective phenomena, and the procedure of analogizing involved in applying the concept of superorganism consequently appears to be misleading for analytical study. The difference in interpretation, it should be remarked, is not one of mere emphasis but of logical procedure in studying the evidence. But fortunately, Wheeler (83) and Emerson as serious students of insect society have not employed the superorganism concept rigidly. Although Emerson (33) finds "a remarkable similarity between insect and human social systems," he also remarks: "In spite of many common analogous attributes, human society shows fundamental differences from insect social organization and these also must be analyzed for a proper perspective." He recognizes important differences in intelligence, leadership, the human use of symbolization, and in the fact that "the human species is the only organism which has developed an additional mechanism supplementing such biological heredity" (*i.e.*, genic transmission). Yet, having noted these differences he returns to emphasize the analogues, as follows:

The development of human social heredity through learned symbols is of such importance that this human attribute would seem to indicate the valid division line between the social and biological sciences. . . . However, even though techniques may differ and phenomena are diverse in numerous instances, I personally believe that scientific method is fundamentally the same whether applied to human social mechanisms or insect social mechanisms or whether applied to the social supraorganism or to the individual organism. In spite of the real differences between the societal types, careful comparison and correlation leads us to the formulation of important principles. (33, pp. 168–169)

But instead of relying upon a method of analogy in studying social levels, stressing apparent but unclear similarities, it is preferable to compare phenomena by looking for the basis of both similarities and differences and endeavoring to emphasize these according to their respective importances. If biological analogues are "similarities in the function or use to the organism" (Boyden, 12), it seems necessary to find how far-reaching or how limited these similarities may be before they can be used in explaining social phenomena. For example, "communication" on the insect and human levels appears to be sufficiently different, both in its mechanisms and in the qualitative consequences of its function in social organization, as to require different conceptual terms in the two instances. In view of the very basic psychological differences which exist between the two processes, it seems preferable to use a term such as "social transmission" for interindividual arousal in insects, reserving the term "communication" for higher levels on which a conceptual process of social transmission is demonstrable. The similarity between these processes appears to have only a minimal and an illustrative, descriptive importance for theory.

A FUNDAMENTAL BIOLOGICAL FACTOR PROMOTING GROUP ORGANIZATION

Thus far our discussion has tended to be particularistic or "reductionistic" (Sloane, 72), in that describing characteristics of similarity or difference serves to emphasize part-processes and not unity of the social group, so far as the group is organized as a unit. Now we must look into the nature of organization process. Wheeler (84) offered an important contribution toward understanding insect social organization in his concept "trophallaxis," which signifies the reciprocal exchange of food or of equivalent tactual and chemical stimulation among individuals in a colony. An approach or "turning-toward" response is established to such stimuli, because of their adequacy to elicit (*i.e.*, to *force* organically) the apppropriate reflexes in dependence upon the conditions of presentation. Thus the queen feeds her first young larvae not because of a "maternal-care instinct," but because of larval secretions which effectively attract her to the larvae and which she licks up readily. The workers, while tending queen, eggs, and brood, lick from the integument of these other individuals fatty exudates and salivary secretions, or regurgitate food as a reflex reaction to the antennal strokings of a nestmate.

Wheeler (82) also presented abundant evidence for the existence of trophallactic relations between the ants and other insects (*e.g.,* various staphylinid beetles) as "guests" in the colony, the latter eliciting a regurgitation of food through tactual stimulation as would nestmates, and attracting the ants through special glandular secretions or exudates. Thus the relation between ant hosts and "symphiles" is insured by much the same factors which are responsible for the existence of basic colony organization.

The process of trophallaxis is based upon the presence of inherited biological factors which insure simple stimulus-response relationships among individuals. In ants the behavior is not inherited in the sense that it is present as an adult pattern when the callow worker appears in the nest. It is probable that a rudimentary conditioning process is involved in the social adaptation of the individual as larva or as callow to the given species-colony chemical which is prevalent in the environment during feeding and equivalent activities. A simple learned basis is thereby provided for more versatile approaches to nestmates, following of colony chemical-trails, and returns to the nest as a foraging adult. This postulation (Schneirla, 66) is based especially upon observations by Heyde (42) of early behavior in young ants, and upon the investigations of Thorpe (75), which demonstrate olfactory conditioning in larval insects as an influence upon adult behavior. As a consequence of the inheritance by its individuals of the stable biological factors underlying trophallaxis, the insect colony acquires a unity which persists throughout the life of the group (Bodenheimer, 11), with disunity occurring only under exceptional conditions involving incidental changes in the trophallactic factors themselves.

It will be recognized that the trophallaxis process depends upon general biological factors somewhat comparable to those involved in the formation of simple and temporary aggregations, the incidental or sub-social groupings ("associations"—Alverdes, 5; Allee, 1) established through the independent approach responses of numerous related individuals to a common external stimulus (*e.g.,* temperature). It is the hereditary specialization and diversification of such responses, as dependent upon particular types of stimuli from other individuals, and their generalization from early reflex-like reactions through simple habituation-learning, which accounts fundamentally for the integrity and persistence of the complex insect social organization. Throughout, the dominance of specific biological factors is apparent; thus the pattern of colony organization retains an essential insect-like constancy almost indefinitely in biological time, unless biological evolution intervenes.

TROPHALLAXIS AS INVOLVED IN HUMAN SOCIALIZATION

A somewhat comparable process of trophallaxis is basically involved in the socialization of the human infant. From the infant the mother receives agreeable stimulation, especially tactual stimulation (Allport, 4), as well as physiological relief exemplified by the effect of suckling, which affords not only agreeable and even erotic sensory effects but also relief from painful tension of the mammary glands and thus is sought by the mother (Ford, 36). Carpenter's (18) discussion of mother-infant relationships in the chimpanzee is enlightening in this connection. In the lower primates and other mammals the mother also receives a chemo-stimulative and physiological gratification through consuming the afterbirth, and licking and handling the young (Tinklepaugh and Hartman, 76, 77; Yerkes and Tomilin, 87). In the primates such intense gratification derives from holding and cuddling the young that in captivity it is sometimes impossible to constrain the mother to release a baby dead for hours or even for days (Lashley and Watson, 46). These and other physiological factors of direct stimulation and sensory gratification to the mother insure maternal orientation toward and psychological attachment to the young in most of the mammals, in which the duration and quality of parent-young association appears to be an important factor contributing to the degree of social organization (Darling, 28; Scott, 69). On the human level of course such developments usually advance to a high degree of elaboration, facilitated by an anticipation of the offspring and a psychological preparation for its arrival.

Because of the human mother's perceptual schema of the process of childbearing and care—typically influenced as it is by a great variety of effects from tradition and social experience—from the beginning the trophallactic relationship has a psychological setting of great importance. This means that by no means can the process of trophallaxis be termed with accuracy "the same process" on the ant and human levels. There is little doubt, however, that in mammals generally the entire maternal process is based upon and energized by the physiological and trophallactic processes, especially since the mother's responsiveness to young generally is intensive even in the primiparous females of lower mammals incapable of psychological preparation for the event and its significance (Wiesner and Sheard, 85; Beach, 7; Cooper, 25).

On the side of the infant human individual the trophallactic processes provide a necessary and effective basis for the development of a psychological affiliation with the mother and, through her, with society. "The child is part of his mother before he becomes an individual for himself and is part of a definite group for a long time before he can enter and join any group actively" (Bühler, 15). We cannot hope to do more here than to sketch the general outlines of the early socialization process deriving from this physiological foundation. Initially the infantile behavior is essentially unpatterned and on a reflexive basis (Pratt, Nelson, and Sun, 60), and those crude differences which may be identified in the infant's overt responses are attributable to the direct physiological *energy* effects of stimulation rather than to the nature of the given situation as adults know it. Under strongly stimulating conditions, as when extreme visceral tension exists or abrupt or intense extrinsic changes come about, vigorous mass action and usually crying result. Two incidental but important characteristics of such reflexive behavior should be emphasized. First, it serves to force the infant and his needs upon the mother and society, at first quite unintentionally of course, later more systematically through learning, thus coming to have a crude "signal-function" in the trophallactic process. Secondly, the mass-action response, although highly variable, also is characterized by a doubling up or flexion of the limbs. Predominance of the flexion response when the limb itself is intensely stimulated locally (Sherman and Sherman, 70) shows the forced reflexive character of the response, and also emphasizes its adaptive character, a getting-away-from the stimulus source (Goldstein, 41). Although at first this is a purely physiological factor, through it in the course of time specialized withdrawal reactions develop to particular objects and situations in which intense stimulation has been encountered (Schneirla, 65). Thus what is initially a purely *biological* part-process provides the basis for learned avoidances and later for negative attitudes in the social situation—a psychological process.[3]

More important for the early trophallaxis process is the fact that weak stimulation, such as the stroking of skin or relieving organic tensions, exerts a soothing effect and relaxes the infant.[4] Not only that, but the reflexive physiological effects under these conditions appear to involve an extension or abduction (i.e., a "getting-out") of the limbs as frequent component. At first reflexive, this response through conditioning becomes transformed into a specialized *reaching toward* or holding to objects or situations which have had soothing or tension-relieving effects. Thus holding out the arms toward the mother indicates *(to her)* that the infant needs her, and is an important objective sign of a growing specialized attachment to the mother and what she represents. *This is the nucleus of the infant's training to approach things, situations and people,—first by reaching out, then by crawling and walking toward, and finally by positive attitude and conceptual expression indicating approval of given aspects of society* (Schneirla, 65). In the early stages of this social-training process the initial

reflexive biological part-process becomes socially specialized and rather completely transformed into an externalized, psychologically organized system. The nature of this system, soon greatly expanded through selective learning rather than simple conditioning as at first, reaches conceptual status and comes to vary according to the relative emphasis of the given social and cultural environment upon things-and-people-and-ideas-to-be-approved.

DIFFERENT ROLES OF TROPHALLAXIS IN ANTS AND MEN

In comparison with this highly specialized psychological outcome of early trophallaxis on the human level, the stereotyped social repertoire appearing in the young ant is rudimentary and limited, dominated throughout as it is by native factors. The social outcome of insect trophallaxis is largely set by hereditary factors; the social outcome of human trophallaxis is highly variable and plastic, in dependence upon the given cultural setting. On the human level the close trophallactic bond first established with the mother, family and immediate social environment expands into diversified and widened social relationships in very different ways according to what social agencies canalize and redirect the organization process. The trophallaxis process, a biological factor which is initially somewhat similar in the insects and in man (more on the human infant's side, and far less on the mother's) has a very different significance in the two cases for the eventual social pattern. In both cases it is a central factor in permitting the rise of a group organization and in molding the individual into the existing group organization, but in highly different ways as to how far the strictly biological processes are modified.

We are led to the conclusion that

---

[3] In thus differentiating biological and psychological processes, it is not intended to suggest that psychological processes are non-organic in the last analysis; rather, that psychological functions represent the elaboration and reorganization of the biological on a qualitatively new and higher level.

[4] Such effects were observed in the early studies at Johns Hopkins and were labeled "love" by Watson (80), although not at all "love" in the adult sense of an emotional attitude.

although important biological processes such as trophallaxis are represented both on lower and higher levels of social organization, their significance may vary greatly according to capacity for learning and in dependence upon factors influencing learning in the social situation. If space permitted, a similar conclusion might be reached from an analysis of individual "drives" or tissue needs in relation to social participation. Their role in insect social behavior is relatively stereotyped; on the human level it is plastic and highly modifiable. In man to be sure they are also fundamental, since tissue needs exist and compel adjustments, but the kinds of adjustments made depend widely upon the particular socialized selective-learning process involved. The socially acceptable satisfactions under given cultural conditions may be very different in nature from those tolerated and approved under other conditions. In all human societies most of the socially acceptable incentives appear to vary more or less widely from objects and processes which directly relieve drive tensions. In the socialization of the individual, his drive-reactions are greatly modified and elaborated through taboo and custom. Thus the relative importance of the principal types of incentives (sustenance, security, acquisitiveness, dominance, and social approval) may vary greatly among human societies and among different sections of the same society (Kornhauser, 44). Maller (52) found American school children from poorer neighborhoods somewhat less acquisitive and more cooperative in classroom situations than were children from well-to-do neighborhoods; and anthropologists (Goldman, 40) find some human societies (e.g., the Zuni) predominantly cooperative in their internal organization, others (e.g., Kwakiutl) predominantly competitive. Essentially similar biological part-processes are involved in these different human instances, but are metamorphosed very differently into psychological relationships according to the predominant pattern. The role of trophallaxis is so different in the socialization of insects and men that, in the final analysis, it cannot be considered the *same* process by any means. On the higher level it has a qualitatively distinct significance as compared with the insect level.

"DOMINANCE" CONCEPTS INADEQUATE FOR STUDYING ORGANIZATION

The role of trophallaxis has not been explored in the general study of vertebrate social organization. Instead, both investigation and theory have featured "dominance" relationships as presumably the most essential factor. As described by Schjelderup-Ebbe (63) in his classical studies with newly assembled groups of barnyard fowl, a social rank or dominance hierarchy is established in time by dint of reciprocal pecking among individuals. After an initial period of fighting and social instability, a ranking is established in which (in the simplest case of a linear peck-order) hen No. 1 pecks all of the others but is not pecked in turn, No. 2 pecks all but No. 1 and is pecked only by No. 1, and so on to the most subordinate member which is pecked by all and may never peck any of the superior members. Dominance orders have been described for a considerable number of bird species and for various other vertebrates including some primates (Allee, 2). Typically, dominance order is considered dependent upon aggressive behavior (Collias, 23); for example Maslow (54) defines the dominant animal as "one whose behavior patterns are carried out without deference to the behavior of his associates."

However, the term does not always have this meaning and in general is not used very consistently. For instance, Noble, Wurm, and Schmidt (57) used the relative height of bill-holding between the members of a heron pair as an indication of dominance, on the assumption that this indirectly resulted from superiority in food-snatching as nestlings. And Maslow, who lays much stress upon dominance in studying primate social relationships, as the above definition suggests, appears to lapse both from consistency and from clarity when he states that "dominance in the chimpanzee is mostly of a friendly kind. . . . The dominant chimpanzee (at least in young animals) is a friend and a protector of the subordinate chimpanzee. . . . They form a close contact group and are dependent on each other. . . ." "Dominance in the macaque is usually brutal in nature, and . . . dominance in the cebus is in the first place tenuous and in the second place relatively non-contactual" (54, pp. 314–315). What these diversified relationships have in common would seem better characterized as "ascendancy," a behavior trait not necessarily dependent upon aggressiveness.

It is highly important to note that describable dominance hierarchies appear under rather special conditions, particularly when groups of birds or primates are confined within a small space, when incentives (i.e., food or drink) are restricted in quantity or in accessibility, or when sexual responsiveness is high. Moreover dominance relationships do not always stand out in grouping lower vertebrates, even in groups of fowl and especially when the groups are large, when social organization displays different characteristics (Fischel, 34). In certain animals studied carefully under field conditions, as for instance the howler monkeys

studied by Carpenter (16), an efficient group organization exists without signs of intra-group dominance or aggressive relationships, whereas in others (e.g., baboons) aggression-dominance is prominent (Carpenter, 17, 19).

We must seriously entertain the possibility that dominance theory is an inadequate basis for the study of vertebrate social behavior. "Dominance" appears to be just one characteristic of social behavior, sometimes outstanding in group behavior and sometimes not, according to circumstances. Since the true "dominance" relationship is one of real or abbreviated aggression and withdrawal (Collias, 23), dominance must be considered a factor promoting the isolation and greater psychological "distance" of individuals, and thus more or less counteractive to factors which hold the group together. Perhaps a dominance situation may be viewed as one in which positive unifying factors are relatively weak but are somewhat artificially reinforced by special conditions, such as food scarcity or sexual receptivity (which serve to heighten reactivity to specific stimuli from other individuals), or by physical confinement of the group.

Factors related to social facilitation (Allee, 2, 3), and probably based upon original trophallactic relationships of one kind or another, may well furnish the major unifying basis in groups of lower vertebrates (e.g., schooling fishes), as seems to be the case in the social insects and in mammals. As an alternative to emphasizing dominance in adult groups, it would be well to examine carefully the nature and persistency of early contacts with one or both parents during the period of incubation in relation to "gregarious tendency" in later life. In the domestic fowl, as Brückner's (14) study shows, an intimate trophallactic relationship first exists between hen and chicks, only to be displaced after a few weeks by a condition in which the hen drives off the young. Comparably in mammals, the age of young when weaning occurs, if this change abruptly enforces separation from family conditions, presumably has an influence upon subsequent readiness to group and upon group behavior (e.g., the prominence of interindividual aggressiveness).

In a real sense, aggressive or dominance reactions are an indication of weak social responsiveness either because of individualistic reactions (e.g., sexual reactions) or an incompletely established group organization. Actually, interindividual adjustments in the formation of a new vertebrate group often pass from a stage of overt aggression to one in which tolerance reactions and social facilitation exist (Taylor, 74; Collias, 23). In different animals, the readiness with which early aggressive reactions change into qualitatively different relationships permitting a closer group unity may be opposed to different extents by the individualizing influence of factors such as sexual responsivenes, but otherwise may depend upon species capacity for modifying behavior. Bard (6) and Collias (23) have remarked that as cerebral cortex increases in the animal series, aggressive dominance relationships appear to drop back as prominent characteristics of social behavior. The importance of interindividual grooming, clearly a trophallactic unifying factor of compelling force in primate groups, has been emphasized by Yerkes (86). An infant chimpanzee first raised in isolation from others of its kind (Jacobsen, Jacobsen, and Yoshioka, 43), was at first aggressive when placed with a strange young chimpanzee, but within a few weeks there developed a pacific relationship of mutual dependence between the two.

Concepts such as dominance hierarchy tend toward a particularistic, static type of thinking about social organization, actually distracting attention from the essential problem of group unity. In studying social organization on any level, a theoretical procedure is desirable which centers around the conception of a dynamic integrative process rather than given characteristics such as dominance which may be sometimes absent. We believe that a more thoroughgoing ontogenetic survey of social behavior in the vertebrates will reveal the prerequisite importance of trophallactic processes for group unity wherever it occurs and whatever its strength.

CONTRASTS IN "COOPERATION" ON DIFFERENT GROUP LEVELS

The probability that trophallactic relationships intimately underlie intragroup approach reactions on all social levels suggests the importance of the principle of cooperativeness, elucidated by Kropotkin (45) as a factor in evolution. But of course the outcome in the group behavior pattern is very different according to the general level on which this factor operates. Some comparisons may be suggested.

Trophallaxis may be represented on its lowest or purely biological level by the common benefiting of individuals in an assemblage of lower invertebrates through the biochemical products of their own activity. Allee (1) cites many examples of such incidental byproducts of sub-social associations. It is clear that the result is an *incidental physiological facilitation of individual* metabolism, without any consequent elaboration of interindividual adjustments through learning. Such an elaboration occurs to a limited extent in the social insects, not only on the basis of hereditary specializations promoting such behavior, but also by virtue of a limited extension of interindividual responsiveness through social conditioning. Group function as a dynamic process thereby acquires a greater persistence and is extended into important functions such as reciprocal feeding and foraging; however, the psychological poverty of the process has been suggested above in our discussion of insect "communication." It is probably more accurate to term the capacity for mutual assistance in social insects *bio-social facilitation,* rather than "cooperation," because of these psychological limitations.

In contrast, *psycho-social cooperativeness* as it is found on the primate level involves an ability to anticipate the social consequences of one's own actions and to modify them in relation to attaining a group goal. The capacity for such behavior does not appear automatically, merely through the possession of cerebral cortex. The basis for some form of "cooperative" behavior in all probability is laid in early trophallactic relationships in family or intimate groups, as suggested above. The social consequences of this foundation may vary according to a variety of circumstances. In the social insects, the predominance of genetic factors in the socialization process assures the appearance of a highly stable "cooperative" pattern of very low psychological calibre. But in the vertebrates, such a pattern is by no means inevitable. The individualization of sexual responsiveness and other genetically contributed or learned factors may reduce its prominence greatly, as shown in dominance patterns.[5] The

evolution of cerebral cortex through the mammalian series admits plastic learning as an increasingly important influence, and makes possible greater versatility in social patterning according to ontogenetic circumstances. Thus Crawford (26) has demonstrated the development through learning of cooperative relationships in the problem-solving and food-sharing of chimpanzees, a process very different in its psychological nature from the relatively stereotyped interindividual facilitation process of social insects, and qualitatively superior to the insect system. But even where cortex is most highly evolved, in man, it must be recognized that the predominance of a trend toward higher-type cooperation patterns is by no means inevitable. It is because man's group behavior depends mainly upon elaborated psycho-social patterns of motives, attitudes and purposes and plans, and is not determined in its *patterning* by biological factors, that his societies are qualitatively distinct from those of insects.

It is frequently stated that, as Novikoff (58, 59) puts it, ". . . higher level phenomena always include phenomena at lower levels. . . ." This statement holds only if it is recognized that phenomena appearing in direct relation to biological processes on lower levels are not represented in the same fashion on higher levels, but there may influence group patterns very differently as they are modified, transformed and elaborated in new and diversified ways. They are not really analogues in a logical sense, for as we have found they do not have the *same* functions on different group levels except in a teleological and descriptive sense.

---

[5] The appearance of dominance behavior in social activities is related by Collias (23) to the function of male sex hormone in particular. However, in a recent study of dominance behavior in chimpanzees, Birch and Clark (10) have been led to a different view on the basis of their finding that not only male hormone but female hormone as well (when in the necessary replacement amounts) (Clark and Birch, 22) can account for the appearance of dominance behavior.

CONCLUDING REMARKS

To sum up, although similar biological factors underlie group unity on various levels of organization, these factors are not identical even in their basic or initial form, and their eventual significance for social patterning varies greatly from level to level. Analogical procedures such as those involved in using the "superorganism" concept are not adequate for studying social levels comparatively, because such procedures become preoccupied with general similarities rather than working toward an understanding of group unity through an evaluation of social similarities and differences.

Underlying the appearance of group unity on all levels are biological factors contributing to the facilitation of primary interindividual stimulative relationships, which, after Wheeler (83), we have termed *trophallaxis*. On different levels these factors are roughly similar in their general significance for grouping. Yet when we compare the insect social level with the simple-aggregation level on the one hand and with the human social level on the other, critically important differences are found in the basic form of trophallaxis and in its functional potentialities for group behavior. In view of qualitatively different consequences evident in the group activities of different animals, the "cooperation" factor as generalized by Kropotkin and Allee is subject to revision. It is necessary to differentiate interindividual relationships as "physiological facilitation" and as "bio-social facilitation" on the levels of simple aggregation and insect colony, respectively, and as "cooperation" on the psycho-social level typified by human society. In the light of this necessary conceptual differentiation, we are prepared to recognize the existence of a qualitatively different process of individual socialization on the human level, influenced very differently by psychological factors according to cultural pattern and social heritage, rather than in dependence upon the direct function of hereditary organic agencies as on the insect level.

The psychological plasticity contributed by human cortex admits the possibility that the role of the basic physiological factors (trophallaxis and drives) may be very different according to circumstances. We have endeavored to sketch a theoretical explanation of the manner in which individual socialization involves a different transformation and elaboration of the physiological factors into psycho-social patterns under different social conditions. As an analytical statement and not simply a value judgment, it may be said that from the standpoint of breadth of social organization and multiplicity of interindividual relationships a cooperative pattern (rather than a "dominance-hierarchy" pattern) represents the fuller attainment of human psychological resources. Dominance factors which emphasize individualistic motivation represent only a partial realization of group resources, on a lower psychological level, on which the clash of different sub-group motivations increases intra-group conflict and promotes tensions which make for social disorganization (Galt, 38). In criticizing "dominance" theory as applied to the vertebrates, we called attention to its reductionistic emphasis upon just those special characteristics which are not essential to group unity but in fact tend to operate against and reduce the psychological level of group integrity. For this reason in particular, "dominance" theory cannot be regarded as adequate for a comparative psychology of social organization.

Although insect-human analogies

may offer an interesting way to introduce social comparisons, as investigation advances we must pass from them to a study of qualitative contrasts, if types of organization are to be understood and their relationships adequately evaluated. Superorganism theory and dominance theory have a common weakness, in that both emphasize particular group characteristics without directing attention toward the nature of group unity and toward conditions which favor or oppose it. Studies with chimpanzees (Crawford, 26; Nissen and Crawford, 56) show that infrahuman primates plastically acquire cooperative relationships under appropriate conditions, working purposively together toward common goals. Under other conditions, as we have seen, "dominance" aspects may be emphasized in interindividual behavior.

Similarly, the studies of Lippitt (47) with boy groups show convincingly that human behavior patterns and attitudes toward cooperative participation differ greatly according to whether the prevailing social climate has encouraged and facilitated the formation of a democratic or an autocratic group organization. Under the former conditions, individual psychological participation is considerably wider in its group references. Such studies assist social scientists to discern and emphasize the procedures which best realize the qualitative superiority of the human social level over sub-human levels. For although an insect society is limited to the bio-social level of organization, its genetically canalized social pattern considered as an adaptive device is obviously superior to a human social climate in which group members are taught not group participation for group goals but sub-group motivation and schismatic attitudes.

BIBLIOGRAPHY

1. ALLEE, W. C. *Animal aggregations.* Chicago: Univ. Chicago Press, 1931.
2. ALLEE, W. C. *The social life of animals.* New York: Norton, 1938.
3. ALLEE, W. C. Social biology of subhuman groups. *Sociometry,* 1945, 8, 21–29.
4. ALLPORT, F. H. *Social psychology.* New York: Houghton Mifflin, 1924.
5. ALVERDES, F. *Social life in the animal world.* New York: Harcourt, Brace, 1927.
6. BARD, P. Neural mechanisms in emotional and sexual behavior. *Psychosomatic Med.,* 1942, 4, 171–172.
7. BEACH, F. A. The neural basis of innate behavior. I. Effects of cortical lesions upon the maternal behavior pattern in the rat. *J. comp. Psychol.,* 1937, 24, 393–439.
8. BENEDICT, R. *Patterns of culture.* New York: Houghton Miffln, 1934.
9. BIERENS DE HAAN, J. A. Animal language in its relation to that of man. *Biol. Rev.,* 1930, 4, 249–268.
10. BIRCH, H., & CLARK, G. Hormonal modifications of social behavior: II. *Psychosomatic Med.* (in press).
11. BODENHEIMER, F. Population problems of social insects. *Biol. Rev.,* 1937, 12, 393–430.
12. BOYDEN, A. Homology and analogy. *Quart. Rev. Biol.,* 1943, 18, 228–241.
13. BROWN, J. F. *Psychology and the social order.* New York: McGraw-Hill, 1936.
14. BRÜCKNER, G. H. Untersuchungen zur Tiersoziologie, insbesondere zur Auflösung der Familie. *Z. Psychol.,* 1933, 128, 1–110.
15. BÜHLER, C. *The social behavior of children.* In *Handbook of child psychology* (Ed. by C. Murchison). Worcester: Clark Univ. Press, 1931. Pp. 374–416.
16. CARPENTER, C. R. A field study of the behavior and social relations of howling monkeys. *Comp. psychol. Monogr.,* 1934, 10, 1–168.
17. CARPENTER, C. R. A field study in Siam of the behavior and social relations of the gibbon (Hylobates lar.). *Comp. psychol. Monogr.,* 1940, 16, 1–202.
18. CARPENTER. C. R. Societies of monkeys and apes. *Biol. Symposia,* 1942, 8, 177–204.
19. CARPENTER, C. R. Concepts and problems of primate sociometry. *Sociometry,* 1945, 8, 56–61.
20. CATTELL, R. B. *The fight for our national intelligence.* London: King, 1937.
21. CHEN, S. C. The leaders and followers among the ants in nest-building. *Physiol. Zool.,* 1937, 10, 437–455.
22. CLARK, G., & BIRCH, H. Hormonal modifications of social behavior. I. The effect of sex-hormone administration on the social status of a male-castrate chimpanzee. *Psychosomatic Med.,* 1945, 7, 321–329.

23. COLLIAS, N. E. Aggressive behavior among vertebrate animals. *Physiol. Zool.*, 1944, 17, 83–123.
24. COMBES, M. Existence probable d'une elite non differenciée d'aspect, constituant les veritables ouvrieres chez les Formica. *C. R. Acad. Sci. (Paris)*, 1937, 204, 1674–1675.
25. COOPER, J. A description of parturition in the domestic cat. *J. comp. Psychol.*, 1944, 37, 71–79.
26. CRAWFORD, M. P. The cooperative solving of problems by young chimpanzees. *Comp. psychol. Monogr.*, 1937, 14, 1–88.
27. CROWELL, M. F. A discussion of human and insect societies. *Psyche*, 1929, 36, 182–189.
28. DARLING, F. F. *A herd of red deer.* London: Oxford Univ. Press, 1937.
29. EIDMAN, H. Das Mitteilungsvermögen der Ameisen. *Die Naturwiss.*, Berlin, 1925, 13, 126–128.
30. EMERSON, A. E. Communication among termites. *IV Int. Cong. Entom.*, II, 1928.
31. EMERSON, A. E. Social coordination and the superorganism. *Amer. Midl. Nat.*, 1939, 21, 182–209.
32. EMERSON, A. E. Biological sociology. *Den. Univ. Bull.*, 1941, 36, 146–155.
33. EMERSON, A. E. Basic comparisons of human and insect societies. *Biol. Symposia*, 1942, 8, 163–176.
34. FISCHEL, W. Beiträge zur Soziologie des Haushuhns. *Biol. Zent.*, 1927, 47, 678–695.
35. FISHER, R. A. *The genetical theory of natural selection.* Oxford: Oxford Univ. Press, 1930.
36. FORD, C. S. *A comparative study of human reproduction.* New Haven: Yale Univ. Press, Publ. Anthrop., 1945.
37. FRISCH, K. v. Über die Sprache der Bienen. *Zool. Jahrb., Zool. Physiol.*, 1923, 20, 1–186.
38. GALT, W. The principle of cooperation in behavior. *Quart. Rev. Biol.*, 1940, 15, 401–410.
39. GERARD, R., & EMERSON, A. E. Extrapolation from the biological to the social. *Science*, 1945, 101, 582–585.
40. GOLDMAN, I. The Kwakiutl Indians of Vancouver Island (Chap. 6), and The Zuni of New Mexico (Chap. 10). In *Cooperation and competition among primitive peoples* (Ed. by Margaret Mead). New York: McGraw-Hill, 1937.
41. GOLDSTEIN, K. *The organism.* New York: American Book Co., 1939.
42. HEYDE, K. Die Entwicklung der psychischen Fähigkeiten bei Ameisen und ihr Verhalten bei abgeänderten biologischen Bedingungen. *Biol. Zent.*, 1924, 44, 624–654.
43. JACOBSEN, C., JACOBSEN, M., & YOSHIOKA, J. G. Development of an infant chimpanzee during her first year. *Comp. psychol. Monogr.*, 1932, 9, 1–94.
44. KORNHAUSER, A. W. Analysis of "class" structure of contemporary American society —psychological bases of class divisions. Chap. II in *Psychology of industrial conflict* (Ed. by G. Hartmann and T. Newcomb). New York: Dryden Press, 1939.
45. KROPOTKIN, P. M. *Mutual Aid. A factor of evolution.* New York: Knopf, 1917.
46. LASHLEY, K. S., & WATSON, J. B. Notes on the development of a young monkey. *J. Anim. Behav.*, 1913, 3, 114–139.
47. LIPPITT, R. An experimental study of the effect of democratic and authoritarian atmospheres. *Univ. Iowa Stud.*, 1940, 16, No. 3.
48. LOEVINGER, J. Intelligence as related to socio-economic factors. *39th Yearbook, Nat. Soc. Stud. Educ.*, vol. I, 1940.
49. MAIDL, F. *Die Lebensgewohnheiten und Instinkte der staatenbildenden Insekten.* Wien, 1933–1934.
50. MAIER, N. R. F., & SCHNEIRLA, T. C. *Principles of animal psychology.* New York: McGraw-Hill, 1935.
51. MALINOWSKI, B. *The dynamics of cultural change.* New Haven: Yale Univ. Press, 1945.
52. MALLER, J. B. Cooperation and competition, an experimental study of motivation. *Tch. Coll. Cont. Educ.*, 1929, No. 384.
53. MANN, C. W. Mental measurements in primitive communities. *Psychol. Bull.*, 1941, 37, 366–395.
54. MASLOW, A. H. Dominance-quality and social behavior in infra-human primates. *J. soc. Psychol.*, 1940, 11, 313–324.
55. NEFF, W. S. Socioeconomic status and intelligence: a critical survey. *Psychol. Bull.*, 1938, 35, 727–757.
56. NISSEN, H. W., & CRAWFORD, M. P. A preliminary study of food-sharing behavior in young chimpanzees. *J. comp. Psychol.*, 1936, 22, 383–419.
57. NOBLE, G. K., WURM, M., & SCHMIDT, A. Social behavior of the black-crowned night heron. *Auk*, 1938, 55, 7–40.
58. NOVIKOFF, A. The concept of integrative levels and biology. *Science*, 1945, 101, 209–215.
59. NOVIKOFF, A. Continuity and discontinuity in evolution. *Science*, 1945, 102, 405–406.
60. PRATT, K. C., NELSON, A. K., & SUN, K. H. The behavior of the new-born infant. *Ohio St. Univ. Stud., Contrib. Psychol.*, 1930, No. 10.
61. REDFIELD, R. (Ed.) Levels of integration in biological and social sciences. *Biol. Symposia*, 1942, 8, 1–26.
62. RÉVÉSZ, G. The language of animals. *J. gen. Psychol.*, 1944, 30, 117–147.
63. SCHJELDERUP-EBBE, T. C. Beiträge zur Sozialpsychologie des Haushuhns. *Z. Psychol.*, 1922, 88, 225–252.

64. SCHNEIRLA, T. C. Learning and orientation in ants. *Comp. psychol. Monogr.*, 1929, 6, 1–143.

65. SCHNEIRLA, T. C. A theoretical consideration of the basis for approach-withdrawal adjustments in behavior. *Psychol. Bull.*, 1939, 37, 501–502.

66. SCHNEIRLA, T. C. Social organization in insects, as related to individual function. *Psychol. Rev.*, 1941, 48, 465–486.

67. SCHNEIRLA, T. C. A unique case of circular milling in ants, considered in relation to trail following and the general problem of orientation. *Amer. Mus. Novitates*, 1944, No. 1253, 1–26.

68. SCHNEIRLA, T. C. Ant learning as a problem in comparative psychology. In *Twentieth century psychology.* New York, 1945. Pp. 276–305.

69. SCOTT, J. P. Group formation determined by social behavior; a comparative study of two mammalian societies. *Sociometry*, 1945, 8, 42–52.

70. SHERMAN, M., & SHERMAN, I. C. Sensorimotor responses in infants. *J. comp. Psychol.*, 1925, 5, 53–68.

71. SIMPSON, G. The role of the individual in evolution. *J. Wash. Acad. Sci.*, 1941, 31, 1–20.

72. SLOANE, E. H. Reductionism. *Psychol. Rev.*, 1945, 52, 214–223.

73. STURTEVANT, A. Essays on evolution. II. On the effects of selection on social insects. *Quart. Rev. Biol.*, 1938, 13, 74–76.

74. TAYLOR, W. S. The gregariousness of pigeons. *J. comp. Psychol.*, 1932, 13, 127–131.

75. THORPE, W. H. Further studies on pre-imaginal olfactory conditioning. *Proc. Roy. Soc. London*, 1939, 127-B, 424–432.

76. TINKLEPAUGH, O. L., & HARTMAN, C. G. Behavioral aspects of parturition in the monkey *(M. rhesus)*. *J. comp. Psychol.*, 1930, 11, 63–98.

77. TINKLEPAUGH, O. L., & HARTMAN, C. G. Behavior and maternal care of the newborn monkey *(M. rhesus)*. *J. genet. Psychol.*, 1932, 40, 257.

78. VEBLEN, T. *The theory of the leisure class.* New York: Viking Press, 1931.

79. WARDEN, F. J., & GALT, W. A study of cooperation, dominance, grooming, and other social factors in monkeys. *J. genet. Psychol.*, 1943, 63, 213–233.

80. WATSON, J. B. *Psychology from the standpoint of a behaviorist.* Philadelphia: Lippincott, 1919.

81. WHEELER, W. M. The ant colony as an organism. *J. Morph.*, 1911, 22, 307–325.

82. WHEELER, W. M. *Social life among the insects.* New York: Harcourt, Brace, 1923.

83. WHEELER, W. M. *Emergent evolution and the development of societies.* New York: Norton, 1928.

84. WHEELER, W. M. *The social insects.* New York: Harcourt, Brace, 1928.

85. WIESNER, B. P., & SHEARD, N. M. *Maternal behavior in the rat.* Edinburgh: Oliver & Boyd, 1933.

86. YERKES, R. M. Genetic aspects of grooming, a socially important primate behavior pattern. *J. soc. Psychol.*, 1933, 4, 3–25.

87. YERKES, R. M., & TOMILIN, M. Mother-infant relations in chimpanzees. *J. comp. Psychol.*, 1935, 20, 321–359.

5

THE ANALYSIS OF SOCIAL ORGANIZATION IN ANIMALS[1]

J. P. SCOTT

Roscoe B. Jackson Memorial Laboratory, Bar Harbor, Maine

INTRODUCTION

Following Darwin's emphasis on the principle of adaptation in organic evolution, biologists became strongly interested in the study of individual animal behavior. Adaptation could only be inferred from fossils, but could be observed directly in living forms. Up till 1900 a great many biologists concentrated their efforts in this field, which was at that time almost as popular as comparative anatomy and embryology. Shortly afterwards two discoveries were made which strongly influenced this type of research. One was the rediscovery of Mendelian heredity, and a great many workers who were interested in general evolutionary problems shifted their attention to genetics. The other was the establishment of general laws of learning following Pavlov's studies of conditioned reflexes, and those workers with psychological interests tended to elaborate these finding with studies of behavior in the rat.

About 1920 a new era in behavioral research began with the description of social organization in birds. Elliot Howard (1920) wrote on the significance of song and territory in birds and Schjelderup-Ebbe (1922) discovered the existence of social dominance in chickens. Shortly afterward Allee appoached the problem from a more general viewpoint with his review on animal aggregations (1927) and later book by the same title (1931). Since that time a large body of information has been accumulating around the problem of social organization, and this paper will briefly describe some general methods and the kinds of results to be expected from its study.

[1] This paper was originally delivered, September 7, 1954, at Gainesville, Florida, as part of a program on "Social organization of Animals" given in honor of Dr. W. C. Allee.

The study of the social organization of a species may be delimited by the natural units of social organization, which are usually species populations and their subgroups. It differs from a general ecological study in which the unit of organization is the animal and plant community, which may be composed of many different sorts of animal and plant populations. It also differs from psychological analysis in which the unit of organization is the individual and where the primary interest lies in internal organization. So defined, the analysis of social organization occupies an intermediate position between the sciences of ecology and psychology, and overlaps with both at many points.

At the present time we have only a few definite standard techniques for the study of social behavior and organization (Scott, 1950). However, there are certain general methods which have achieved wide use and which can be recommended to anyone working in the field. The analysis of social organization in any animal should begin with a thorough descriptive study in which the seasonal and daily cycles of behavior are thoroughly surveyed. An essential part of this study is the identification of individuals, since without this information the details of social organization can only be surmised. Descriptive study should also include a study of the development of social behavior and organization from birth to maturity, since behavior and organization change with age. This systematic descriptive information naturally leads to certain hypotheses which can be studied experimentally, and it is found that many types of factors can affect social organization. These may be described as ecological, psychological, physiological, and genetic; and appropriate techniques used for the examination of each.

CLASSIFICATION OF GENERAL BEHAVIOR PAT-
TERNS AND THEIR COMPARATIVE STUDY

In relating genetic factors to social organization it has been found that each species has certain characteristic ways of adapting to the environment and that these "patterns of behavior" in part determine the nature of social organization which can be developed in the species. As a guide to the study of characteristic patterns of behavior the author has developed a general scheme of classification. This scheme is intended to include the kinds of behavior generally recognized by students of animal behavior rather than to conflict with them, and has been found useful in making general descriptive studies of the type described above (Scott 1945, 1950a). It was developed originally by attempting to classify the behavior patterns described in several species of animals by other authors, as in Allen's (1911-1914) study of the blackbird and Dean's (1896) study of the river dogfish.

The chief rule for any good scheme of classification is that it should be natural and conform to discontinuity which exists in nature and which can be recognized by independent observers. The scheme of classification should be logical, and one which includes all related phenomena. Terms in common usage may be employed, provided their meanings are clear and describe the observed phenomenon correctly, but they should not be redefined with new meanings as this leads to great confusion. Finally, terms should be descriptive and not imply some theoretical interpretation of the facts.

These rules for classification and terminology conform to general usage in biological science. In addition, the author feels that one should be conservative about forming new terms inasmuch as they add to the labor of acquiring a scientific vocabulary, but that one should not hesitate to form new ones where necessary. Those which are useful will stand the test of time and those which are not useful will tend to be discarded.

The following classification includes all the major patterns of social and semi-social adaptation which have so far been described.

Contactual Behavior. This may be defined as simply maintaining bodily contact and, as Allee (1931) has shown, the formation of simple aggregations through behavior of this sort occurs very widely throughout the animal kingdom. The adaptive significance of the behavior may vary a great deal. A group of mammals may huddle together for warmth whereas a group of Paramecia may form because the bodies of their fellows afford protection against unfavorable chemical conditions. This extremely simple type of social behavior affords a possible basis for the evolution of higher types of behavior.

Ingestive Behavior. This may be defined as behavior concerned with the taking of solids and liquids into the digestive tract and is found very widely although not universally throughout the animal kingdom. It may have an important social significance in animals which feed their young, and becomes highly social in the nursing behavior of mammals.

Eliminative Behavior. This is defined as behavior associated with the elimination of urine and feces from the body. Special behavior is rarely seen in aquatic animals but highly elaborate patterns may be developed in terrestrial species which build nests or lairs. In such forms as wolves and the prong-horned antelope, it may acquire considerable social significance.

Sexual Behavior. This may be defined as behavior connected with the fertilization process and includes the usual courtship and copulation behavior of animals. It occurs very widely though not universally in the animal kingdom and is undoubtedly one of the most primitive forms of social behavior.

Epimeletic Behavior (Gr. *epimeleteon,* caregiving). This may be defined as the giving of care or attention. It has been called maternal behavior but is also found in males in animals like the ostrich which incubates the eggs; and in many other animals where there is biparental care of the young. It could be called parental behavior except for the fact that in many species it is done by animals other than the parents as, for example, the mutual grooming of adult primates, and the care of the young by worker females in the social insects. This behavior includes what has been called by the more specific terms of attentive behavior and nurturance.

Et-epimeletic Behavior (Gr. *aeteo,* beg, + epimeletic). This is defined as calling and signaling for care and attention and is very widely found in animals which give some care to the young. The behavior may be vocal, as in infant mammals, or simply be some sort of movement, as in the larvae of bees and ants. This behavior could be called infantile except that it also occurs in adult animals. In most cases it is used as a substitute for direct adaptation by an individual which is itself helpless or unable to adapt.

Agonistic Behavior (Gr. *agonistikos,* combative). This is defined as any behavior associated with conflict or fighting between two individuals. The term fighting behavior was originally used, but it was found that patterns of behavior involving escape or passivity were very closely related and could not be included under the narrow term of fighting. This type of behavior occurs principally in the arthropod and vetebrate phyla.

Allelomimetic Behavior (Gr. *allelo,* mutual, + *mimetikos,* imitative). This is defined as any behavior in which animals do the same thing with some degree of mutual stimulation and consequent coordination. It is seen developed to a high degree in schools of fishes, flocks of birds and herds of mammals. It could be called imitative behavior except that to most people this implies some degree of learning, which is not necessarily involved, and the idea of a model and a mimic rather than mutual stimulation. Such behavior in birds has been described as mimesis by Armstrong (1952) and as contagious or infectious behavior by other authors. The two latter terms appear to be somewhat undesirable in that they suggest that the behavior is transmitted in the manner of a disease.

Investigative Behavior. This may be defined as sensory inspection of the environment. This has been called exploratory behavior in the rat, where the animal actively explores the environment with nose and whiskers. However, in an animal with highly developed eyes, such behavior may consist merely of glancing around without movement of the whole body. The more general term of investigation appears preferable.

When the above classification is used to organize descriptive data the result is a detailed list of activities under each category which gives the characteristic ways in which a species responds in relation to major behavioral functions (Scott 1950a). Many of these exist in playful or immature forms as well as adult patterns.

It is also possible that patterns of behavior grouped in the above or similar categories may reflect an underlying nervous organization. For example, the behavior of an inexperienced male mouse of the C57 Black strain attacked by a superior fighter follows a regular and predictable sequence (Scott and Fredericson 1951). He first fights back, and when this is not successful he attemps to escape. If he fails in this he adopts a defensive posture, and if cornered he may lie on his back with feet in the air. We may eventually come to think of "systems of behavior" organized around a particular function, but general evidence still needs to be obtained on the point.

The categories may also be used to check the completeness of a descriptive study, leading to the discovery of behavior which has passed unnoticed. Sometimes a major category of behavior is almost or entirely absent, and this affords an opportunity for characterizing the social life of the species.

For example, allelomimetic behavior is highly developed in the sheep and almost entirely absent in the mouse, which results in two very different types of social organization (Scott 1945a).

Species differences can also be found within a category of behavior. For example, the agonistic behavior of sheep and goats consists largely of butting. Sheep back off and run together head on, but goats typically rear up and butt with a sideways thrust of the head. A scheme of classification can be used as a systematic framework for such comparative studies of species differences in behavior, as well as making it easier to do a complete descriptive job. A high standard of scholarship requires such a complete systematic method, which has too often been missing in past studies of comparative behavior.

Comparative studies lead to the conclusion that the presence or absence of a given type of social behavior affects the type of social organization developed by the species. Stated more specifically, the presence of a given behavior pattern defines the types of social relationships which may be developed from it.

SOCIAL RELATIONSHIPS

Social organization may be analyzed in terms of social relationships. Such a relationship is defined as regular and repeatable behavior between two or more individuals, and it in turn may be described in terms of the patterns of behavior exhibited by the individuals taking part. The classical example of a social relationship is the peck order in hens in which one individual regularly threatens or pecks the other which just as regularly dodges or submits to pecking. Social behavior is not identical with social organization. In the case of two strange hens, behavior at first consists of unorganized fighting, and it is only after several encounters that behavior is organized through learning and habit formation into a regular dominance-subordination relationship. Even in the social insects, where heredity seems to play a relatively stronger role, social organization is not automatic. Experience in early life determines the species with which social relationships are formed by larvae taken by slave-raiding ants. Social behavior may therefore be considered an important but not the sole determiner of social relationships, and this should be kept in mind in making analyses of the two phenomena.

A great many types of social relationships are theoretically possible if all types of social behavior are considered. The number of possible combinations of behavior patterns in a species which shows all of the 9 types of behavior classified above may be theoretically calculated as follows. There are 9 possible combinations where both members of the relationship exhibit the same type of behavior and 36 possible combinations where the individuals exhibit unlike behavior, making a total of 45 (Scott 1953). Some of these combinations are frequently seen, such as the dominance order in which behavior of both individuals is agonistic. Some of them may be commonly overlooked as, for example, the combination of sexual and agonistic behavior exhibited by male and female rodents when the female is not in heat, and many of them are yet to be described. It is possible, of course, that certain ones are only theoretical and do not occur in nature. Some of them which have been widely observed are described below, using the names commonly applied to them.

Simple Aggregations. In this type of relationship the behavior of both animals is contactual. As Allee has shown, such relationships exist very widely in the animal kingdom, but their occurrence is apt to be irregular and nonspecific because of their dependency on environmental conditions.

Dominance-Subordination Relationships. In this type of relationship the behavior of both individuals may be described as agonistic. That of one individual consists of a threat or actual fighting while the other individual remains passive or attempts to escape. Evidence gathered by Allee (1950) and his students and summarized by Collias (1944) shows that this type of relationship is widespread in the vertebrates and occurs in at least some arthropods. It does not occur in many of the lower animals which are incapable of fighting.

Leader-Follower Relationships. The behavior of both individuals may be classed as allelomimetic, but there is an unequal degree of stimulation so that one tends to lead and the other follow. This has been described in sheep (Scott 1945), deer (Darling 1937) and ducks (Allee et al. 1947) and deserves more extensive study. It should not be confused with cases in which a dominant animal drives a group before him, as in the case of a stag and does in the rutting season.

Sexual Relationships. The behavior of both individuals is sexual. Although sex behavior has been extensively described in many species, very few analytic studies of the resulting social relationships have been made. Dominance and sexual relationships may be interdependent in both chickens (Guhl and Warren 1946) and chimpanzees (Nowlis 1942).

Care-Dependency Relationships. In this case the behavior of one individual is epimeletic while the behavior of the other individual may be one of several different types. An infant animal may exhibit et-epimeletic behavior, as when a young lamb is separated from the flock, or it may exhibit ingestive behavior in the process of nursing. In animals like dogs which regularly clean the young, the latter may exhibit eliminative behavior. This type of relationship has been widely described but subjected to very little experimental analysis in animals.

Mutual Care. In this case the behavior of both individuals is epimeletic, and an example is seen in the mutual grooming of primates. In spite of its theoretical importance for considerations of basic human sociology, such relationships have been little studied in animals.

Trophallaxis. This is a complex relationship described by Wheeler (1923) in the social insects. Both individuals may exhibit investigative behavior, and one usually exhibits epimeletic behavior in providing food from the crop while the other ingests the food. On subsequent occasions the roles may be reversed.

Mutual Defense. This is another complex relationship in which the members exhibit both agonistic and allelomimetic behavior. It has been described in such animals as muskoxen, wolves, baboons and many kinds of birds in reaction to hawks, but has not been subjected to extensive analysis.

It would appear from the above presentation that the complete analysis of the social organization of a species can be an extremely complex affair, and that the simple description of the dominance-subordination relationships does not give the whole story by any means. However, the task is not hopeless since in any actual case the number of important relationships turns out to be relatively small.

BIOLOGICAL CLASSIFICATION OF SOCIAL RELATIONSHIPS

As described above, analysis of the social organization of an animal species consists of a systematic description of the basic patterns of behavior and their organization into social relationships. If an animal gave the same responses to all individuals the relationships would be simple to describe and a low degree of social organization would result. As indicated above, the behavior of individuals is frequently unlike and the concept of differentiation (Tinbergen 1953) becomes a useful one. Behavior may be differentiated by biological factors such as age and sex on the one hand and by psychological factors involving learning on the other. As Carpenter (1934) has pointed out, there are three biologically determined types of individuals in mammals: males, females and young.

When these are combined in all like and unlike combinations, a total of six basic relationships can be established. These are essentially super-categories under which those relationships mentioned in the previous section may be grouped. For example, male-male relationships can include dominance-subordination and mutual care relationships as well as many others. This general scheme can be applied to other animals as well as primates and is particularly useful in studying groups of wild animals where it is difficult to distinguish individuals except by age and sex.

In the case of the social insects which have biologically differentiated castes, the scheme may be amplified to include greater numbers of basic relationships. A similar extension must be made when social relationships are developed between two different species, as commonly occurs in domestication (Scott 1953).

ANALYSIS OF SPECIFIC SOCIAL RELATIONSHIPS

So far, the analysis of social organization has been considered in terms of behavior of the whole organism and relationships between organisms. It is also possible to investigate the internal processes which in part determine behavior and through it influence social organization. Since the methods used are well known to physiologists and psychologists, the scope of such investigations will be described only briefly here.

One basic field of study is the analysis of the complex network of physiological factors which is associated with any major pattern of behavior (Scott and Fredericson 1951). Available information of this sort tends to be scattered and unequal. The physiology of sexual and ingestive behavior has been widely studied, but our knowledge of agonistic behavior comes largely from the domestic cat. The physiology of certain types of behavior, such as investigative and allelomimetic, is almost completely unknown.

A related field of investigation is concerned with the physiological genetics of social behavior and social relationships. As Collias has shown (1943), dominance-subordination relationships may in part be determined by such physiological factors as weight and hormones, which in turn may depend on the genetic differentiation of sex. Further genetic variability in physiological factors affecting social behavior introduces an individual specificity into social relationships. Those of the genetic variant will fall into the usual broad categories but will also show certain unique and specific differences.

In animals which are capable of learning, social behavior becomes differentiated on the basis of mutual adaptation and habit formation as well as on the basis of biological differences. As shown by Ginsburg and Allee (1942) the formation of a dominance order is at least in part related to the psychological principles of learning. Once such a relationship is formed and firmly established by habit, it may be extremely difficult to upset it by altering biological factors, as shown by Beeman and Allee (1945). Among the higher animals it is probable that all social relationships are affected by the psychological processes of problem solving (or adaptation) and habit formation. Neglect of these factors in favor of an assumption of purely genetic or instinctive determination of behavior may lead to serious errors of interpretation (Lehrman 1953).

As with physiological factors, psychological processes tend to introduce specificity into social relationships. In a study of social organization of sheep it was found that any given lamb had a general leader-follower relationship toward all adults in its tendency to follow them, although this tendency was strongest with regard to the mother. Its care-dependency relationships, on the other hand, were extremely specific and developed only with its own mother (Scott 1945).

In any species which has the psychological capacity to discriminate between individuals, such specific relationships will be expected to occur. It likewise follows that unless the scientific observer develops a similar ability to identify individuals, specific social relationships will pass unnoticed, and with them a large part of social organization. There is a vast difference between the observation that hens tend to peck each other (a general relationship) and the knowledge that within each pair there is one which always pecks and one which always dodges (a peck order consisting of a group of specific relationships).

If social relationships were entirely determined by biological factors it would be expected that all female-female relationships in a given species would be the same except as affected by chance genetic variability. Such is obviously not the case in the peck-order studies, and it is probable that a similar specificity will be found in many other types of social relationships when studied in detail. The general analytic method is to identify all individuals in the social group and then bring them together in pairs in a situation which will elicit the kind of behavior being studied (agonistic, sexual, etc.). A well-designed experiment includes results on all possible combinations of pairs, observed in an order which controls factors such as fatigue, habit formation, etc. The total possible number of combintions in a group is given by the formula $\frac{n(n-1)}{2}$ (Carpenter 1940), so that such a study becomes very lengthy if n (the total number of individuals in the group) is large. The study should be supplemented with observations on the whole group, as it is possible that pairs may react differently when in the presence of others.

Many models for this type of study may be found in the literature (Collias 1951), particularly those done with dominance-subordination relationships, and the methods need not be described in greater detail here. Guhl (1953) has written an excellent summary of methods used for the study of specific dominance and sexual relationships in the domestic fowl. As indicated above, there is a need for the extension of these methods to the study of other sorts of social relationships in order to obtain a more complete picture of social organization.

The importance of specificity in social relationships is likewise apparent when analysis is made from a developmental viewpoint, and this has led to the concept of socialization.

SOCIALIZATION

Experiments which modify the social environment have tended to bring out the general principles of socialization. Any highly social animal that has been so far studied has behavioral mechanisms whereby, early in development, an individual forms positive social relationships with its own kind and usually with particular individuals of its kind. At the same time, other behavioral mechanisms operate to prevent such relationships being formed with other species and, to a lesser extent, with other groups of its own kind. Raising an individual with any other species including human beings usually produces an individual which is socialized to the strange species. Essential patterns of social behavior remain unaltered and are given in response to the strange species to the extent that such behavior is compatible.

The process of socialization has been shown to take place within limited periods of time by Lorenz (1935) in birds and also in such mammals as the sheep (Collias 1953) and dog (Scott and Marston 1950). Lorenz has been chiefly concerned with the positive mechanisms to which he has given the general term "imprinting," but, in the case of the mammals mentioned, some information is also present regarding the negative mechanisms which normally prevent social attachment to other species. In the dog and wolf, for example, agonistic behavior may be elicited by strangers at a very early age, effectively preventing positive contact as the animal grows older (Scott 1953a).

These effects have led to the concept of critical periods in the process of socialization (Scott and Marston 1950). There are points in development where it is very easy to change the individuals to which an animal becomes socialized, and other points later or earlier in development when this is difficult or impossible. A study of the process of socialization thus forms an important part of the developmental analysis of social organization.

An excellent review of experiments on the general effects of modifying early experience has been made by Beach and Jaynes (1954). As applied to the problem of social organization the most useful techniques may be outlined as follows. The first step is a detailed descriptive study of development, paying particular attention to social behavior patterns and changes in sensory, motor and psychological capacities. We have attempted to set up a model outline for this type of study in the case of two species of mammals, the dog and mouse (Scott and Marston 1950; Williams and Scott 1953).

A developmental study may be supplemented by two sorts of simple experimental procedures. One is to take male and female individuals at birth or hatching and have them reared by another species, human or otherwise. This method has produced brilliant results in the hands of Lorenz (1935), and may be extended further to the rearing of individuals in semi-isolation (Thompson and Heron 1954). The other is to castrate male and female individuals as early as possible in development, which gives some indication of the importance of hormonal changes and sexual behavior in social development.

Observations on the time of development of various capacities and patterns of behavior require experimental verification. On the basis of present evidence, the development of many species falls into natural periods based on beginnings or changes of important social relationships. This gives a basis for the experimental analysis of the factor of time in early social experience. Modification of social experience may be begun at different points in development and continued for appropriate periods. Controls for physiological and structural changes, genetic differences, random environmental factors and the psychological factors of learning are important in the design of these experiments (King and Gurney 1954).

SOCIAL ORGANIZATION BETWEEN GROUPS

Most of the work on analysis of social organization has been done within groups, and correctly so. The most frequent aspects of social life in animals appear to be concerned with what are called by the human social psychologists "face-to-face contacts in small groups." However, some organization between groups apparently does exist, as reported in Carpenter's (1934, 1940) studies of howling monkeys and gibbons. Likewise, the territorial organization of passerine birds during the breeding season may be considered as organization between small family groups. Obviously, such organization is most likely to occur under natural conditions, and its existence is one of the possibilities which should be looked for in field studies.

The evidence for the existence of territoriality in small mammals has been summarized by Blair (1953), who finds that instances of defense of definite boundaries are relatively rare. As with birds, group territorial defense in mammals may be described simply in terms of a general dominance-subordination relationship, in which the group or individuals within the group tend to be dominant over the individual trespassing on the territory (Murie 1944; King 1955). However, in the case of howling monkeys Carpenter (1934) reports that groups maintain regular daily contacts with each other by howling at dawn, and this is possibly also the case in birds such as quail which vocalize regularly. The "wars" between colonies of ants appear to involve a general relationship in which all strange individuals are attacked.

The relationship between separate social groups is a problem of considerable theoretical interest and should be investigated wherever possible. Analysis should include the means by which new groups are formed, and the fate of isolated individuals as well as territoriality. King (1955) has found a definite group territorial system in prairie dogs, with a tendency for adults to move out and colonize new areas after raising the young, which are left to occupy the old territory. Carpenter's (1940, 1934) studies of the gibbon and howling monkey are excellent models for the analysis of this and other aspects of social organization.

CONCLUSION

It will be seen from the foregoing outline that the complete analysis of the social organization of even a single species is an enormous task. It is not likely that the job will ever be entirely completed, though such a point is being approached in such favorite objects of study as the chicken, mouse, stickleback, honeybee, army ant and termite. The task must be divided into sections, and emphasis placed upon important points of fact and theory.

Collecting observations on behavior patterns and their organization into basic biologically determined social relationships can often be done with a study of a year or two. Much material is already available in publications written for other purposes, particularly in natural history and ecological life history studies. A great deal of information on European species is contained in studies of comparative behavior patterns done by the "ethologists," although they have frequently concerned themselves with the nature of stimulation ("releasers") rather than social organization. The systematic classification of material from these sources often saves a great deal of time as well as providing confirmatory evidence for original study.

The analysis of socialization and specific social relationships is more complicated and time-consuming, and is difficult to do except in laboratory or semi-natural environments, although it has been shown that social dominance does exist under natural conditions (J. W. Scott 1942). Indeed, the more basic the analysis, the slower it becomes, and a division of labor among scientists is necessary when physiological and genetic factors are studied. This brings up the problem of coordination of effort.

The study of the social organization of behavior tends to unify knowledge. Behavior is usually defined as activity of an entire organism and thus is affected by factors customarily divided up in all the conventional subdivisions of zoology and psychology. When behavior is considered in relation to social organization the subject matter is widened still further to include certain aspects of general social science. It is obviously impossible for anyone to have a thorough and complete knowledge of all these fields, but the specialist in any one of them should at least have a general knowledge of the others and their relation to his own. The ecologist, as a last surviving type of general biologist, should have a particular advantage in this respect. Meetings such as this one, in which the speakers have training in such diverse fields as genetics, psychology, and ecology, can do much to promote the interchange of ideas and mutual comprehension.

A final word may be said concerning the necessity of maintaining a high standard of scientific work. As stated in the beginning of this article, we still have very few specific standardized techniques for the analysis of social organization, and to insist upon the enforcement of standards prematurely would do much to discourage new and creative lines of research. The harm that can be done by publishing inconclusive pilot experiments or generalizing from superficial descriptions based on study of a few individuals is equally great. We can and should insist on general standards of thoroughness and repeatability, on controls for the wide variety of genetic, physiological, developmental, psychological, ecological, and social factors known to affect behavior, and on the necessity for final experimental verification of any theory.

The resulting well-established facts and theories of social organization in animals have widespread significance. Such information has great practical importance in wildlife management and in the care of animals under domestication. For example, if it is planned to recolonize a vacant territory with social animals, the new individuals must have been socialized toward each other or otherwise the group will disintegrate and fail to survive.

A knowledge of social organization is an essential part of understanding the general ecology of the species. Populations of animals do not exist in nature as randomly distributed individuals but as socially organized groups. Enough has been said in the foregoing article to indicate that an immense amount of information is still to be gathered about the social organization of animals, and even some of our common and familiar species are far from completely understood.

Finally, the collection of knowledge regarding social organization in animals and the resulting generalizations which will be made should prove to be of great help in understanding certain human problems. A great many problems of human maladjustment can be attributed to interference with the process of socialization, whose study should be one of the primary concerns of child psychology. A more thorough knowledge of animal societies should contribute greatly to our understanding of the biological and psychological bases of human social organization.

SUMMARY

An attempt has been made in this article to present a systematic general outline for the analysis of social organization in animals. Methods include systematic description of the important behavior patterns of the species based on studies of the daily and seasonal cycles of behavior and upon the development of behavior in the individual from birth until maturity and old age. These basic patterns of behavior determine to a large extent the general types of social relationships which are possible in the species.

Further analysis of specific social relationships depends upon identification of every individual in the social group and should be verified by experimental combinations of every possible pair.

Study of the process of socialization should include a systematic descriptive study of changes in the basic behavioral capacities: sensory, motor, psychological, and patterns of social behavior, with particular reference to relative timing and its consequences. Some remarks are made concerning general application of the method, and the significance of the results.

REFERENCES

Allee, W. C. 1927. Animal aggregations. Quart. Rev. Biol., 2 : 367-398.
——. 1931. Animal aggregations. A study in general sociology. Chicago : Univ. Chicago Press.
——. 1950. Extrapolation in comparative sociology. Scientia, 43 : 135-142.
——, M. N. Allee, F. Ritchey, and E. W. Castle. 1947. Leadership in a flock of White Pekin ducks. Ecology, 28 : 310-15.
Allen, A. A. 1911-14. The red-winged blackbird : a study in the ecology of a cat-tail marsh. Abs. Proc. Linnean Soc. N. Y., No. 24-25.
Armstrong, E. A. 1952. Behavior is contagious, too. Animal Kingdom, 55 : 88-91.
Beach, F. A., and J. Jaynes. 1954. Effects of early experience upon the behavior of animals. Psych. Bull., 51 : 239-263.
Beeman, E. A., and W. C. Allee. 1945. Some effects of thiamin on the winning of social contacts in mice. Physiol. Zool., 18 : 195-221.
Blair, W. F. 1953. Population dynamics of rodents and other small mammals. Rec. Ad. Genetics, 5 : 1-41.
Carpenter, C. R. 1934. A field study of the behavior and social relations of howling monkeys. Comp. Psych. Monogr. No. 48, 10(2) : 1-168.
——. 1940. A field study of the behavior and social relations of the gibbon (Hylobates Lar). Comp. Psych. Monogr. No. 84, 16(5) : 1-212.
Collias, N. 1943. Statistical analysis of factors which make for success in initial encounters between hens. Am. Nat., 77 : 519-538.
——. 1944. Aggressive behavior among vertebrate animals. Physiol. Zool., 17 : 83-123.
——. 1951. Social life and the individual among vertebrate animals. Ann. N. Y. Acad. Sci., 51 : 1074-1092.
——. 1953. Some factors in maternal rejection in sheep and goats. Bull. Ecol. Soc. Amer., 34 : 78.
Darling, F. F. 1937. A herd of red deer. London : Oxford Univ. Press.
Dean, B. 1896. The early development of Amia. Quart. J. Mic. Sci., 38 : 413-444.
Ginsburg, B. and W. C. Allee. 1942. Some effects of conditioning on social dominance and subordination in inbred strains of mice. Physiol. Zool., 15 : 485-506.
Guhl, A. M. 1953. Social behavior of the domestic fowl. Kansas State College Ag. Exp. Sta. Technical Bull. No. 73.
——, and D. C. Warren. 1946. Number of offspring sired by cockerels related to social dominance in chickens. Poultry Sci., 25 : 460-472.
Howard, H. E. 1920. Territory in bird life. London : Murray.
King, J. A. 1955. Social behavior, social organization and population dynamics in a black-tailed prairiedog town in the Black Hills of South Dakota. Contr. Lab. Vert. Biol., Univ. Mich., No. 67, 123 pp.
——, and N. L. Gurney. 1954. Effect of early social experience on adult aggressive behavior in C57BL/10 Mice. J. Comp. and Physiol. Psych., 47 : 326-330.
Lehrman, D. S. 1953. A critique of Konrad Lorenz's theory of instinctive behavior. Quart. Rev. Biol., 28 : 337-363.

Lorenz, K. 1935. Der Kumpan in der Umwelt des Vogels. Journal für Ornithologie, **83** : 137-213, 289-413. For English summary see : The companion in the bird's world. Auk, **54** : 245-273, 1937.

Murie, A. 1944. The wolves of Mt. McKinley. U.S.D.I. Fauna Series No. 5.

Nowlis, V. 1942. Sexual status and degree of hunger in chimpanzee competitive interaction. J. Comp. Psych., **34** : 185-194.

Schjelderup-Ebbe, T. 1922. Beiträge zur sozial-psychologie des Haushuhns. Zeit. Psych., **88** : 225-252.

Scott, J. P. 1945. Social behavior, organization and leadership in a small flock of domestic sheep. Comp. Psych. Monograph No. 96, **18**(4) : 1-29.

————. 1945a. Group formation determined by social behavior; a comparative study of two mammalian societies. Sociometry, **8** : 42-52.

————. 1950 (ed.) Methodology and techniques for the study of animal societies. Ann. N. Y. Acad. Sci., **51** : 1001-1122.

————. 1950a. The social behavior of dogs and wolves : an illustration of sociobiological systematics. Ann. N. Y. Acad. Sci., **51** : 1009-1021.

————. 1953. Implications of infra-human social behavior for problems of human relations. *In:* Group Relations at the Crossroads, M. Sherif and W. A. Wilson, eds. New York : Harper.

————. 1953a. The process of socialization in higher animals. *In:* Interrelations between the social environment and psychiatric disorders. New York : Milbank Memorial Fund.

————, and M. Marston. 1950. Critical periods affecting the development of normal and mal-adjustive social behavior of puppies. Jour. Genet. Psych., **77** : 25-60.

————. and E. Fredericson. 1951. The causes of fighting in mice and rats. Physiol. Zool., **24** : 273-309.

Scott, J. W. 1942. Mating behavior of the sage grouse. Auk, **59** : 477-498.

Thompson, W. R. and W. Heron. 1954. The effects of early restriction on activity in dogs. J. Comp. and Physiol. Psych., **47** : 77-82.

Tinbergen, N. 1953. Social behaviour in animals. New York : John Wiley.

Wheeler, W. M. 1923. Social life among the insects. New York : Harcourt, Brace.

Williams, E., and J. P. Scott. 1953. The development of social behavior patterns in the mouse, in relation to natural periods. Behaviour, **6** : 35-64.

6

Reprinted from *U. of Queensland Fac. Papers Vet. Sci.,* 1(2):75-110 (1964)

A GENERAL THEORY OF SOCIAL
ORGANIZATION AND BEHAVIOUR

G. McBride

I. INTRODUCTION

There are many patterns of social organization in animals, almost as many, in fact as there are gregarious animal species. Though patterns differ so widely, there are also many similarities. It is in these similarities that one looks for order. It is unlikely that a universal theory of social behaviour can yet emerge from the order apparent at this time. Nevertheless there is enough order discernible now for an attempt at a general theory of social behaviour and organization. Time will modify, improve, or replace such a theory, but the process can only be initiated by the presentation of such an attempt; this is the object of this paper.

Animals cannot be clearly classified as social or non-social, but all levels of social contact can be observed. In the extremely non-social species, individuals come together briefly to mate or to fight: they show well-developed spacing mechanisms which maintain their isolation from others of the same species. Coming down the scale, one finds patterns which add parental behaviour with one or both sexes participating in temporary or permanent associations. Further, one finds larger aggregations of animals of the same sex, then of both sexes, in permanent or temporary associations.

At the social extreme, one finds societies which remain aggregated throughout the year. Such species are described as gregarious. It is difficult, however, to define any limits associated with the concept of gregariousness. Nevertheless we recognize that in aggregated groups of all types there is a large category of intraspecific behaviour that we refer to as social behaviour. It is this behaviour which must be analyzed into components in order to find the similarities which lead to order.

It is apparent from the start that to consider aggregated or non-aggregated species we are automatically discussing two-dimensional spatial relationships between animals. It seems probable that these spatial relationships between animals are also basic in animal organization, since most social (or non-social) behaviour can be expressed in terms of the distances between animals and their arrangement in space. Close social organization must also mean small physical distances between individuals.

To get at the factors underlying the spatial organization of animals, we must consider what may be called a third dimension, a social dimension. This social dimension is expressed in a wide range of animal behaviour. The expression of this dimension is social dominance, which appears to be a basic component of the behaviour of social animals. This social dimension may be called "social mass". In territorial species of animals, high social dominance is reflected directly in large and well-situated territories, that is in two-dimensional space. It is this spatial effect of the social dimension which will provide the basis of the theory of social organization to be presented.

The last important factor to be considered is that animal groups are not uniform in type, but are divided into castes such as adult males, adult females, young males, and immature animals. One knows empirically that animals of one caste behave towards other members of the same caste differently from the way in which they behave towards members of other castes. Intercaste behaviour includes such important categories as courtship, mating, and parental behaviour, while agonistic behaviour (social dominance contests) is generally better organized on an intracaste basis.

The problem, then, is the basis of organization behaviour in animals which range along a spectrum from extremely non-social species to extremely gregarious species. The problem should also include that of society classification or taxonomy.

The approach to the problem is concerned with analysis of social behaviour, consideration of the spatial organization and its underlying social dimension, and the interactions of this system with the caste subgroups within species.

The proposed analysis of social behaviour aims to define categories of behaviour in terms of the effect that they have on the spatial relationships between animals, that is, their effect upon individual distances between neighbouring animals. These distances may be maintained visually or be formalized in territories. Basically, the categories include behaviour which causes distances between animals to be increased and behaviour which allows these distances to be reduced. The most complete analysis is best carried out on the behaviour of aggregated groups, since only in these groups can one observe the behaviour components which reduce distances between animals as well as those which increase them. The spatial organization of a group of animals at a given time will reflect the balance between these opposing types of behaviour.

The social dimension is concerned with the behavioural organization between and within castes. The category of behaviour associated with intraspecific aggressiveness is regulated by social organization within castes. Social dominance and/or the territorial system are the most common forms of organization. These systems regulate the supply of such facilities as food and sex in competitive situations which arise when the supply of these facilities is limited. It will be suggested that the basic expression of the social dimension is a type of "extension of self" in space to create a field of social force associated with an animal. This theory of social force is based on the work of McBride, James & Shoffner (1963). These social force fields are expressed between individuals in a group by releasers or stimuli for intraspecific aggressiveness. The stimuli are the species and caste morphological or behavioural characteristics. In groups organized by territories, the territories represent social force fields fixed in space. Here the stimuli are the territorial marking mechanisms which release avoidance and spacing behaviour or alternatively aggressive behaviour.

The force fields associated with animals reflect their dominance status within castes. They also vary from caste to caste, and here are controlled by the mechanisms, generally hormonal, determining caste status. The effect of force fields differs within from that between castes since caste as well as species characteristics provide the social force stimuli. The stimuli have the greatest agonistic effect within castes, and thus the primary organization of animal groups occurs in this situation. Other types of regulation of force fields, particularly diurnal, will be discussed.

When only intracaste organization occurs, a simple society occurs. Where more than one caste is present, complex societies develop, and these show the primary organizations within castes as well as organized intercaste behaviour.

II. Analysis of Social Behaviour

(1) Gregarious Behaviour and the Social Bond

The basic feature of gregarious behaviour is the mutual attraction which draws animals together. This has been considered by a number of authors, and is apparently controlled by "social releasers" (Tinbergen, 1951). More recently Chance (1962) has referred to this behaviour as a "social bond" and has considered the nature of this bond. Wynne-Edwards (1962) has described it as a centripetal social force.

Its existence is an observed fact, and it is responsible for the coherent nature of social groups as well as the tendency for groups to organize around a central dominant animal. That is, the attraction is not only to the other animals, but more specifically to the dominant animal. In this sense, other animals, and particularly the dominant animal itself, become a type of positive response situation (P.R.S.) (Calhoun, 1962). A P.R.S. is defined by Calhoun as a "stationary place whose characteristics are such as to lead to securing a reward by the individual who responds there".

Chance (1962) has considered the nature of this social bond in the social primates, drawing particularly upon the evidence of Kummer (1957). Unweaned baboons seek the protection of their dams when threatened or frightened; after weaning, they seek out the animal of highest possible rank. The situation may be summed up in his quotation of Kummer (1957): "It seems that with advancing age the elements of behaviour of the young in the relationship with their mother do not vanish but project themselves upon ever higher ranking individuals." The source of fear or threat may be from outside the group, e.g. a predator, or from within the group, when one individual is threatened by another. A frightened animal is attracted to the highest ranking animal, even when the latter is the cause of the fear (Kummer, 1957). This situation can also be observed in flocks of domestic hens.

This attraction between animals is responsible for what Hediger (1962) calls a "social distance". This is the maximum distance that individuals of one society will move from one another. Once again, in mammalian societies this has its origin in parent-offspring behaviour, when the young will not move far from their dams.

In territorial animals a similar pattern develops, with the dominant animal often occupying the central territory. Here the point of attraction is not so obviously the other animals and the dominant animal in particular. The P.R.S. is more reasonably a spatial one. In both systems, the animals on the fringes of the group area are more subject to predation, and one of the important functions of animal aggregation is the greater protection afforded from predators. Presumably this is one of the factors responsible for the role of the central space as a social P.R.S. through the availability of facilities for competition (e.g. Bartholomew, 1952).

In the evolution of the aggregated group type of social organization from a territorial type of situation, the social P.R.S. has changed from a central area occupied by the dominant animal to the dominant animal itself occupying the central position. The selective forces favouring the development of a social bond seem to be that any animal leaving a social group must re-enter it or another as a stranger. This usually leads to low social status. To the extent that low social status is associated with low reproductive performance, selection will operate to maintain this centripetal social force generated by the social bond.

Though the role of social dominance has been, and will continue to be, emphasized, there is a category of gregarious behaviour which involves negligible intragroup aggressiveness. In its simplest form, animals may aggregate because of uneven distribution of suitable environmental conditions, physical or nutritional. The phenomenon is best described as schooling. The level of organization in schools is more difficult to determine than when social dominance is present, though Keenleyside (1955) has described some of the conditions determining the organization of spatial relationships between fishes.

The school pattern can be observed in various forms: animals may spend their whole lives in schools; they may show schooling when juvenile, develop aggressiveness and take up territories during the breeding season, and revert to schools during the non-breeding season; alternatively the schooling may be associated with group territories, and show elements of dominance behaviour as in the hymenoptera.

In its simplest forms, schooling seems to be a primitive or juvenile type of gregarious behaviour. Even among the gregarious species showing strong dominance behaviour, there is generally an early period of aggregation without intraspecific aggressive behaviour. This can be well seen in the domestic fowl, where there is strong attraction between individuals and following behaviour, before agonistic behaviour develops.

(2) *Intracaste Behaviour*

Analysis of intracaste social behaviour may be carried out at any level, from the neurophysiological through individual behaviour patterns to categories of behaviour grouped in terms of their function in the society. It is this last type of analysis which will be attempted: thus each category will contain a wide range of individual behaviour patterns, though all have a common element in terms of the role they play in their particular society. The proposed components are as follows:

  (i) Intraspecific aggressiveness.

  (ii) Submission.

  (iii) Socializing behaviour.

For reasons given above, the range of social behaviour to be discussed can only be seen in closely aggregated single-caste societies. Probably the best examples of such societies are found in the domestic animals where husbandry methods keep the different castes separate and the species have evolved towards closely aggregated societies under conditions of restraint on movements. The peck order is the ideal type of closely aggregated society. The same categories of behaviour have been described in non-domestic species where strong flock- or herd-forming tendencies are normal, particularly during those seasons where the castes form separate societies (or a single society).

(i) *Intraspecific aggressiveness.*   Intraspecific aggressiveness has long been recognized as an important category of animal behaviour, and Darwin (1883) considered the subject at great length, particularly in relation to sexual selection.

The exact factors involved in intraspecific aggressiveness are complex. Species recognition is certainly involved, as is also caste recognition. Hormones are involved both in their effects on secondary sex characteristics and through a direct effect on aggressiveness. Guhl (1961) showed that the use of testosterone propionate caused hens to respond aggressively to their penmates, leading to increases in their status in the normally stable peck order. Bennett (1940) showed that the same hormone caused ringdoves to increase the size of their territories.

Spatial factors are also involved; the level of aggressiveness of an animal increases as strangers either approach or cross some defined territorial boundary. In closely aggregated animals, aggressiveness also varies with the orientation of the heads of neighbours. This observation suggested that every animal treated as its own the space in the immediate vicinity of its head, and exerted a social force in this area.

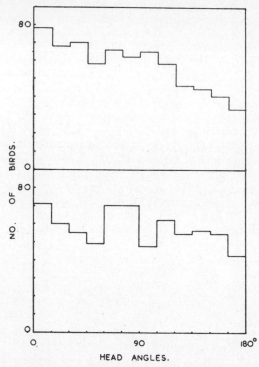

FIG. 1.—The frequency distribution of head angles (angles through which an animal must turn to face a neighbour) for nearest (a) and second nearest (b) neighbours in a flock of 9 week old chickens. (McBride, G., Parer, I. & Shoffner, R. N.; unpublished.)

(a) A theory of fields of social force:

As animals of the same species approach each other, they react to each other. This is so even in a well-integrated group where the reactions are not strong but nevertheless may be detected in various ways. For example, if we look at the frequency distributions of angles through which birds must turn to face their nearest and second nearest neighbours (head angles), we find quite different distributions (Figure 1); that is, birds react differently to their nearest neighbour from the way in which they react to their second nearest neighbour. This can be seen more clearly in adult hens, when only pairs which are mutual nearest neighbours are considered (Figure 2).

The pattern of behaviour of fowls towards their neighbours was described by McBride, James, & Shoffner (1963). They photographed adult hens from above,

G. McBRIDE

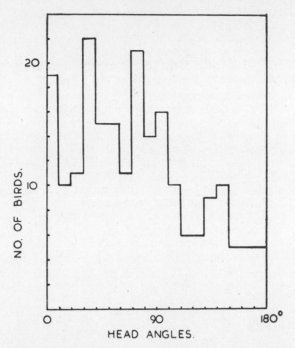

FIG. 2.—The frequency distributions of head angles between mutual nearest neighbours in adult hens (McBride, G., James, J. W. & Shoffner, R. N.; unpublished.)

drew arrows through the heads of each bird with the arrowhead at the beak, and analyzed the relationships in terms of the model shown in Figure 3, using the partial regression equation $y = b_1x_1 + b_2x_2$, y being the head angle of the bird under consideration (A), x_1 the distance to the neighbour (B), and x_2 the head angle of B. The first, second, and third nearest neighbours were considered, but significant effects arose primarily from the nearest neighbours. In general, b_1 was positive and b_2 negative. Linearity was not considered. Thus at any constant distance apart, birds tended to avoid the face to face situation; if A looks at B, B turns aside. At any constant head angle of B, as the distance from A descreases, A turns to face B.

These authors also looked at the distribution of heads in the fields of their photographs. The heads were considered as points; this seemed reasonable, since they are higher than the bodies and were often observed above the bodies of neighbours. The test for randomness of the scatter of such points was that of Clarke & Evans

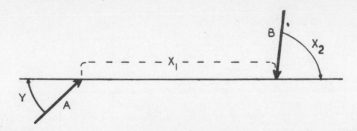

FIG. 3.—The model used for the analysis of the relationships of head orientation and spacing between neighbouring adult hens. The arrows represent the heads of birds A and B: B is the nearest neighbour of A; their head angles to face each other are Y and X_2 and the distance between them is X_1.

91

(1954). A highly significant tendency towards uniform non-random spacing was obtained. Such a result suggests that birds maintain maximum distances from their neighbours.

Since the photographs were taken at constant time intervals, regardless of the contents of the field, the behaviour of the hens is one of constant adjustment of spacing and head orientation relative to their neighbours.

If adjustment of head orientation alone is used as a measure of this behaviour, it can be shown that the behaviour pattern is not constant over the whole range of distances. The model in Figure 3 was used to analyze the behaviour of birds at different distances apart: the result is shown in Figure 4. The fact that animals behave differently at different approach distances was observed by Hediger (1950), who referred to flight distance as the approach distance at which an animal would take avoidance action, and fight distance as the distance at which an animal would fight when unable to move away. Though these distances were concerned with interspecific situations, the points marked A and B on Figure 4 may be analogous to fight and flight distances respectively in the intraspecific situation in the domestic hen. These distances are approximately 16 and 30 inches respectively.

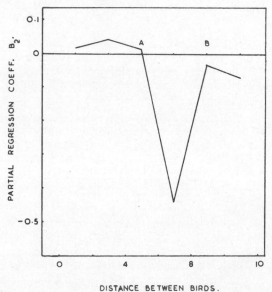

FIG. 4.—The relationship between b_2 and the distance between neighbouring adult hens, where b_2 is the partial regression of the head angle of the bird under consideration on the head angle of its nearest neighbour. (McBride, G., James, J. W. & Shoffner, R. N.; unpublished.) (Units of distance are 3.19 inches.)

From Figure 4, it seems that neighbouring birds more than 30 inches apart are relatively unconcerned with each other or show a slight tendency to turn to avoid each other's faces. As they approach each other between 16 and 30 inches, this tendency is very marked. However if they approach within 16 inches, the behaviour changes and they turn face to face, or, more accurately, beak to beak in an offensive or defensive position.

In adult turkey males, a similar pattern of behaviour has been observed. The dominant bird may turn towards a subordinate who turns and moves away but never vice versa. In addition, the intensity of the reaction can be varied by specific postures. If the dominant male adopts a threat posture, the avoidance action of the subordinate is quite violent.

Other circumstances can affect the behaviour of animals towards their neighbours. Scott (1948) found that delayed feeding increased aggressiveness in a herd of goats. After feeding, there was a decreased aggressiveness of dominant animals. When a well-distributed and ample feed supply was available, fighting and obvious dominance activity was eliminated.

From these observations, a theory of social force fields is proposed. The behaviour of animals towards neighbours arises as a response to stimuli provided by the neighbours. These stimuli are the species and caste characteristics, morphological and behavioural, of the animals. They appear to have two effects: one is centripetal or mutually attractive, due to the so-called "social releasers" (Tinbergen, 1951); and the other is centrifugal or repulsive, which maintains spacing between animals. This intraspecific spacing results in the individual distances observed by Hediger (1950).

The reaction to species and caste traits is not independent of the distance between animals, and in this the animal behaves as though the area in its immediate vicinity, particularly in front of its face, is its own "territory". In and over this area, it may be said to exert a social force. These social force fields are repulsive and directional. Their intensity is related to the social dominance of the animal. That is, animals higher in the social order maintain more intense and presumably larger force fields than do animals lower in the dominance hierarchy. The dominance order governs the priority in keeping these field areas clear of neighbours. Animals low in the social order attempt to avoid the social force fields of dominant neighbours: if they are unable to avoid the dominant neighbour's force field, they adopt a submissive posture while within this area. The intensity of the force fields can be controlled: it increases as threat is communicated and decreases when an animal submits.

It is suggested that this social force is the primary manifestation of intraspecific aggressiveness. It affects individuals of the same species, and particularly of the same caste, which intrude into a particular area in front of an animal. The effect of these forces is to control the spacing between animals, thus giving rise to the individual distances of Hediger (1950). The mutual attraction between animals (controlled by social releasers) brings animals to the limits of the force fields of their neighbours. At this range, the movement pattern of animals is concerned with minimizing the social forces operating on them.

Animals differ in the intensities and magnitude of their force fields, and these are learned by other members of the group by agonistic contacts (King, 1963).

Males appear to have greater force fields than females. All show increases with age, and most vary with season, usually reaching a maximum in the breeding season. There appears to be a diurnal rhythm in force field intensity: many animals aggregate closely at night with little agonistic behaviour. In the domestic fowl and turkey, the forces appear to be maximum in the morning and evening and lower in the heat of the day, and during group activities such as feeding and moving.

The caste characters which control the force fields are under hormonal control and often show graded series related to the dominance hierarchy. For example, McAlister (1958) found a correlation between social rank in *Gambusia hurtadoi* and the intensity of the yellow markings on and at the base of the caudal fin and on the ventral portion of the caudal peduncle.

The term force is borrowed from Physics: it suggests something measurable with the possibility of a quantitative approach to social relationships. Work is proceeding in this problem. The measurement of social forces is more likely to be on a relative than on an absolute scale. In other words a matrix of forces will operate between the $\frac{1}{2}n(n - 1)$ combinations of animal pairs within a group of n individuals. The magnitude of the forces will only have significance relative to the other social forces operating between pairs within the group.

The force normally operating between a pair of indíviduals in a group appears to be a function of:

a. the social position of the dominant;
b. the social position of the subordinate;
c. the difference in the social positions;
d. the frequency of agonistic contact between the pair;
e. the size of the group.

The relative agonistic contact frequency is the operational component of the function.

Earlier, the field of social force exerted by an animal was described as a "territory". This was done deliberately, since the parallel between normal territories of animals and these social force fields is close. This will be considered more fully below.

(b) Formal regulation of intraspecific aggressiveness:

Most social organization is concerned with the regulation of aggressiveness within a group. The two main systems are the social dominance hierarchy and territorial patterns of organization. The systems are not mutually exclusive in any way, and many animal species change from one to another and back again each year.

During the formation of either system, animals adjust to the changing force fields which are under hormone control. In the early stages of this adjustment considerable fighting occurs, and this is gradually reduced and takes on a formal expression as the organization develops in either system. Both systems are organized within castes, though in some species, in the non-breeding season, the males and females constitute a single asexual caste and form a common dominance hierarchy; for example, the ring-necked pheasants in winter (Collias & Taber, 1951).

The dominance hierarchy system has been most thoroughly studied in the domestic fowl, where it has been described by Guhl (1953). The socializing mechanisms involved in the regulation of intraspecific aggressiveness into a dominance hierarchy will be discussed below. However, the sequence of aggression followed by submission is common to both systems. This sequence is concerned in each specific dominance relationship between pairs of animals. In dominance hierarchies, these may be uni-directional and based upon "peck right" in the fowl (Allee, 1952); or they may also be bidirectional (Allee, 1952), as in the ringdove (Guhl, 1961). In the territorial system, they are bidirectional, but this is regulated by territory and the bidirectional system appears to be associated with territory (Castoro & Guhl, 1958). The territorial system was described by Carpenter (1958) as a system reducing stress, pugnacity, and non-adaptive energy expenditure, and this description could equally well be applied to the dominance hierarchy.

Social force fields generate individual distances, ranging from small areas associated with the face in aggregated groups to quite large areas in some species during the breeding season. While the small force fields move with the animals in aggregated groups, the force fields are stabilized in space under the territorial system. This occurs when the territories are large, as they are when they must provide food or nesting sites, or when they fix in space a competitive situation which occurs when some areas are superior to others: an example is the beach-front mating territories of the elephant seal (Bartholomew, 1952).

Dominance and territorial systems both regulate the supply of such essential facilities as mates, nesting spaces, or feeding space. Where the availability of these varies in different areas, the territorial system regulates their supply by regulating the control of space. The dominance hierarchy, on the other hand, is a more flexible system, in that it regulates facility or commodity supplies in whatever competitive situation they occur. For example, in a flock of hens, dominance gives priority at the feeders, waterers, nests, roosts, and dust baths; and for males, availability of mates (Guhl, 1953).

Many methods are used by animals for marking territorial areas; scent from bodily secretions or excreta is used by many species, e.g. by urine in dogs, by chin glands in the rabbit (Mykytowycz, 1962), and by scapular glands in the flying-fox (J. Nelson, pers. comm.). Sound marking is used by song birds. The territorial marking has the same effect on a stranger as does the presence of the controlling animal. Aggressiveness increases as an individual enters either the territory or the force field of another.

While the territories are an expression of force field, their stabilization by marking serves, with time, to reduce the amount of contact between animals and thus the amount of aggression.

When territorial rights are respected, aggression and submission take on the ritualized patterns associated with low levels of aggressiveness. These are also observed in dominance hierarchies.

Both territory and social dominance have been shown to be increased by injections of testosterone propionate (Bennett, 1940; Guhl, 1961).

(ii) *Submission.* Lorenz (1952) pointed out that when intraspecific aggressiveness was present in a gregarious species, associated as it often is with powerful weapons of intraspecific offence, there must be some mechanism to prevent its operation from harming the individuals of the species. That is, there should be some mechanism to stop an act of aggression when one individual has been defeated. This mechanism is that of submission.

(a) Flight:

The simplest and one of the most common forms of submissive behaviour is flight. However, if flight is not to be followed by pursuit and continued aggression, flight itself must be an acceptable act of submission; that is, it must dissipate the aggressive drive of the victor. This can be seen in domestic animals, which are normally restrained so that effective flight is not possible. Here, a slight avoidance movement appears to have the effect of restraining the victor, though the latter may continue to give a formal nip on the neck in pigs or a peck on the comb in fowls. Probably the acceptance of such a formal flight pattern instead of complete flight represents an adaptation to high density conditions. Such adaptations have a special significance in domestic animals. In terms of the force field model, flight operates by removing the vanquished animal from the force field of the victor. This can be achieved by retreat, turning aside, or a combination of the two.

(b) Force altering behaviour:

Lorenz (1952) described the motionless exposure of the neck of a submitting wolf after a fight. In the domestic fowl, in the baboon, and in the rhesus monkey, the submission behaviour signal seems to be basically sexual in origin, though formal versions of these signals are more common in integrated groups. In the hen, the intense version of the submissive signal is a sexual crouch. This can occasionally be seen after a conflict between two birds. More rarely the dominant bird grasps the comb of, mounts, and attempts to treat the submitting hen. One can demonstrate this sexual submission signal very easily by walking quietly among a flock of domestic hens and giving a slight but sudden movement of the foot beside a hen. It will normally respond with a sexual crouch. The formal equivalent is a lowering of the head for a peck on the comb. The use of sexual presentation as a submission signal between two males or two females has been commonly described in apes. Zuckerman (1932) has described this pattern in baboons under the abnormal restrained conditions of a zoo, and Mason (1961) and others have also described this pattern in the chimpanzee. Morris (1955) referred to this as pseudofemale behaviour.

The requirement of submissive behaviour is that it shall prevent continued aggression. The sexual type in the fowl and the baboon illustrates this well. After a conflict, the loser presents sexually. The dominant animal thus receives a signal which converts its aggressive drive to a sexual drive. It attempts to mate—abortively—thus inhibiting the sexual drive. The pertinent point is that aggression has ceased.

If the reaction to aggression involves certain types of sexual or infantile displacement behaviour, these may provide the basis for the evolution of submissive behaviour patterns. This type of behaviour is basically intercaste and therefore not subject to normal intracaste force fields.

In the pig, a modification of the distress squeal of a young pig is the intense form of submissive behaviour. Possibly the significance of the neck in submission in wolves (Lorenz, 1952) is that cubs are normally carried by the neck. These both appear to be of infantile origin.

A more subtle form of submissive behaviour can often be observed in well-integrated groups, particularly of cattle and pigs. In cattle, when two animals are standing together, the subordinate beast generally has its eyelids lowered and its neck and head angles nearer to horizontal than its dominant neighbour. By such a posture it is continually communicating submission and presumably lowering its own force field.

(c) "Acceptance" of submission:

It is apparent that the dominant animal is involved in the "acceptance" of submission.* Any form of displacement behaviour activity as a response to aggression is not likely to be effective while the vanquished animal remains within the area of the force field of the victor and is itself exerting a social force. Thus the submissive behaviour must remove this force field of the vanquished animal and the stimulus to aggressiveness it provides to the victor. The "acceptance" occurs when the stimulus to aggressiveness is removed by flight or when the aggressive drive is inhibited or transferred to some other drive as a result of the stimulus provided by some intercaste behaviour of the vanquished animal.

(iii) *Socializing behaviour.* While submissive signals operate to eliminate aggressive drives in specific pair encounters, there is a group of mechanisms which together bring about socialization. These lower the general level of aggressiveness between individuals in a social group, thus allowing aggregation to occur. The factors to be described have this in common, they all allow individual spacing distances or force fields to be reduced.

(a) Recognition:

Non-social animals maintain their spacing by epideictic display (Wynne-Edwards, 1962). When pairs meet there is a contest, sometimes involving a fight, sometimes highly formalized. Aggregated animals on the other hand meet regularly, yet generally only settle their dominance relationships once, and these are learned and reinforced by a series of repeated threats and submissions. The result is that individuals within an aggregated group learn to recognize each other. The dominance type of social organization involving close aggregation could not exist without individual recognition.

Recognition has three components, individuality, identification, and memory (learning).

In cross-fertilized animals there is no problem of identity, since each individual is genetically unique (with the exception of identical twins). This genetic individuality shows in all aspects of an animal's phenotype, in both morphological and biochemical traits. Kalmus (1955) showed that trained police dogs were able to distinguish between the scent of unrelated individuals but did not normally distinguish the scents of

---

*The use of the term "appeasement behaviour" (Kikkawa, 1961) illustrates this point.

identical twins. These body scents are the normal means of identification in dogs, and they recognize each other by sniffing each other's anal glands. The evidence of Kalmus (1955) suggests that each individual has its own unique identity in this character. That animals have morphological individuality is clear to anyone who has ever observed them.

The senses that animals use for identification are sight, sound, and smell. Guhl (1953) studied the means by which domestic fowls identified each other, using attacks by penmates as the criterion of non-recognition. He found the recognition was visual and that alterations to the head and comb caused more non-recognition than alterations to other parts of the body.

Bartholomew & Collias (1962) described the use of sound for the location of their pups by elephant seal females when they return from feeding trips. Very young pigs respond immediately to the feeding call of their own dam but not to the feeding call of a sow in an adjoining pen.

Very many animals use scent for individual identification; dogs are probably the best example because of their practice of sniffing each other's anal glands. Pigs will gather and sniff a newcomer to a pen for some minutes before they attack it.

Memory, or length of unreinforced recall, has not been widely studied, but Guhl (1953) reported that fowls failed to recognize each other after a period of approximately two weeks: once again he used attack by previous penmates as his criterion of non-recognition. The author has observed attacks on nine weeks old littermates, in pigs, after a separation of only one week.

The number of individuals which an animal can recognize is important, since it will determine either the size of a group or the size of a subgroup and the degree of isolation of subgroups. McBride & Foenander (1962) found that domestic hens in flocks of eighty tended to restrict their areas of movement to one-third of the available area so that they would normally meet about two-thirds of the birds in the flock. It was postulated that the area of movement would reflect the number of birds a hen could remember. This number is probably of the order of fifty in the domestic fowl. A good indication of the presence of unrecognized individuals in a flock of hens housed together was reported by Morgan & Bonzer (1959), who observed many pairs of birds fighting after a flock was moved to a new pen.

In domestic animals, recognition of other individuals often appears to operate at two levels; that is, subgroup formation occurs in some groups. When McBride & Foenander (1962) removed the fence between two flocks of the same inbred line of domestic hens, fights occurred between individuals of the two flocks, and an open area or "no man's land" developed along the old fence line. Few birds crossed this line during three weeks' observations. The behaviour of birds towards individuals of the same original pen was quite different from that towards individuals of a different original flock. Hale (1956) showed that there was a tendency for birds of any breed to generalize their dominance reactions towards individuals of a different breed after an initial encounter with a member of the second breed. An unpublished study showed that when individuals of four different breeds of chickens were intermingled at either a few hours or ten weeks of age, the level of interaction between members of the same breed was quite different from their level of interaction with individuals of the other breeds. In both cases the behaviour of birds towards others of the same breed was different from their behaviour towards individuals of the other breeds. The experimental conditions were not such as to eliminate the type of self-recognition implied in the "species learning" aspect of imprinting proposed by Lorenz (1952).

Though members of the different breeds in the studies reported above did not behave towards each other as would strangers, the basic element of discrimination necessary for subgroup formation was present. Recognition, not only of individuals but of breed type, was necessary to provide the basis of the discriminatory behaviour observed.

Recognition operates in aggregated groups by allowing formal dominance relationships to exist between recognized pairs of individuals. Thus its basic role is socialization since it allows individual distances to be reduced.

Another type of recognition may be associated with the schooling type of aggregation. This involves the use of a hive or colony scent on all individuals of a group, e.g. in the ants. This allows recognition of large numbers of individuals within a group. Aggressiveness is only displayed between individuals of different colonies, and the colony is the unit of lowered aggressiveness and thus of lowered approach distances.

(b) Asocial behaviour:

What may be described as asocial behaviour is often observed in a flock of domestic fowl. What appears to happen is that birds can engage in some forms of behaviour which "switch" off their force fields. Feeding, self-grooming, apparent sleep or "eye-shutting" are examples.

One can often observe two hens approach each other in an erect posture beside a feed trough. They stand motionless for up to thirty seconds; then, as though by mutual consent, they lower their heads, turn to the trough and commence feeding. Submission is involved by the subordinate bird's remaining motionless, and the dominant bird breaks off the interaction and commences feeding first.

Again a hen can often be observed to approach another in a threatening posture; the bird approached will either turn and start grooming or squat and close its eyes. In all of these cases the birds appear to have "switched" out of the social field of forces. Self-grooming and grazing can often be observed in cattle in similar situations. Fowls often have their heads almost touching when feeding. If one raises its head so as to bring its beak close to the other, the latter may make a swift avoidance movement; alternatively it may raise its head and cause the first bird to avoid it. In other cases, both birds hold their heads raised a few inches for perhaps five to ten seconds and then resume feeding.

Another example of asocial behaviour was observed in turkey males in a pen which was halved in area at ten day intervals. Birds were observed standing along the fences facing outwards pecking at the wires.

Once again, asocial behaviour involves socialization, since it allows individual distances to be reduced.

(c) Play in animals:

Play has often been described by observers in many species. It takes several forms, though agonistic behaviour is very common. For example, in pigs, the initiator will usually scamper around the pen before running up to another animal, often a socially dominant pig, and biting the latter on the neck. A typical mouth to neck struggle will develop, often lasting for some time, with the subordinate animal commonly initiating each renewal of the struggle. One seldom observes an agonistic type defeat and retreat in such play. In dogs, play is initiated by a wagging of the tails after normal recognition formalities.

The justification for including play in the socializing behaviour category is that it appears to relax the social force fields completely for the duration of the play and reduce them for some time after.

(d) "Contact" behaviour:

Some socialization mechanisms involve physical contact between animals within a group. Zuckerman (1932) appears to have been the first to suggest that mutual grooming among baboons was not exclusively concerned with the removal of ecto-parasites and foreign bodies. He suggested (p. 58) that "it is perhaps legitimate to regard the picking reaction and the stimulus of hair as factors involved in the main-tenance of a social group in sub-human primates". One often observes, in cattle, mutual grooming of parts of the body which could easily be licked by the animal itself. There is a clear communication pattern based on agonistic behaviour involved in the initiation of grooming. The animal to be groomed gives a neighbour a slight bunt with its head and turns aside to bring the area to be groomed into close proximity to the mouth of the neighbour. The latter generally responds by licking the former, though sometimes the communication pattern is repeated several times before the licking commences. The sequence of bunt followed by licking (parental and intercaste behaviour) is strongly reminiscent of threat and submission.

Pigs are described as "contact" animals (Hediger, 1950), because they move and rest in physical contact with each other. This physical contact may operate as a socializing mechanism in pigs.

Since contact brings animals together, it is classified as socialization.

(e) Group activities (Allelomimetic behaviour):

It seems to the author that many of the activities carried out together by a group of animals are socializing (by definition) in function. For example, after the breeding season, many migratory birds come together; that is, their individual distances are reduced. During this period they spend hours flying together in circles. While this presumably prepares them for flying together in a group while migrating, it also seems likely that this enables them to adjust their behaviour by bringing about low individual distances and force fields until they can operate as an integrated group. The "come fly with me" call described by Lorenz (1952) in his jackdaws seems to be an example of this type of behaviour.

The group of activities described as social facilitation falls into this category. Social facilitation refers to the tendency of animals to "join in" when they observe another performing some activity such as feeding. The sight of an animal feeding is a stimulus to cause other animals to commence feeding. It seems to be this tendency which is partly responsible for the diurnal rhythmic patterns often observed in animals, e.g. Wood-Gush (1959).

This type of phenomenon may have the effect of keeping the force fields of a group of animals synchronized. Because different activities are associated with differ-ent individual distances, joint activities ensure that, at any given time, the force fields of the majority of animals are approximately equal.

(f) Ritualization of behaviour:

Ritualization, particularly of agonistic behaviour, brings about milder formal communication patterns, particularly of threat behaviour which increases social tensions. Agonistic communication patterns are not observed only in intense and mild forms. Instead, there is a complete range varying in intensity from one extreme to the other, depending on the specific set of circumstances.

The process of integration can be observed for a week or so after a group of strange fowls are brought together in a flock. During this period there is a change in the pattern of agonistic behaviour from the extreme form, involving fighting, down to the highly ritualized forms observed in the integrated flock. Formal threat behaviour involves milder force fields and thus smaller individual distances. It also requires only mild submission signals.

(g) Schooling behaviour:

Schooling behaviour in fish seems to be a non-aggressive aggregation of the animals, with elements of the other socializing behaviour mechanisms, i.e. asocial and group activities. Animals aggregate without overt aggressiveness, though their spatial relationships are well organized (Keenleyside, 1955). Sticklebacks and rudds aggregate closely in new quarters, when disturbed, or when feeding (social facilitation). Dispersal increases when animals become familiar with their environment and increases further when they are hungry. Male sticklebacks show greater individual distances than females, and these increase with the onset of sexual activity, the males taking up territories.

These schools seem to represent a juvenile type of social organization, involving close aggregation without overt aggressiveness. Keenleyside (1955) suggests that schooling behaviour may be similar to resting or sleeping. He quotes supporting conclusions from the studies of Holzapfel (1940) and Craig (1918). He suggests that schooling behaviour is appetitive and that, as in rest or sleep, the "doing nothing" states actually have consummatory value.

This interpretation of schooling behaviour would fit the situation in the asocial behaviour category, which could perhaps account for the extremely low intragroup aggressiveness. Nevertheless, the patterns of varying spacing relationships suggest that social force fields are still present, though at an intensity at which overt aggressiveness does not occur.

(iv) *Discussion.* The mutual attraction of the social bond is aided by socializing mechanisms and opposed by force fields resulting from intraspecific aggressiveness; this constitutes an ambivalent situation. It is the author's belief that intraspecific aggressiveness, which is subject to intragroup natural selection, is the basic component in non-schooling social behaviour and that the social organization is an adaptation to the existence of this component. The situation may be represented in the model shown in Figure 5.

The model suggests that animals are mutually attracted by the social bond and repelled by the social force fields. The balance of this ambivalent situation is modified by the socializing mechanisms operating on the force fields. Acts of aggression are terminated by submissive behaviour and result in learned dominance relationships.

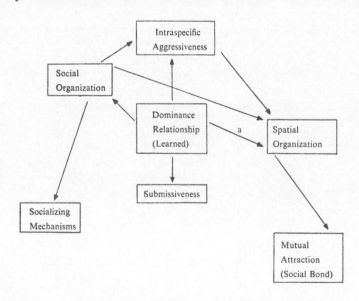

(a) learned by agonistic contact frequency

FIG. 5.—A model showing the relationship between the various components of social and spatial organization in animals.

A matrix of dominance relationships results in the social organization of the group which is learned by individuals. The relative frequency of agonistic contact between any pair of individuals results in the learning of approach distances to each dominant animal by its subordinates. This gives rise to spatial organization within a group.

During observations on turkey males, two pairs of birds were generally observed together. In each pair the dominant member of the pair seldom pecked or threatened the subordinate. Pairs of birds in which the dominant male pecked or threatened the subordinate frequently showed spacing distances greater than average, and these were maintained by the subordinate bird.

Another interesting pattern of behaviour was observed in these turkeys, which were restrained in relatively small areas. On two occasions after fights, birds dropped considerably in social position. The vanquished birds were persistently attacked, and spent over a day running almost continuously. Though they each only fought with one bird, after their defeat they were attacked persistently by all of their social superiors, and, during this period, some of their subordinates achieved dominance over them while they were persistently in a submissive posture. If the vanquished bird tried to attack these subordinates by adopting a threat posture, a well-established dominant male would immediately attack it. On the second day, the situation became less violent and the social force fields operated in the opposite way to protect the vanquished bird. In this situation, if any but the alpha bird threatened the defeated male, one of its social superiors would immediately threaten the aggressor. Sometimes an even higher bird in the social order would then join in. The chain reaction led to a moving queue or chain with the vanquished bird leading, followed by its aggressor, followed in turn by the latter's social superior, and so on. All except the first bird were in threatening postures, though these did not usually involve full male display. In a restricted area only the alpha bird normally showed full male display.

These observations suggest that, at least in restricted areas, the force fields operated directly to lower the level of overt aggression in the group. In force field terms, any except the alpha bird adopting a threat posture in a restricted area was within the force field of a dominant neighbour and would stimulate a dominant to attack it. This situation has also been regularly observed in flocks of domestic hens.

It is generally considered that a dominance hierarchy is made up of a series of independent pair dominance relationships. However if this were so, the probability of achieving a linear peck order would be extremely low.* Under these conditions, the probability of achieving a linear peck order in a flock of ten fowls would be quite negligible. Two factors could alter this situation. The first would occur if the birds formed an ordered series in aggressiveness, fighting ability, and all other factors which enter into the probability of success in winning dominance encounters. This is unlikely because of the existence of the "peck lag" effect (McBride, 1958), which operates to cause a correlation between the results of the succeeding encounters of each bird during the formation of a peck order. The second factor is a lack of independence in the results of pair encounters. This would occur if an animal's probability of winning a dominance encounter were influenced by observation of the other bird's dominance relationships with other flockmates of known dominance relationship.

Linear social orders are so common in small groups, yet it is highly unlikely that they could be formed by the occurrence of independent pair dominance relationships; this appears to have been overlooked in past studies.

$$* \quad N!^{\frac{1}{2}} \left[\frac{N(N-1)}{2} \right]$$

Since a lack of independence in pair dominance relationships implies a higher level of organization than is usually attributed to animals, evidence on this point would be desirable.

Guhl (1961) showed that aggressiveness and submissiveness, as well as the male and female sex drive, are mediated respectively by androgen and oestrogen. Presumably the intensity and size of the force fields in animals are regulated by a balance between these hormones. This balance, or the sensitivity of the tissues to these hormones, varies throughout the year, as do the sizes of the force fields: these reach a maximum during the breeding season. They also vary with age, reaching full development with sexual maturity. Thus each caste has its own balance, which may be expressed in individual distances in a social situation. Here, however, there may be some qualitative differences in the nature of the stimuli provoking aggressiveness within a force field, as well as differences in field size. These stimuli would be concerned with secondary sex (or other caste) characteristics, both morphological and behavioural.

Independent of this basic pattern, changes in force fields occur throughout the day and are associated with specific activities. Examples are movement, rest, and feeding. When animals are hungry, they commonly disperse and increase their movement activity (Keenleyside, 1955). Sexually receptive females show similar behaviour (Young, 1961). The increased aggressiveness in hungry goats (Scott, 1948) suggests that the tendency towards dispersal with hunger may be regulated by the change in force field size.

Other types of organization in space will be considered later (spatial organization).

(3) Intercaste Behaviour

(i) *Courtship and mating.* In most animals, mating is preceded by a series of behaviour patterns collectively referred to as courtship behaviour. Courtship, leading as it does to intraspecific mating, has many aspects. However, animals generally are subject to a pattern of spacing behaviour which is regulated by the fields of social forces in which they move. One would therefore expect that part of the behaviour leading up to mating would be concerned with the lowering of individual distances to zero. One would also expect this aspect of courtship to be regulated by agonistic behaviour.

When one considers a non-social animal like the cat, the courtship precedes a mating which Scott (1958) describes as follows: "Even in mating, the male and female will yowl, spit, and claw at each other in a way which is difficult to distinguish from conflict." It appears as though courtship in this species reduced social forces to a threshold sufficient only to allow the pair to approach and mate. Even in social species the female maintains an individual distance from the male, unless it is in oestrus or in a period of sexual receptivity.

A number of workers have studied the agonistic aspects of courtship and mating and its effect on approach distances between the sexes. Marler (1956) has made some comparisons between the normal approach distances in chaffinches. Dominant females allowed subordinate females to approach closer than other females, and males allowed females to approach more closely than other males. Tinbergen (1953) suggested that the prenuptial behaviour of females appeased the males and suppressed the escape behaviour of females.

The agonistic component of mating can be seen in the domestic fowl, where dominant hens are less receptive sexually than subordinates (Guhl, 1961). This author suggests that the submissive attitude is a component of sexual receptivity: it is, however, possible that sexual behaviour has evolved a secondary function in submission.

Hinde (1953) found that in chaffinches the males were dominant in winter and the females were dominant in spring. The male display occurred in situations in which there was a balance between the approach tendency (courtship) and the tendency to flee from the female. Again he suggested that the attempts to copulate were unsuccessful unless the sex drive was adequate to inhibit aggressiveness.

Peterson (1955) observed that pairing in migratory bank swallows resulted from the persistent returning by a female to an area in the face of aggressive attacks by the male. After a pair-bond formed, both mates shared in attacks on other birds.

This tendency towards lowered aggressiveness between mates during pair formation has the normal elements of aggregated social group formation. Courtship leading to mating is still necessary within the pair. This can be observed in paired flying foxes in the breeding season. In this species mating occurs frequently each day at the height of the season, but considerable courtship, often in the face of strong aggressive responses by the female, is necessary before copulation can occur towards the end of the receptive period (Nelson, 1963). Collias & Taber (1951) found that ring-necked pheasants formed a single-caste social order in winter. In the breeding season, aggressiveness increased within sexes and decreased between sexes. There is probably a general inverse relationship between the aggressiveness within and between groups which is also present in these intercaste mating pairs. Moynihan (1958) also observed that pairing in gulls was followed by a reduction in intersexual hostility and the emergence of sexual behaviour.

The breakdown of individual distances during courtship leading to mating seems to suggest that external stimuli from the male lead to the crossing of a threshold beyond which sexual behaviour replaces repulsive intersex aggressive behaviour. The level of this threshold determines the amount of courtship necessary to cross it or even the possibility of crossing it. This level is under hormone control. Kislack & Beach (1955), working with hamsters, found that oestrogen increased aggressiveness and that progesterone alone did not reduce aggressiveness. However, oestrogen followed by progesterone eliminated fighting and reduced other forms of aggression. "The ovarian hormone that renders the female receptive also inhibits her tendency to attack the male."

An example of a courtship pattern leading to mating was described by Guhl & Craig (1962), in the domestic fowl. The sequence is as follows:

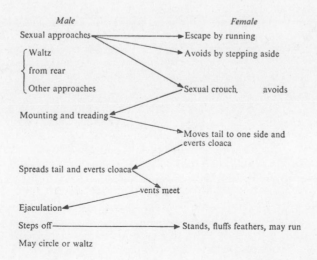

The ability of the male's courtship to stimulate the female to the mating threshold is determined in part by the sexual approach. Guhl (1962) showed that the "rear" approach was more effective than the "waltz" or "wing flutter" approaches. Guhl (1962) considered that the mating approach also contains an element of dominance; "cocks need sufficient aggressiveness to dominate the females. . . . .".

In force field theory, the female responds to the approach of the male by submitting in a sexual crouch or by flight, which is similar to a normal intracaste situation. By crouching, she provides the stimulus for the male to mount; if he does not, the sequence ends by the hen moving out of the male's field. The crouch, however, involves an elimination of her individual distance with regard to the male, so allowing the sequence to continue.

(ii) *Parental behaviour.* Scott (1958), has described two types of behaviour associated with parenthood, care-soliciting and care-giving behaviour. In parent-offspring relationships, care-giving is generally a response to specific stimuli associated with care-soliciting behaviour. However, this situation depends on the existence of a specific parent-offspring bond. In many animals, a parent will normally respond quite aggressively to young animals which are not her own (Mykytowycz, 1959). Thus once again the basic relationship between adult and young is agonistic, with some combination of recognition and care-soliciting behaviour responsible for the low level of aggressiveness within a parent-offspring bond. Once again it is extremely clear that the formation of this bond and its associated behaviour is strongly under hormone control (Lehrman, 1961).

Even within the specific parent-offspring relationship, there is an agonistic element which generally increases until the offspring reaches maturity. The rate of increase is not even throughout but it is subject to sudden increases at certain times, for example at weaning in mammals. This increase in agonistic behaviour is expressed in the consequent development of spacing tendencies. In the pig, the first expression of a spacing mechanism occurs when the sow first stands up after farrowing (McBride, 1963). She gradually rocks herself into an upright position and then slowly rises to her feet. Throughout, she gives a specific avoidance call or snort, with strong exhaling. This call appears to have the effect of causing the piglets to stand well away from her, huddled in a group. If one approaches her, she will nudge it firmly away; if it persists, she may take it in her mouth and throw it back quite violently. Her behaviour reinforces their tendency to avoid her. Her call sounds very similar to the threat call of a sow to another sow.

There are also several attracting calls at this age. The distress squeal or the feeding squeal (not well differentiated at this age) will cause the sow to give a feeding call and roll over to give the piglets better access to the udder. Her feeding call attracts the piglets to the udder. When a piglet wanders away from the sow to defaecate or urinate, it gives a location call. The sow immediately responds with her location call which attracts the other piglets to her snout.

As the piglets grow and become more persistent with their hunger squeals, the sow responds less frequently with her feeding call and more frequently with an avoidance call which inhibits the ingestive drive of the piglets. This avoidance call also resembles the sow's threat call. If the piglets persist, she again nudges them away and often gives one a sharp nip. The spatial balance in this ambivalent situation moves from close proximity at birth to the regular maintenance of individual distances at weaning. Nevertheless, during movement or sleeping, the pigs show typical contact behaviour and when feeding or moving they maintain contact with location calls.

Altmann (1960) has also considered this situation in the young moose and elk. In the moose, the first change in dam-offspring relationship occurs at weaning, where the individual distances are increased. The second change occurs just before the next calf is born, when the dam "drives the juvenile away from her". Up to this point the offspring has ranked socially with its dam; it must now rank very low socially. The ambivalent situation of attraction to and rejection by the dam is resolved in terms of an individual distance from the dam. This is found by the offspring by trial and error. Two other alternatives are observed; one involves attachment to an older bull moose and the other the formation of a caste group of yearlings; both involve lower individual distances than is possible with the dam. As sexual maturity approaches during the rutting season, the young males are forced to maintain great individual distances, as indeed are all males. The young female is also rejected by other females, now paired with bulls. In the elk, only the male goes through the extreme spacing procedure as it reaches sexual maturity; the females are incorporated into the harem groups.

(iii) *Flight Distance.* An interesting parallel exists between the behaviour of animals within the individual distance of a social dominant and their behaviour when approached by an animal of another species. In both cases the animals take avoidance action when approached closer than a specific distance. The interspecific situation has been described by Hediger (1950) as the flight distance of an animal; that is the minimum distance of approach to an animal before it flees.

This parallel suggests that flight distance and individual distance may be two aspects of the one phenomenon, the former operating between species and the latter within species and subject to caste differences.

(iv) *Discussion.* The main caste groups in animals are mature males, mature females, young males, and immature animals. Young animals characteristically show negligible individual distances. These increase with the onset of agonistic behaviour. In the pig this begins early, at about twenty-four hours of age. The development occurs slowly and is generally independent of sex until secondary sex characteristics begin to develop. From this stage the social situation develops more on an intrasex basis, with the males developing greater individual distances than the females. The females continue to increase the intensity and specificity of their force fields until sexual maturity. Males on the other hand suffer a check as they approach sexual maturity, due to the presence of other adult males.

At times in some species an anoestrus or asexual type situation develops, and this is associated to some extent with changes in secondary sex characteristics, especially in birds. At this phase the species may form unisexual groups, for example cattle and red deer; and, in others, the sexes may form a single group, almost homogeneous and asexual in organization, for example, in some migratory birds and the ring-necked pheasant (Collias & Taber, 1951).

The castes are primarily sexual, though at different seasons they may become more or less discrete subgroups within the general society organization. Age castes are not clearly discrete subgroups. Rather there is a continuous pattern of change, though the rate of change increases sharply at critical growth stages, particularly the onset of adolescence and sexual maturity.

The intercaste behaviour patterns associated with immature animals seem to be primarily agonistic, with the older animals dominating the younger ones. In this sense they are relatively simple. However, the two main types of intercaste situations, male-female and parent-offspring are qualitatively different and associated with specific hormonal states. Nevertheless, there is a strong element of agonistic behaviour in their expression. To this extent only, the force field theory can be evoked to account for their spacing and mutual approach behaviour. Obviously there is also a wide range of other specific behaviour patterns associated with these hormonal situations, though these have not been discussed.

(4) *Inter- and Intra-caste Behaviour*

(i) *Spatial organization.* Spatial organization in animals, in so far as it reflects individual distances peculiar to a caste at any season, must be considered separately for each caste in a society of animals. In addition, the spatial relationships of the castes, one to another, must also be considered. Most spatial organization is concerned with adjustment to force fields, both in aggregated groups and in widely spaced and territorial situations. Nevertheless other types of spatial patterns are associated with certain behavioural activities such as movement, feeding, and resting. On the whole, little is known of these patterns, but it is likely to be a rewarding field of study.

(a) Aggregated groups:

In aggregated groups, spatial organization can arise in many situations. The basic type of spatial organization seems to be concerned with maintenance of complex patterns of individual distances. In turkey males, the dominant animals are generally the most widely separated in a pen. A similar pattern was observed by Beilharz (1963), in penned dairy cows.

In turkey males of lower social status, spacing distances seem to be related to relative peck frequencies, particularly in the extreme case of animals usually found together. In such pairs, pecks are relatively rare.

Social dominance confers advantages in many ways, one of which is flight distance (Hediger, 1950). Beilharz (1963) reported that, in pens of dairy cows, the dominant animals retreated to the far fence when a man entered the pen. As he approached, the subordinate animals were those which were nearest, and these tried to break past him. The tendency of dominant animals to maintain central territories or central positions in a group may be a reflection of this tendency to maintain the maximum flight distance from possible predators. In a number of activities, the dominant animal occupies a position in the centre of a group. This is found in domestic fowls roosting at night (Guhl, 1953), and in pigs at feeding, where the fodder is strewn on the floor of the pen. In a group of cattle being worked in yards, the dominant animals are generally in the centre of the group. This is common in various territorial situations, for example the strutting grounds of the sage grouse (Scott, 1958).

Crofton (1958) has described an interesting pattern of grazing behaviour in British sheep. The flock breaks down into sub-groups of approximately six. The orientation of the animals with their heads down grazing is such that each animal keeps another two sheep each at an angle of 55° on either side of its head and body axis. This pattern seems to be concerned with maintaining their location relative to each other and to physical features of the locality, since Crofton found too that trees and other conspicuous features of the landscape were also incorporated into this pattern.

McBride & Foenander (1962) reported a structure, based upon territorial type behaviour, in domestic fowls penned intensively. This structure seemed to be an adaptation to large flock conditions by birds which can recognize only a limited number of their penmates.

There have been a number of cases of organized movement reported in animals. One interesting example is described by Washburn & De Vore (1962), in baboons. When groups of baboons are moving, the dominant males, females, and young are in the centre, while two groups of subordinate males form an advance guard and rear-guard. If the group is attacked from behind, the advance guard and central body accelerate while the rear-guard does not (Washburn, pers. comm.)

These spatial arrangements concerned with normal activity situations take many forms. Three important types have been described, one concerned with memory of individuals, one with location relative to other animals and physical features of the landscape, and the third with the tendency of dominant animals to keep to the centre of a group. The last is probably related to priority in flight distances or advantages in a specific location, if it is formalized by territory formation.

(b) Territory:

Territories in animals are fixed areas which are maintained, marked, and defended by individuals of a group against intrusion by other members of (usually) the same caste and species. Defence is generally carried out by the male, and sometimes by the female or both sexes. Territories may be permanent or seasonal, and usually some territories are "better" than others in terms of the facilities they provide.

The hypothesis suggested here is that territories are force fields stabilized in space. They appear to have all of the properties of force fields, and their stabilization seems to be equivalent to a socializing mechanism. This is not in the sense of lowering individual approach distances, though the distances animals can approach on their own territories without conflict is reduced. They are equivalent to socializing mechanisms only in the sense that they formalize the force fields in space and lower the degree of agonistic behaviour in a group.

Territories in social animals are chiefly concerned with mating, feeding, or nesting. At these periods one also expects force fields to be maximum in size. Territories may be for mating only, for example the lekking areas of the sage grouse (Scott, 1942). In addition, they may serve to include the females with or without young, e.g. the elephant seal territories described by Bartholomew (1952). Mating territories are generally associated with polygamy.

Nesting territories may or may not be associated with feeding areas, and Mykytowycz (1960) separated feeding and shelter territories in the rabbit. The word nesting is used in a broad sense to include all associations for the purpose of rearing young—there may be no fixed nest-building behaviour. The flying fox (Nelson, 1963) provides an example of nesting territories, and the rabbit or prairie dog (King, 1955) provide examples of nesting, mating, and feeding territories.

All territories do not have equal values, and considerable competition occurs for favoured situations and large-sized territories. Mykytowycz (1959) showed clearly that the dominant female in a rabbit group maintained control of the only burrow, while subordinate females lived and bred in unprepared surface shelter sites. This led to considerable differences between the reproductive rates of the dominant and subordinate females. Bartholomew (1952) showed that dominant elephant seals maintained mating territories suitably placed on the beaches, while subordinate and young males had territories on the outskirts of the mating areas, and these were seldom observed with females.

It would appear that populations are regulated partly by this territorial system (Ratcliffe, 1958). Animals able to secure suitable territories appear to suffer less predation than animals forced to occupy fringe territories, such as those in single caste (usually male) aggregated on the outskirts of the group (Mykytowycz, 1959; Bartholomew, 1952). Ratcliffe (1958) suggests this provides a mild population regulating system. Further population regulation seems to be associated with the social stress involved in territorial and society organization breakdown, e.g. Ratcliffe (1958), Deevey (1960).

There are other territorial situations beside the two major groups described. The most common of these are found in sedentary or semisedentary animals, for example limpets (Scott, 1958), and aggregated group territories, for example the howling monkeys described by Carpenter (1934), Altmann (1959), and the wolves described by Scott (1958). In these two cases, the group moves around the territory, which is marked by sound (monkeys) and scent (wolves).

Wynne-Edwards (1962) has drawn attention to the scale on which territorial size and desirability must be measured. The scale is in terms of availability of the facilities for which the territory operates; in feeding territories, the size will be related to the long-term harvesting of an area without depletion of resources.

(c) Breakdown of spatial organization:

Force fields lead to spatial distances between animals, and organization occurs as a result of these. However, if density increases beyond the level which will allow this spatial organization, one obtains abnormal behaviour associated with what Calhoun (1962) has described as a "behavioural sink". This may occur as a result of either restraint on movements or increased numbers. The situation described by Calhoun (1962), in rats, and Zuckerman (1932), in baboons in a zoo, arose as a result of physical restraint. The situations described by Deevey (1960) in migrating lemmings under conditions of very high density represent a complete breakdown in normal spatial organization, while the breakdown under conditions of restraint is not quite so complete. Such situations do not normally occur in domestic animals, even at extremely high densities.

The greater the spacing tendencies operating at the season of plague in rabbits or lemmings, the more complete is the breakdown of the spacing behaviour. Heape (1931) says that, in the emigration that occurs under these conditions in lemmings and springbucks, they become "fatalistic", losing all timidity and courting danger. They have no fear of man (Elton, 1942); rabbits moved from their own territories appear to behave similarly (Ratcliffe, pers. comm.)

Social stress under these conditions is extremely harmful to the animals and is generally associated with hypertrophy of the adrenal cortex (reviewed by Deevey, 1960), and such animals die easily. If a domestic cockerel is placed in a standard commercial laying cage for single birds with wire partitions, between two well-established males of the same age, it will generally die within three weeks. Probably a more common situation is that the additional stress of reproduction (Deevey, 1960) is responsible for many of the deaths of adults. A lower fecundity of the female, due to resorption of foetuses and loss of young, is the important factor associated with social stress. The effect of this mortality on population density control has been referred to above.

(ii) *Communication.* Communication is an integral part of all aspects of social behaviour. It has evolved primarily to serve the needs of the social group. Thus the level and range of communication is likely to be correlated with the complexity of the social organization. It is involved in all of the behavioural components referred to above. Since more of these are found in closely aggregated groups than in semi-social species, the range of communication is maximum in aggregated groups comprising males, females, and young, and is observed in all inter- and intra-caste situations.

It is the author's view that communication is basically emotional in character. That is, it is concerned with the communication or stimulation of emotional states. As such it is not purposeful. This can be seen by its use in interspecific situations, where it usually has no significance. This is in contrast to communication as we understand it, where our communication is concerned with both emotional and non-emotional information. Further, we have the ability to make a purposeful use of emotional communication, such as the smile. Even here, the smile can occur as a response to an appropriate stimulus, even in a non-social situation.

The model proposed here for this type of communication in animals is as follows:

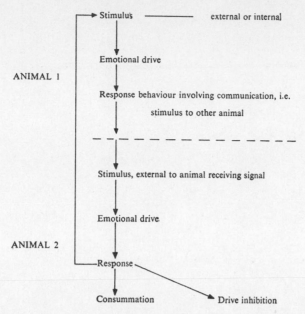

For example, when a large bird flies over a flock of domestic hens, the sequence is as follows:

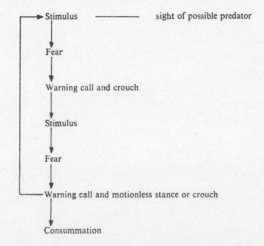

The warning call in a flock of fowls is taken up almost simultaneously by all birds. It is a low drone which rises and falls slowly. This has non-locating properties, since the consummatory response is to remain motionless. It is extremely difficult to determine which bird started it. The motionless response is rather ineffective in domestic fowls, which do not have the protective coloration of their ancestors.

Communication can involve sight, sound, smell, and touch; probably visual communication is the most common, especially in aggregated groups. Since so much of the movement in animals is related to the presence of other animals in the vicinity, almost every movement is the stimulus for a response by another animal; this response may be to move or to adopt some specific posture, perhaps threat or submission. Continuous communication of submission by posture may occur in stationary cattle.

109

Vocal signals are more widely used in some species than others. Their use as threat calls in song birds, or in the "bugling" of male elk, is concerned with maintaining greatly extended force fields often formalized by territories. Vocal signals are also used in the maintenance of group territories (Altmann, 1959). Sound is also used by animals with poor sight, such as pigs, which have a higher proportion of sound signals than most domestic mammals. Sound is commonly used for location (social distance), when animals lose visual contact with one another. The author observed the following pattern in pigs at night. The pigs (approximately six weeks of age) approached the sow and gave their feeding squeals. The sow failed to respond. After a while the piglets left the sow one by one to walk about 20 yards to a self-feeder. As each piglet reached a distance of about 4 to 6 feet from the sow, it started to give its location grunts, which ceased when it arrived at the feeder and commenced feeding with its litter mates.

Scent is widely used for communication of territorial information. Many animals, particularly the males, have well-developed secretory glands and spend a considerable proportion of their time marking territory boundaries, e.g. rabbits (Mykytowycz, 1962), and flying foxes (Nelson, 1963). Scent is commonly used in mammals for communication of oestrus information, e.g. female dogs.

Touch is often used in the communication sequence leading to mating, usually by the male muzzling the female's genitals. When piglets are born they locate the teats by pressing their snouts against the sow and moving "down hair". The hair orientation pattern is such that the piglet is led to the midline of the udder, thence outwards past the nipples. Thus the hair pattern of the sow contains information on the location of the teats. This is transmitted to the piglets by touch (McBride, 1963).

The stimulus-response aspect of communication can be observed in interspecific situations. The hen will give a sexual crouch (submission), and the pig a submissive squeal, when suddenly disturbed by a man. These have little communication value. An important exception is the human-dog situation, where familiarity has enabled individuals of both species to learn something of the meaning of each other's communication signals.

III. Synthesis: The Organization of Societies

(i) *Hypothesis*

(a) Simple Societies:

Spatial relationships within animal castes (simple societies) are the result of an equilibrium between intraspecific aggressiveness, which increases spacing, and socializing mechanisms, which decrease it. The equilibrium is under hormone control; it varies with age and season. In aggregated societies, the spacing also follows a diurnal rhythm, with maximum individual distances usually occurring at dawn and dusk.

The spacing between animals is a reflection of the social force fields, and is learnt and reinforced by agonistic contact between individuals. The spacing is proportional to the frequency and intensity of these contacts. Variations in spacing organization give rise to a spectrum of society types, ranging from extreme spacing of non-social animals through territorial social animals to aggregated groups showing well-developed socializing behaviour patterns. The situation is represented in Figure 6.

This model suggests that caste formation is under hormone control, which leads to the development of caste characteristics, both morphological and behavioural. Though the main castes are male, female, immature males, and young, at times the castes may join to form a homogeneous single caste, usually organized as a dominance hierarchy or school.

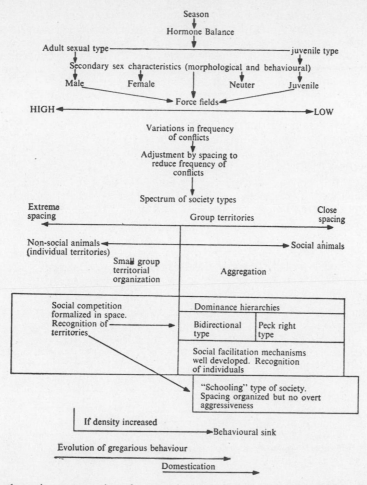

FIG. 6.—A schematic representation of a general model of adult animal dispersion and aggregation.

Caste characteristics are releasers for intracaste aggressiveness expressed in the development of force fields in which these releasers operate. Agonistic encounters enable animals to learn the individual spacing distances dictated by these force fields. The scale of these distances varies between species, but is related to availability and long-term harvesting of facilities (Wynne-Edwards, 1962).

These spacings within a çaste (simple society) give rise to a spatial organization of a society. All behaviour not directly concerned with spacing must operate within and adjust to the matrix of force fields responsible for the spacing. The spacing is generally extreme in the breeding season, and may be formalized at this time (or throughout the year in some species) by the formation and defence of territories, marked and fixed in space. Territories appear to be fixed force fields which regulate in space (size and suitability) the dominance situation among individuals.

The territories may be maintained by individuals or by groups of various size. The organization of animals within these group territories is generally a dominance hierarchy. This may arise from the territorial type of organization when animals are aggregated—either by man or by the development of a more intense social bond. Perhaps the mechanism described by Calhoun (1962) as a "behavioural sink" may be an example.

111

Schools may develop as primitive aggregations without aggressiveness; they may precede the sexually mature aggressiveness situation, which may revert to the school, or may develop aggregation in the "adult" form with aggressiveness, i.e. a dominance order. Alternatively schools may arise from and within the territorial system by an alternative recognition pattern, the "hive scent" of the hymenoptera. Schools often show spatial organization which is generally greater in males than females and varies with different activities. The spacing is maintained without overt aggressiveness.

Within the simple territory situation, agonistic behaviour is regulated and formalized by the location of animals relative to their own territories. In aggregated groups, agonistic behaviour is modified by a series of socializing mechanisms which lower individual distances.

It is suggested that the dominance hierarchy system is a more flexible system, allowing the priorities associated with social dominance to be applied to a wide range of competitive situations which cannot all be expressed in fixed spatial terms.

If, by density increases or by restraint, animals are forced to intrude into each other's force fields, whether territorial or free, social stress develops in individuals. A hypertrophy of the adrenal cortex follows with an associated change in the hormonal environment of the animals. This leads to increased mortality and decreased female reproductive rate. Animals fail to avoid man and other predators, and all social organization appears to break down.

The social organizations of various species range along this spectrum, which represents an evolutionary sequence towards or against gregariousness. Such an evolutionary movement has occurred during the process of animal domestication, since domestic animals do not develop serious abnormal behaviour at high densities. Associated with this movement there has also developed a tendency to include man in the social organization of the domestic species. This is most apparent in the dog (Guhl, 1962).

Within a single species, there can also be seasonal movement along this spectrum. This may merely involve greater spacing within an aggregated group during the breeding season, or a change from aggregate to territorial structure. Generally, the spacing and the level of intense agonistic behaviour is maximum in the breeding season.

(b) Complex societies:

Complex societies are those involving two or more castes. Each caste maintains its own separate organization, which is reflected in its spatial organization. The simple societies overlap each other in space, so that intercaste behaviour is present.

In aggregated groups, males and females each form their own separate dominance hierarchy (e.g. Guhl, 1953; Mykytowycz, 1958). In territorial species this also occurs. For example, in the rabbit, Myers & Poole (1959, 1961) reported that male home ranges were larger than female home ranges and the home range of dominant animals larger than those of their subordinates of the same caste, except under extremely crowded conditions. In this case territorial breakdown occurred, and all animals in the group moved over the same area irrespective of dominance or sex. This work and that of Mykytowycz (1958, 1959) show that rabbits form breeding groups of two or three males and several females. Social hierarchies develop within each sex, and females set up their own territorial system with the dominant female in possession of the only burrow.

In monogamous species where aggregation occurs, both members of a pair often hold the same social rank (Lorenz, 1952): in territorial situations the pairs operate as social territorial groups and defend their territory together (Peterson, 1955). This differs from the usual situation where territorial defence is generally a responsibility of males.

Intercaste behaviour is found in complex societies. Though functionally this is not primarily concerned with spatial organization, it operates in a matrix of intra-specific social force fields. Thus it involves adjustment to the force field situation when animals of different castes adjust spatial relationships to one another. Several authors have drawn attention to this aspect of courtship and mating behaviour (reviewed by Guhl, 1961). Parental behaviour involves stimuli given by the young which allow close approach by its own parent. However, spacing increases as the young approach maturity and change caste (Altmann, 1960).

The intracaste organizations in complex societies alter separately in response to seasonal changes.

(ii) *The classification of societies.* Any logical classification of animal societies should rest on evolutionary considerations. While these will be dealt with more fully in the next section, the spectrum of simple society types along with the concepts of simple and complex societies provide an adequate basis for society classification. The spectrum model provides a plausible pathway for evolutionary change, yet it says nothing of evolutionary direction in any group. One may argue that evolutionary changes in terms of increasing complexity and level of organization have been the general pattern of social evolution: nevertheless, within any group, changes probably occur in either direction along this spectrum and may precede, accompany, or follow speciation.

In addition, evolutionary changes along a relatively constant spectrum have occurred independently in a large number of animal groups. Thus the basis for taxonomy within any particular group must be in terms of such a general basic evolutionary pathway, rather than through a specific evolutionary hierarchy as is normal in phylogenetic taxonomy.

Since relatively few animal social organizations have been adequately described, the proposed basis for classification must be tentative only. Developments in this field will occur in response to a demand for an adequate taxonomy and with increased interest in the field of society evolution, which seems to be as important a problem as species evolution.

The system of classification proposed is hierarchical and descriptive at each level of classification. The criteria for description are as follows:

A   Society either stable throughout the year or subject to seasonal change.

A1   If subject to seasonal changes, the number of phases of the society, e.g. asexual phase, mating phase, or reproductive phase, should be listed. The relationship between phases should be stated, i.e. migratory at beginning and end of asexual phase. The periods of the phases should also be stated.

B   Caste structure for the stable society for each phase of the seasonal type society. The caste structure should include the number of castes, whether they form a simple society, and/or a statement of the castes which form a complex society. In the case of complex societies at mating, polygamy or monogamy should be indicated as well as the numbers of individuals of each caste which form social groups or subgroups.

C   Spectrum values for each caste at each season. The spectrum values suggested are: non-aggregated, territorial, bidirectional hierarchy, "peck-right" hier-archy, "school" type. Further classification is desirable in the case of territorial and dominance hierarchial systems and, though these are treated separately, organization may consist of a combination of these systems.

C1   Where territorial behaviour is involved, the type of territory or territorial function should be indicated, i.e. mating, mating and harem, nesting, nesting and feeding, or aggregated group territory.

The territory-marking mechanism and spacing distances should be stated.

C2   Where the society is a dominance hierarchy, the degree of aggregation (individual distances), the size of the group, and, where possible, the types of socializing mechanisms should be stated.

The following is a proposed classification of the European wild rabbit social organization in Australia, based on the observations of Myers & Poole (1959, 1961), and Mykytowycz (1958, 1959, 1960, 1961). This example illustrates most aspects of the proposed descriptive classification, though some information is not available, e.g. individual distances.

A   Society seasonal.

A1   Phases: Asexual (dry summer), prebreeding (autumn), reproductive (winter and spring).

B1   Asexual phase: Group feeding—male, female, and young castes. (Plague or high density phase: Group behaviour which shows no apparent organization).

C   Spectrum male and female peck-right hierarchies—Location preference in squats, burrows and logs. Submission by flight. Social facilitation, feeding.

B2   Prebreeding phase: castes; territorial males, females, and surplus males.
Subgroups: complex, two to three males and three to four females.
Surplus males: separate, simple society.
Polygamy, though dominant female attracts dominant male.

C   Spectrum: Breeding males and females: territorial—peck-right combination.
Surplus males: non-territorial, school—young and subordinate males.
Dominant male: defends and marks territory subgroup by scent (chin glands).
Territory for mating (priority over other males), nesting, and feeding.
Dominant female: defends burrow—nesting.
Subgroup males and females form separate peck-right dominance hierarchies.
Submission: flight or prone posture.
Socializing: by social facilitation and mutual grooming.

B3   Reproductive phase: As for B2, except for complex societies including young caste; social facilitation by contact and play.

While it is possible to describe the social behaviour of rabbits much more fully, the descriptive classification presented includes all essential features of the social organization of this species of rabbits in Australia.

IV. Evolutionary Considerations

It was pointed out that evolution along the social spectrum, probably in either direction, has occurred separately within each major taxonomic group of animals. Though the range of social animal species is large, the conformity to the spectrum is extraordinary. Social evolution in this sense is probably the most comprehensive example of iterative or convergent evolution ever reported. This is an extremely powerful argument for the very early evolution of a basic component of social behaviour. This would appear to be intra-specific aggressiveness, which is probably expressed best in terms of social force field theory and consequent spacing behaviour.

It is probable that evolutionary change has been predominantly in the direction of increased organizational complexity, with consequent increased range of social behaviour. This increase in organization, which occurs at the aggregated group end of the spectrum, represents the behavioural adaptation to intraspecific aggressiveness.

Many authors from Darwin (1883) to Wynne-Edwards (1962) have considered the evolution of intraspecific aggressiveness. Darwin placed great emphasis on the role of sexual selection in the development of this (and other characters). He seemed quite clear in his understanding of what he meant by that form of selection which he designated sexual selection and was quite scathing in his criticism of those who would over-emphasize the role of female choice in sexual selection (Footnote, page 210). Nevertheless, this type of criticism is still applied, e.g. by Wynne-Edwards (1962), to sexual selection.

Selection in any form can only be expressed in terms of differences in reproductive rate associated with specific characteristics of animals. If there is additive genetic variation in these characteristics, selective changes will occur. Differences in reproductive rates can be associated with viability, longevity, mating success, fertility, embryo viability, ovulation rate, and so on. Darwin defined sexual selection as being concerned with mating success, "depending on the will, choice and rivalry of individuals of either sex". There is ample evidence that sexual selection for factors associated with social dominance does occur, e.g. in domestic cocks (Guhl, 1953), elephant seals (Bartholomew, 1952), and rabbits (Mykytowycz, 1959). However intraspecific aggressiveness may be subject both to sexual selection, especially in polygamous species, and also to other forms of natural selection. For example, it plays a part in the suitability of feeding territories of monogamous species, giving a reproductive advantage in rearing numerous healthy young. Wynne-Edwards (1962) was probably justified in shifting the emphasis to "social selection" for this type of character.

Two main types of selection were recognized by Darwin (1883), intra- and inter- group. Aggressiveness is primarily subject to intragroup selection, while the mechanisms which regulate aggressiveness, submission, territorial, and socializing behaviour, have probably been subject to intergroup selection.

Intragroup selection for aggressiveness operating through social dominance is probably relatively inefficient, since there are several important non-genetic factors involved in social dominance. Age is an extremely important component of social dominance, while many social influences are also observed. Lorenz (1952) observed that female jackdaws took the social rank of their mates. The study of Myers & Poole (1961) suggests that the dominant female initiates territorial activities by defending a burrow, while the dominant or younger males compete for the dominant female. Altmann (1960) found that offspring took the social rank of their dams, and the study of Mykytowycz (1961) suggests that an extension of this system may be responsible for a tendency for the dominant males to be produced by the dominant females. Here, however, this could be confounded by genetic, health, and age factors.

In addition selective changes are subject to selective limits or genetic plateaux (e.g. Lerner, 1958), which prevent indefinite responses to selection.

In spite of these considerations, intraspecific aggressiveness is a component of social dominance and, in any stable population, it is presumably maintained at a high genetic plateau or at an equilibrium level, balanced chiefly by intergroup selection for socializing mechanisms.

Wynne-Edwards (1962) has placed the basis for intergroup selection on a sound ecological foundation in terms of the efficient long-term harvesting of the food supply of an area. Well-integrated groups, in terms of lowered intragroup aggressiveness, are able to devote more of their energies to obtaining food.

Nevertheless intergroup selection does not appear to be a readily acceptable phenomenon to many biologists, e.g. Elton (1963). Probably the strongest case for intergroup selection was the work of Wright (summarized, 1951) who showed that a large population divided into small, semi-isolated subgroups provided optimum conditions for evolution. The mechanism of intergroup selection did not involve replacement of inferior subgroups by superior ones any more than intragroup selection means "the survival of the fittest". The emphasis was on migration of individuals from genetically superior subgroups to neighbouring subgroups, so providing the basis for genetic change within the latter.

Genetic changes arise primarily by the operation of the four forces of selection, migration, mutation, and chance. Intragroup genetic change is due to mutation, selection, and chance while intergroup changes are controlled primarily by migration, intergroup selection, and chance. If mutation and chance are ignored and populations are divided into subgroups, genetic differences which develop in the efficiency of subgroups lead to increased population size. This in turn leads to emigration of individuals and to the splitting of the subgroups (e.g., Southwick, 1962). The latter is clearly intergroup selection by a form of group "budding", while migration leading to genetic changes in neighbouring groups is also indistinguishable from selection.

Social groups are an important form of population subgroups in animals. The available evidence suggests that migration between social groups occurs, though is not common, e.g. Washburn & De Vore (1962). However when it does occur and is associated with reproductive success, the offspring from immigrants may well show some degree of heterosis, which could in turn lead to social and reproductive advantages. Evidence for this type of intergroup selection would be extremely difficult to obtain.

The work of Guhl (1950) and Castoro & Guhl (1958) suggests other mechanisms for the selective limitation of aggressiveness by the greater readiness to mate by subordinate females. Dominant females generally took longer to pair than subordinate females. Where any limitation of nesting sites is present, this slower pairing behaviour of dominant females could lead to selection for aggressiveness in the males and against aggressiveness in the females. It could also be responsible for the lower aggressiveness of females than males.

It is extremely important to recognize that intraspecific aggressiveness is only regulated within social groups, either within territories or within aggregated groups. *Strangers are normally attacked.* In fact there may be an inverse relationship between the levels of intra- and inter- group aggressiveness, e.g. Myers & Poole (1961, page 21).

This form of regulation of aggressiveness places an extremely high premium on mutual attractiveness (social releasers). Individuals leaving social groups are usually subordinate, young, or displaced dominants and generally remain subordinates in other groups which they join. They are thus discriminated against reproductively, especially if males (Myers & Poole, 1961; Mykytowycz, 1959). This is an important consequence of the regulation of aggressiveness within social groups. Animals low in the dominance hierarchy either in a group or in a territorial system, are also subject to increased predation (Ratcliffe, 1958). This also applies to outcast groups, such as surplus males in polygamous species.

Evolution is generally thought of in morphological terms. On the whole, structure is a reflection of behaviour, which is subject to evolutionary change, and these changes presumably precede morphological changes. In this discussion we are concerned only with certain types of social behaviour. Behaviour patterns, like morphological patterns, probably seldom arise *de novo* in evolution but arise from pre-existing patterns.

It is probable that displacement behaviour plays a fundamental role in the evolution of behaviour. The evolution of secondary roles for intercaste behaviour such as sexual and infantile patterns as submission signals is of interest, since both operate to reduce force fields.

A similar situation can be seen in a model proposed for the evolution of communication in the pig. The pig communicates mainly by sound, and these calls can be readily observed in appropriate social situations. The sounds can be grouped into squeals and grunts. The following diagram sets out the calls and their possible interrelations in terms of an evolutionary model.

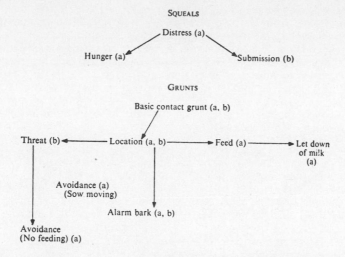

The calls marked "a" are for intercaste and those marked "b" are for intracaste communication. The interesting groups are the use of squeals in inter- and intra-caste (submission) communication, and the relationships between the threat and avoidance calls. These three calls are snorts with exhalation (McBride, 1963). The first is given by the sow when she stands for the first time after farrowing. She rocks herself slowly into an upright position, then slowly stands. Throughout she gives this avoidance call. The piglets stand away, huddled in a group. If one approaches her, she nudges it away; if it persists, she may pick it up in her mouth and throw it away. The second is given when the piglets are older and pester her for a feed. Once again, if they persist, she may reinforce the call with a nip.

These examples of similar calls having similar functions in inter- and intra-caste communication are presumably examples of evolutionary sequences, in the same sense that sequential morphological adaptations can be traced from species to species within a genus. Considerably more study is necessary to locate good evolutionary sequences in behaviour similar to those which are available in morphological terms.

It has been suggested that when mating pairs develop, the process is similar to that of group formation. The pair, or even polygamous groups, show low intra-group and high extragroup aggressiveness. However the process of welding individuals of different castes together also involves socializing mechanisms. Etkin (1954) has used the term "dominance modification" for this phenomenon which permits closer socialization of male and female at the time of oestrus. He suggests that the extension of oestrus enables the "development of a diffuse sexual activity between male and female as a type of socializing behaviour"; thus sexual behaviour becomes part of the socializing mechanism. This seems to be also true of other species during the mating season, e.g. the flying fox (Nelson, 1963).

It would appear that the best examples of evolution along the social spectrum can be found within those species which show seasonal changes in their social organization. Here the social organization at the mating season is generally associated with high levels of aggressiveness and extremely rigid regulation by territory. This may well be the basic situation, with increased socialization and a return to intra-caste organization or elimination of castes during the rest of the year. Evolution here has involved a hormonal change, and generally organization changes towards the aggregated group situation of dominance hierarchy, or in extreme situations, the school pattern.

No consideration of the evolution of behaviour, particularly social behaviour, would be complete without some comment on the development processes associated with its ontogeny, that is, the problem of instinct or learning. In this field, biometrical geneticists have long ignored any attempt at separation of genetic and environmental factors in the development of animal characteristics. It is known empirically that both genetic and environmental factors have contributed to the variation in most traits. It would appear reasonable for behaviourists to adopt a similar approach to this problem. The learning of avoidance behaviour in the piglet is a good example. The piglets respond to the call and behaviour of the sow by standing aside when the sow is standing. However the "instinctive" aspect of the behaviour is not the whole story, since they often approach the sow and "learn" as the sow thrusts them roughly aside.

The situation seems to be that evolutionary processes can develop "programmes" (in the electronic computer sense) for behaviour patterns, and these will be incorporated in the genetic code. However these "programmes" may or may not be complete. In this case, "blank programmes" appear to be available for learning particular aspects of behaviour. The phenomenon of imprinting (Lorenz, 1937) seems to associate specific periods early in life to learning particular types of information, that is, filling particular "blank programmes", or, more commonly, incomplete "programmes".

In conclusion, if advances are to be made in the study of social organization, far more work is needed in comparative animal sociology. The biggest weakness in this type of study is the paucity of information on specific types of social organization. Society structure has been studied in only a few species. The type of analysis of social behaviour proposed needs to be examined in a far greater range of animal societies. The study of social organization within general taxonomic groups should provide the evidence necessary to examine the proposed social spectrum, the types of inter- and intra-caste behaviour, and the validity of the force field theory.

V. Discussion and Summary

What has been presented is an attempt at a general theory of social organization and behaviour. It attempts to explain the organization and social behaviour of animals in terms of the spatial relationships between animals. These spatial relationships or individual distances, represent a balance between intraspecific aggressiveness and the socializing and territorial factors regulating it. Since the basic social aggressiveness force operating between animals is directional, repulsive, and restricted in area, a theory of social force fields has been presented by anology with force fields in physics.

An adequate case could readily be made for skipping the force field theory and dealing with all animal spatial adjustments in terms of stimulus and response, species and caste characteristics providing the stimuli. Individual distances thus are learned as the distances at which aggressive stimuli are negligible. Nevertheless, the force field theory provides a good descriptive picture of the behaviour of animals towards one another. Accordingly it has been used throughout this study.

It should, however, be clearly recognized that the general theory of animals' social behaviour and organization can be developed either in terms of force field theory or in terms of the individual distances between animals.

The general theory consists of the analysis of social behaviour in aggregated groups into components based upon their effects in increasing or decreasing individual distances or stabilizing them in territorial situations. The existence of castes in animal societies leads to the examination of simple (single caste) and complex societies and allows behaviour to be analyzed in terms of intra- and inter-caste components.

Intracaste society organization shows a surprising uniformity between a wide range of animal species when considered in terms of an organizational spectrum. This is considered as probably the most widespread example of iterative evolution, and it is suggested that this is best explained in terms of social organization arising in response to a single component of behaviour found widely in animal groups, intraspecific aggressiveness.

Though intercaste behaviour is not primarily concerned with spatial adjustment, it operates in a milieu of social force fields and thus is partly concerned with adjustment to these force fields. Only these aspects of intercaste behaviour have been discussed.

An attempt has been made to synthesize the various components of social behaviour into a general theory which attempts to systematize our knowledge of animal societies on to a basic frame or spectrum on which all societies can be described and classified.

In the light of our present limited knowledge of comparative animal sociology, this theory must be considered tentative. The justification for its presentation, if such is needed, must be that of Darwin (1883): "False facts are highly injurious to the progress of science, for they often endure long; but, false views, if supported by some evidence, do little harm, for everyone takes a salutary pleasure in proving their falseness; and when this is done, one path towards error is closed and the road to truth is often at the same time opened."

References

ALLEE, W. C. (1952). Dominance and hierarchy in societies of vertebrates. *Structure et Physiologie des Sociétés Animales*, pp. 157–81. Colloques Internationaux, 34, ed. P. P. GRASSE. Paris: Centre National de la Recherche Scientifique.

ALTMANN, MARGARET (1960). The role of juvenile elk and moose in the social dynamics of their species. *Zoologica, N.Y.* **45**: 35–39.

ALTMANN, STUART, A. (1959). Field observations on a howling monkey society. *J. Mammal.* **40**: 317–30.

BARTHOLOMEW, G. A. (1952). Reproductive and social behaviour of the northern elephant seal. *Univ. Calif. Publ. Zool.* **47**: 369–472.

BARTHOLOMEW, G. A. & COLLIAS, N. E. (1962). The role of vocalization in the social behaviour in the northern elephant seal. *Anim. Behav.* **10**: 7–14.

BEILHARZ, R. (1963). Social position and behaviour of dairy heifers in yards. *Anim. Behav.* (in press).

BENNETT, M. A. (1940). The social hierarchy in ringdoves. II. The effect of treatment with testosterone propionate. *Ecology* **21**: 148–65.

CALHOUN, J. B. (1962). "A 'behavioural' sink." Chapter 22. *Roots of Behaviour*, ed. E. L. BLISS. New York: Harper.

CARPENTER, C. R. (1934). A field study of the behaviour and social relations of howling monkeys. *Comp. Psychol. Monogr.* **10**: 1–168.

CARPENTER, C. R. (1958). *Territoriality: a Review of Concepts and Problems in Behaviour and Evolution*, ed. A. RAE & G. G. SIMPSON, pp. 224–50. New Haven: Yale University Press.

CASTORO, PAUL L. & GUHL, A. M. (1958). Pairing behaviour of pigeons related to aggressiveness and territory. *Wilson Bull.* **70**: 57–69.

CHANCE, M. R. A. (1962). "The nature and special features of the instinctive social bond of primates", in *The Social Life of Early Man*, ed. S. L. WASHBURN, pp. 17–33. London: Methuen.

CLARKE, P. J. & EVANS, F. C. (1954). Distance to nearest neighbour as a measure of spatial relationships in populations. *Ecology* **35**: 445–53.

COLLIAS, N. E. & TABER, R. D. (1951). A field study of some grouping and dominance relationships in ring-necked pheasants. *Condor* **53**: 265–75.

CRAIG, W. (1918). Appetites and aversions as constituents of instincts. *Biol. Bull.* **34**: 24–27 (not seen).

CROFTON, H. D. (1958). Nematode parasite populations in sheep on lowland farms. VI. Sheep behaviour and nematode infections. *Parasitology* **48**: 251–60.

DARWIN, CHARLES, (1883). *The Descent of Man and Selection in Relation to Sex*, 2nd ed. London: John Murray.

DEEVEY, E. S. (1960). The hare and the haruspex, a cautionary tale. *Amer. Scient.* 48: 415–43.

ELTON, C. S. (1942). *Voles, Mice and Lemmings.* Oxford: Clarendon P.

ELTON, C. S. (1963). Self regulation of animal populations. *Nature, Lond.* 197: 634.

ETKIN, WILLIAM (1954). Social behaviour and the evolution of man's mental faculties. *Amer. Nat.* 88: 129–42.

GUHL, A. M. (1950). Social dominance and receptivity in the domestic fowl. *Physiol. Zool.* 23: 361–66.

GUHL, A. M. (1953). Social behaviour of the domestic fowl. *Tech. Bull. Kans. agric. Exp. Sta.* 73.

GUHL, A. M. (1961). "Gonadal hormones and social behaviour in infra-human vertebrates", in E. ALLEN, *Sex and Internal Secretions*, 3rd ed., ed. W. C. YOUNG, vol. 2, pp. 1240–67. London: Baillière, Tindall & Cox.

GUHL, A. M. (1962). "The social environment and behaviour", in *The Behaviour of Domestic Animals*, ed. E. S. E. HAFEZ. London: Baillière, Tindall & Cox.

GUHL, A. M. & CRAIG, J. V. (1962). Genetics and sociobiology of chickens. *Proc. 12th World's Poult. Congr., Sydney* 2: 105–10.

HALE, E. B. (1956). Breed recognitions in the social interactions of domestic fowl. *Behaviour* 10: 240–54.

HEAPE, W. E. (1931). *Emigration, Migration and Nomadism.* London. Quoted by WYNNE-EDWARDS, 1962.

HEDIGER, H. (1950). *Wild Animals in Captivity.* London: Butterworth.

HEDIGER, H. (1955). *Psychology and Behaviour of Captive Animals in Zoos and Circuses.* New York: Criterion Books.

HEDIGER, HENRI, P. (1962). "The evolution of territorial behaviour", in *The Social Life of Early Man*, ed. S. L. WASHBURN, pp. 34–57. London: Methuen.

HINDE, R. A. (1953). The conflict between drives in the courtship and copulation in the chaffinch. *Behaviour* 5: 1–31.

HOLZAPFEL, M. (1940). Triebbedingte Ruhezzustande als Ziel von appetenzhandlungen. *Naturwissenschaften* 28: 273–80 (not seen).

KALMUS, H. (1955). The discrimination by the nose of the dog of individual animal odours and in particular the odours of twins. *Brit. J. anim. Behav.* 3: 25–31.

KEENLEYSIDE, MILES H. A. (1955). Some aspects of the schooling behaviour of fish. *Behaviour* 8: 183–248.

KIKKAWA, JIRO (1961). Social behaviour of the White Eye *Zosterops lateralis* in winter flocks. *Ibis* 103a: 428–42.

KING, J. A. (1955). Social behaviour, social organisation and population dynamics in a black-tailed Prairie Dog town in the Black Hills of South Dakota. *Contr. Lab. Vertebr. Biol. Univ. Mich.* 67.

KING, M. G. (1963). Peck frequency and minimal approach distance in the domestic fowl. *J. genet. Psychol.* (in press).

KISLACK, J. W. & BEACH, E. A. (1955). Inhibition of aggressiveness by ovarian hormones. *Endocrinology* 56: 684–92.

KUMMER, H. (1957). Sociales Verhalten einer Mantelpaviangruppe. *Schweiz. Z. Psychol.* 33. Quoted by CHANCE, 1962.

LEHRMAN, D. S. (1961). "Gonadal hormones and parental behaviour in birds and infrahuman mammals" in E. ALLEN, *Sex and Internal Secretions*, 3rd ed., ed. W. C. YOUNG, vol. 2, pp. 1268–362. London: Baillière, Tindall & Cox.

LERNER, I. M. (1958). *The Genetic Basis of Selection.* New York: Wiley.

LORENZ, K. (1937). "Companionship in bird life" in *Instinctive Behaviour*, ed. C. H. SCHILLER. New York: International Universities Press.

LORENZ, K. (1952). *King Solomon's Ring.* London: Methuen.

MARLER, P. (1956). Studies of fighting in chaffinches. III. Proximity as a cause of aggression. *Brit. J. anim. Behav.* 4: 23–30.

MASON, W. A. (1961). The effects of social restriction on the behaviour of Rhesus monkeys. II. Tests of gregariousness. *J. comp. physiol. Psychol.* 54: 287–90.

MCALISTER, W. H. (1958). The correlation of coloration with social rank in *Gambusia hurtadoi*. *Ecology* 39: 477–82.

MCBRIDE, G. (1958). The measurement of aggressiveness in the domestic fowl. *Anim. Behav.* 6: 87–91.

MCBRIDE, G. (1963). The teat order and communication in young pigs. *Anim. Behav.* 11: 53–56.

MCBRIDE, G. & FOENANDER, F. (1962). Territorial behaviour in the domestic fowl. *Nature, Lond.* 194: 102.

McBride, G., James, J. W., & Shoffner, R. N. (1963) Social forces determining spacing and head orientation in a flock of domestic hens. *Nature, Lond.* **197**: 1272–73.

Morris, D. (1955). The causation of pseudofemale and pseudomale behaviour; a further comment. *Behaviour* **8**: 46–56.

Morgan, W. C. & Bonzer, B. J. (1959). Stress associated with moving cage layers to floor pens. *Poult. Sci.* **38**: 603–6.

Moynihan, M. (1958). Notes on the behaviour of some North American gulls. III. Pairing behaviour. *Behaviour* **13**: 112–30.

Myers, K. & Poole, W. E. (1959). A study of the biology of the wild rabbit, *Oryctolagus cuniculus* (L) in confined populations. I. The effects of density on the home range and the formation of breeding groups. *C.S.I.R.O. Wildl. Res.* **4**: 14–26.

Myers, K. & Poole, W. E. (1961). A study of the biology of the wild rabbit, *Oryctolagus cuniculus* (L) in confined populations. II. The effects of season and population increase on behaviour. *Wildl. Res.* **6**: 1–41.

Mykytowycz, R. (1958). Social behaviour of an experimental colony of wild rabbits *Oryctolagus cuniculus* (L). I. Establishment of the colony. *C.S.I.R.O. Wildl. Res.* **3**: 7–25.

Mykytowycz, R. (1959). Social behaviour of an experimental colony of wild rabbits *Oryctolagus cuniculus* (L). II. First breeding season. *C.S.I.R.O. Wildl. Res.* **4**: 1–18.

Mykytowycz, R. (1960). Social behaviour of an experimental colony of wild rabbits *Oryctolagus cuniculus* (L). III. Second breeding season. *C.S.I.R.O. Wildl. Res.* **5**: 1–20.

Mykytowycz, R. (1961). Social behaviour of an experimental colony of wild rabbits *Oryctolagus cuniculus* (L). IV. Conclusion: outbreak of myxomatosis, third breeding season and starvation. *C.S.I.R.O. Wildl. Res.* **6**: 142–55.

Mykytowycz, R. (1962). Territorial function of chin gland secretion in the rabbit, *Oryctolagus cuniculus* (L). *Nature, Lond.* **193**: 799.

Nelson, J. E. (1963). "The biology of the flying fox, *Genus Pteropus*, in south east Queensland." Unpublished Ph.D. thesis, University of Queensland.

Peterson, A. J. (1955). The breeding cycle of the bank swallow. *Wilson Bull.* **67**: 235–86.

Ratcliffe, F. N. (1958). Factors involved in the regulation of mammal and bird populations. *Aust. J. Sci.* **21**: 79–87.

Scott, J. P. (1948). Dominance and the frustration-aggression hypothesis. *Physiol. Zool.* **21**: 31–39.

Scott, J. P. (1958). *Animal Behaviour*. 281 pp. Chicago: University of Chicago Press.

Scott, J. W. (1942). Mating behaviour of the sage grouse. *Auk* **59**: 477–98.

Southwick, Charles H. (1962). Patterns of intergroup social behaviour in primates, with special reference to rhesus and howling monkeys. *Ann. N.Y. Acad. Sci.* **102**: 436-54.

Tinbergen, N. (1951). *The Study of Instinct*. London: Oxford University Press.

Tinbergen, N. (1953). *Social Behaviour in Animals*. London: Methuen.

Washburn, S. L. & De Vore, Irven (1962). "Social behaviour of baboons and early man", in *The Social Life of Early Man*, ed. S. L. Washburn, pp. 91–105, London: Methuen.

Wood-Gush, D. G. M. (1959). A technique for studying diurnal rhythms. *Physiol. Zool.* **32**: 272–83.

Wright, S. (1951). The genetical structure of populations. *Ann. Eugen. Lond.* **15**: 323–54.

Wynne-Edwards, V. C. (1962). *Animal Dispersion in Relation to Social Behaviour*, 653 pp. Edinburgh: Oliver & Boyd.

Young, W. C. (1961). "The hormones and mating behaviour", in E. Allen, *Sex and Internal Secretions*, 3rd ed., ed. W. C. Young, vol. 2. London: Baillière, Tindall & Cox.

Zuckerman, S. (1932). *The Social Life of Monkeys and Apes*. London: Kegan Paul, French & Trubner.

Part II

STUDIES OF SOCIAL SPACING

Editor's Comments
on Papers 7 Through 11

One of the most elementary examples of social spacing is that of schooling behavior in fishes. In spite of the relative ease with which this behavior can be observed, there is still considerable debate regarding the adaptive significance of the pattern. Many hypotheses impute to schooling behavior the advantage of providing protection for school members. There has appeared a range of suggestions as to how protection from predation comes about. These include the theory that masses of individuals comprising a school tend to confuse a potential predator; targetting on an individual is alleged to be more difficult because of the number of school members. Another idea is that the school, itself, mimics a large organism and hence frightens a potential predator. The fact that the multitude of sensors of school members permits more careful monitoring of the environment and detection of an approaching predator has also been suggested as a protective feature. These all have in common the assumption that natural selection has operated on the school as an entity, i.e., that group selection has occurred in the evolution of this kind of social group.

It has recently been reiterated by Alexander (1974) that one of the few advantages of group living is to provide individuals with a place to hide (within the mass of the school, herd, flock, etc.), and thereby to reduce their vulnerability to predation. This view was proposed earlier by Williams in "Measurements of Consociation Among Fishes . . ." (Paper 7). In essense, Williams proposed that schooling behavior evolved from an initial phase in which conspecifics were attracted to a particular area, perhaps a momentary source of food. If members of a species exhibited a tendency to tolerate close proximity, a second stage would ensue in which those individuals having a greater tendency to place themselves within, rather than on the periphery, would enhance their prospects to escape predation and to reproduce. Natural selection would act to canalize the evolution of this adaptive trait by favoring those individuals that literally used the school as a cover. This argument was based on several lines of evidence, e.g., the observation that pelagic fishes are more apt to be schoolers than near-shore or stream forms, where other aspects of the environment such as rocks or emergent vegetation serve as hides. The experiments conducted by Williams and described in Paper 7 also tend to support his view.

This problem of social spacing was also the focus of Breder in his paper "On the Survival Value of Fish Schools" (Paper 8). Breder reviews the literature on schooling behavior and presents an array of definitions of this phenomenon. He also examines the then-current literature on the proximal causes of spacing in schools, i.e., on the sensory modalities used in the maintenance of spacing and geometry of fish schools. In addressing the question of survival value of schooling behavior, Breder favors a much broader spectrum of factors than does Williams. He disagrees with some, but not all, of the latter's premises and comments critically on the experimental part of Williams' paper. Breder accounts for the variations of school morphology and function in terms of behavioral homeostasis and eschews the one-factor hypothesis advanced by Williams as being unrealistic and too narrow in scope.

The idea that aggregations evolved by virtue of natural selection operating on the cover-seeking behavior of individuals is further fostered by Hamilton in his essay on the "Geometry for a Selfish Herd" (Paper 9). Hamilton has made significant contributions to the development of a firm mathematical–evolutionary basis for sociality. In the present selection, Hamilton develops a model that shows how aggregations may have evolved as a function of predator pressure acting on individual evasive behavior. The major premise of this hypothesis is that predators concentrate their activity on marginal members of a group. To avoid predation, a potential prey individual must situate itself within, rather than on the periphery of, the group. This is, of course, the same argument proposed by Williams. In fact, Hamilton points out that the nucleus of this concept was conceived by Francis

Galton in 1871 in an article dealing with gregarious behavior in cattle and man. Individual survival, according to Hamilton, necessarily leads to the formation of aggregations such as fish schools or ungulate herds in those species occupying environments relatively devoid of natural refuges. Although the assumptions explicit in Hamilton's models are quite simplified when contrasted with natural situations, his arguments are supported by known responses of real animals to predators and serve to render any group-selection hypothesis highly unlikely.

Another common form of sociality in vertebrates is flocking behavior in birds. Paper 10, by Emlen, a leading student of sociobiology, represents an early attempt to evaluate a variety of factors that appear to be influential in the manifestation of flocking behavior. A flock is defined by Emlen as "any aggregation of homogeneous individuals, regardless of size or density." Not included are such special groupings of age and sex as occur in breeding and parent–young assemblies. Gregariousness of birds is related to positive and negative reactions of individuals. Emlen uses the terms "centrifugal" and "centripetal" forces with much the same connotation as does McBride in Paper 6. It is also interesting that a brief discussion of Emlen's observations on cliff swallow groups makes the point that early in the assembly of a loafing group on a span of telephone wire, the birds appear to concentrate their efforts in the central portion of the perching area. This observation could be interpreted as evidence for Hamilton's model presented above.

The main emphasis in Emlen's paper is on causal factors of flocking, which are now referred to as proximal factors. That is, the roles of hormones, photo-period, season, climate, and the like are evaluated as mechanisms that promote or inhibit the formation of bird flocks. There is little discussion of the adaptive significance or ultimate causality of this form of sociality.

A recent paper by Cody (1971) exemplifies a more current trend in the analysis of flocking behavior directed to understanding the ecological impact on sociality. It reflects the influence of optimization theory on ecological and behavioral processes. Cody deals with the question of what advantages accrue to large numbers of individuals banding together for feeding over solitary feeding. The species he has studied are all fruit or seed eaters. The putative advantages of flocking behavior in terms of an appetite for close contact, sorting out of dominance-subordinance relationships, and protection from predation are all discounted in the case of the finch species being considered. Cody develops several hypotheses that relate benefits of flocking to feeding efficiency on nutrients that are differentially abundant at different times of the year.

To this point, the papers in this section have dealt with various aspects of rather rudimentary forms of social spacing. There is a voluminous literature covering more complex systems, such as territoriality and dominance hierarchies. These two topics have received attention from Stokes (1974) and Schein (1975). It is nonetheless appropriate to introduce some of the major concepts underlying territoriality and dominance hierarchies in the present volume. For this purpose, the review article by Marler of Rockefeller University "On Aggression in Birds" has been included. In Paper 11, Marler reviews his earlier work on social spacing in the chaffinch; and enters the sensitive debate concerning the motivational basis for aggression in vertebrates.

REFERENCES

Alexander, R. D. 1974. The evolution of social behavior. *Ann. Rev. Ecol. System* **5**: 325–384.

Cody, M. L. 1971. Finch flocks in the Mohave Desert. *Theor. Pop. Biol.* **2**: 142–158.

Schein, M. W. 1975. Social Hierarchy and Dominance. Dowden, Hutchinson & Ross. Stroudsburg, Pa. 401 pp.

Stokes, A. W. 1974. Territory. Dowden, Hutchinson & Ross. Stroudsburg, Pa. 398 pp.

7

Reprinted from *Publ. Mus. Mich. State Univ., Biol. Ser.,* 2(7):351–382 (1964)

Measurement of Consociation Among Fishes and Comments on the Evolution of Schooling[1]

George C. Williams

State University of New York, Stony Brook

The schooling phenomenon presents the student of animal behavior with a paradox. On the one hand, schooling is superficially a simple phenomenon and would seem to lend itself readily to quantification and casual analysis. On the other hand, there has been a notable lack of success in relating schooling to general biological principles, and there are no really convincing ethological, ecological, or evolutionary explanations. I believe that this deficiency can be ascribed to the absence of any immediately apparent "purpose" for schooling. There is no vital function to which it seems to make an efficient contribution, and it can not be immediately assigned to reproductive, defensive, or any other category of adaptive behavior.

Certain school-like groups, such as that formed by several male guppies chasing the same female, can readily be discussed in functional terms, but for most schools that would be recognized as such, there have been only some rather weakly supported explanations, which I will attempt to evaluate later in this paper. The explanation which suggested the experiments described here is one that relates schooling to the category of defensive hiding. It proposes that a fish places itself among others of its kind to make itself less conspicuous and to place the others between itself and sources of danger. According to this view, schooling behavior (the individual reaction) is adaptive, but a school (the statistical consequence) is not.

In the design of experiments to bear on this possibility, the central consideration is that the vicinity of conspecific individuals constitutes a hiding place and that such usage can be demonstrated by removing all other hiding places. Fishes of open-water habitats in nature have nowhere to hide effectively from predators. This might furnish a partial explanation for the pre-

[1]This study was supported by grants from the All-University Research Fund at Michigan State University, where the experiments were conducted. The author expresses his gratitude to Dr. C. M. Breder, of the American Museum of Natural History; to Dr. Sol Kramer, of the State University of New York; and to Dr. James W. Atz of the New York Aquarium for their comments on the manuscript. Help was provided on mathematical problems by Dr. Philip J. Clark, of Michigan State University, and by Dr. William G. Lister, Dr. Abram V. Martin, and Dr. Donald Malm of the State University of New York. This study is a contribution from the Department of Biological Sciences, State University of New York, Stony Brook; this institution also provided financial support for publication.

valence of schooling in pelagic waters. Fishes that inhabit localities with abundant cover can be placed experimentally in cover-deficient environments, and if schooling is a form of cover-seeking, such treatment might induce schooling even among species that do not show this behavior in their usual habitats. The experiments described here gave a clear demonstration of this expected effect. They were also designed to provide an objective measurement of the intensity of schooling behavior in homotypic groups of various size, and in heterotypic pairs.

There are other theoretical standpoints from which schooling may be studied, besides that of ecological adaptation. Breder (1959) analyzed schooling from the point of view of information theory, and reviewed several aspects of physiological causation. He also discussed the theory proposed by Parr (1927), that schooling results from a force of social attraction that is replaced by repulsion when individuals approach to within a critical distance. This distance would be a point of stable equilibrium at which there would be neither attraction nor repulsion. This concept is useful in explaining some of my observations.

TERMINOLOGY

Animals often convene because each independently seeks some localized conditions. A gathering of people beneath a shade tree on a hot day is a good example. Other groups result from mutual attractions among the members themselves. Spooner (1931) proposed that the term *school* be used for any socially motivated fish grouping, but not for groups formed by forces other than social attractions. Nikolsky (1963) indicated that current Soviet usage recognizes two kinds of social groupings, called *school* and *shoal* in the Birkett translation. A school is defined as a group that is sufficiently small for every fish to be aware of every other fish. A shoal would be a larger group in which a fish could sense and be attracted only to the individuals in its neighborhood, often only a small proportion of the whole. The word shoal is more common than school in the European literature and is seldom used by Americans. Breder (1929, 1959) and Atz (1953) confined the term school to social groupings in which all the individuals are oriented in one direction. They used the term aggregation for disoriented social groups and did not propose a terminological distinction between social and non-social groups.

I have been unable to discern, either in my own work or in the reports of others, any importance of such orientation in a school, other than indicating that it is in motion, relative to the water. The locomotor machinery of almost all fishes is such that to move in a given direction they must point their bodies in that direction. I have observed that whenever a moving (and oriented) school stops moving, the unanimity of orientation ceases, and then reappears when a school again moves away. It may be of some value to distinguish between these two phases of activity, but the difference between social and non-social groupings is in greater need of a terminological distinction. I will therefore use the term *school* to refer to any contagious distribution that owes its persistence to social (but not sexual) forces. *Aggregation* will refer to groups that arise by individuals independently seeking the same localized conditions. This use of the term aggregation is widespread, as is its use to cover both social and non-social groupings (see various general texts on ecology and behavior).

A school, therefore, is to be recognized on the basis of its cause rather than its appearance. There are situations in which appearances may reliably indicate cause. For instance, when one observes a dense concentration of fish of the same species moving about in a pelagic or other uniform habitat, and can be reasonably sure that the fish are not all chasing the same moving object, he is probably safe in calling it a school. Social attraction would be the most reasonable explanation for the cohesion of the group. Schools and aggregations can not be as confidently distinguished in heterogeneous environments, and it must often happen that groups are formed that owe their cohesiveness to both schooling and aggregation in mutual reinforcement (*heterogeneous summation* of Tinbergen, 1951).

METHODS

To demonstrate schooling one must show that a species has a contagious distribution in the absence of any factor that could cause aggregation. I believe that this requirement can be met by satisfying the following conditions:

1. The experimental environment should be visually and topographically uniform.

2. An experimental fish should verify this uniformity by its behavior. Singly or with companions, it should not show any tendency to prefer one part of the tank to another. The amount of time spent in any region should be in direct proportion to its area.

3. In groups (two or more) the fish should show a contagious distribution in the experimental enviroment.

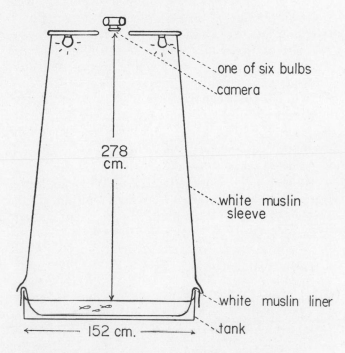

Figure 1. Essential features of the experimental apparatus.

To satisfy the first requirement I used a cylindrical steel tank 153 cm in diameter and 30 cm deep, filled to a depth of 17 cm (Figure 1). The inside of the tank was painted white and coated with paraffin, with a thick fill of paraffin to eliminate the corner between the side and bottom of the tank. When algal and fungal growths on the paraffin made it difficult to maintain as a uniform white field, I provided the tank with precisely fitting, bowl-shaped liners of bleached cotton muslin, which were easily removed and laundered after each use. A wooden scaffolding supported an overhead platform with a hole through which observations and photographs could be taken. The sleeve of white muslin hung around the tank below the platform to complete the visual uniformity for the fish inside. A ring of six 200-watt bulbs hung from the white underside of the platform. A fish in the tank could see nothing but nearly uniform whiteness in any directon, except for the light bulbs and the hole in the ceiling directly overhead.

The experimental fish came from various sources, as indicated in the discussion of each species. I kept most of them in 10- to 30-gallon aquaria when they were not under investigation, and fed them commercial, dehydrated fish food, with occasional meals of live or fresh foods. After an experiment, I put the fish into a different aquarium from the one they had previously occupied, so that the post-experimental behavior and condition of each specimen could be observed, and to avoid the too-frequent use of the same individuals. To minimize the effect of their previous learning, I allowed at least three weeks between experiments with the same specimens.

When there was a temperature difference between the experimental tank and the aquarium from which the fish were taken, I allowed time for acclimation by floating the fish in aquarium water in a glass jar in the experimental tank with the lights turned on. This gave the fish an opportunity to adjust to the visual environment for several minutes before being released in the tank. The tank was on a floor where, most of the year, it kept a temperature of about 18°C. For tropical species, I heated the water with electric heaters to at least 22°C. before introducing the fish. The heaters had to be removed for the experiment, but the overhead lights kept water from cooling appreciably, and would raise the temperature from 18° to more than 20° in six hours. No experiments were conducted at the higher temperatures of the summer months.

After releasing the fish in the experimental tank, I allowed a few minutes of additional time for their adjustment to the unusual conditions before starting to photograph. I usually used a 36-frame roll of 35-mm. film and took pictures, nine consecutively, at 30-second intervals. I continued in this pattern until all 36 frames were exposed, and then left the fish in the illuminated tank for from five to eight hours (usually about six) and exposed another roll of film in the same way as the first. In this way each group of fish provided two series of 36 observations.

By comparing the two series I could note any effect of the amount of time spent in the experimental tank.

Most of the data came from photographs taken through the observation port. Moving pictures were used initially, but snapshots proved adequate, and only the three native minnows, the first species studied, were investigated with moving pictures.

The camera (Kodak Cine-Special or Ansco Memar) was on a wooden frame, permanently fixed to the overhead platform so that the relationship of film to tank was the same for every frame. The shallowness of the water assured that the fish were always about the same distance from the camera, and that distances measured on the film bore a nearly linear relationship to horizontal distances in the tank. Also, there were detectable landmarks in each picture, the circular water-edge, the seams in the tank-liner, and the reflections of the six light bulbs and of the observation hole, which showed as a small circle in the center of the field. I used the middle of this circle as the origin for measuring distances of fish from the center of the tank. The fixed position of both camera and tank meant that lines on a photograph drawn through the central circle parallel to the sides of the picture defined geometric quadrants that corresponded to fixed equal areas in the tank.

Measurements on moving-picture records were made with a meter-stick on single-frame projections, and on snapshot negatives by an optical micrometer in a low-power microscope. Photographs of a meter-stick floating in the experimental tank served to indicate conversion factors for obtaining tank distances from the measurements.

The intensity of schooling is determined by comparing the size of a group formed by a certain number of individuals with the size that would have been shown if each individual had moved at random. Area would undoubtedly be a suitable measure of group size, but I believe that the perimeter is more convenient and useful. Perimeters of irregular polygons, which small groups of fish usually form, are easily calculated as simple sums of individual measurements. The calculation of areas of such figures is tedious. Since the study involved recording several thousand school sizes, ease of calculation was of prime importance. The perimeter has other advantages over area, however, besides that of convenience. The perimeter of a school of a few individuals corresponds more often to an intuitive judgment of group size than the areal measurement would (Figure 2). Moreover, the perimeter is a linear measurement of distances between individuals, while area is a power function of such measurements, and therefore much more variable. The perimeter is, in fact, a special case of the traditional distance-to-nearest-neighbor measurements. It merely requires the restriction of such measurements to peripheral neighbors.

The measure of central tendency used throughout the work is the median. This usage is also justified partly on the basis of convenience. The median of a series of measurements is more rapidly calculated than is any kind of mean. It is superior to the mode in that it is a precise value for any distribution, while a modal value depends partly on an arbitrary decision as to what class limits should be recognized for the independent variable. The school-perimeter frequency distribution for a group of schooling fish is often highly skewed. Most of the time the fish will school, and show a small group-perimeter. Occasionally, however, one or more individuals will leave the school and move about independently, thereby greatly and variably augmenting the perimeter. So the frequency distribution shows a low modal value but includes some values of several times the mode. A number of statisticians (e. g. Yule & Kendall, 1950:115-6) have urged the advantage of the median over a mean as a measure of central tendency in such asymmetrical distributions, of which annual per-capita income is the usual example.

For graphical and mathematical comparisons of species it proved convenient to express all median perimeters as a proportion of the maximum that a group of fish could possibly show in the tank. This maximum is the perimeter of an inscribed regular polygon.

$$2N \sin \frac{\pi}{N}$$

where N is the number of fish.

THE PARAMETERS OF RANDOM MOVEMENT

I take random movement in the experimental tank to mean that a fish shows no tendency to prefer one region to another and that the movements of a fish have no influence on those of its fellows. Bias for certain regions could be on any pattern or scale: a preference for the northern half over the southern; a preference for the point of introduction and avoidance of areas a few centimeters away; a tendency to move along a set itinerary; etc. The observed movements of the fish were such, however, as to rule out any small-scale bias. The fish could and often did travel through widely separate regions in a few seconds. I assume that counts of occurrences in large areas are a sufficient test of regional bias. Constant geometric quadrants could be defined on the photographs, and if a fish moves at random it should be found about equally often in each of the four quadrants. It should also occur about equally often in equal central and peripheral areas. In other words, its median distance from the center should be about $1/\sqrt{2} = .707$ of the tank radius.

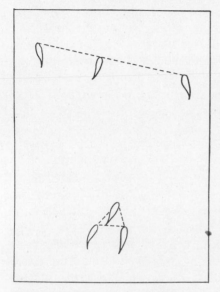

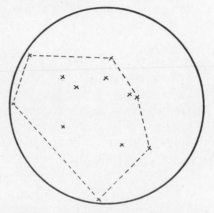

Figure 3. Perimeter of a group of twelve points scattered at random in a circle.

Figure 2. Area vs. perimeter as a measure of schooling intensity. Of the two triangular schools, the upper has a much greater perimenter but lesser area than the lower. The perimeter measurement accords with the intuitive impression that the lower group is a more cohesive school. Note also that a slight movement of any fish in the upper school would greatly increase the area but have slight effect on the perimeter or apparent cohesiveness of the group.

The distribution of a group of fish with no regional bias and with independent movement should be the same as that of a numerically similar group of points scattered at random within a circle. Descriptions of distributions of points in terms of deviations from randomness, either in the direction of clumping or of overdispersal, are available (Clark and Evans, 1955). They are not, however, appropriate to the present study. They relate to single distributions of large numbers of individuals in large spaces, for example, trees in a forest. The treatment of schooling required a statistical summary, the median perimeter, of the smallest convex polygons circumscribed about small numbers of points repeatedly scattered in a circular area.

I have been unable to deduce the medians of the perimeters of such polygons, and have therefore estimated them empirically. The technique is to draw a circle of unit radius on graph paper, and then to use a random number table to choose x and y coordinates of two decimal places for a point within the circle (disregard the point if it falls outside the circle). When the requisite number of points are thus located, the perimeter of the smallest enclosing polygon can be measured. The median of a large number of such perimeters then serves as the value for random distribution (Table 1).

The perimeter for two points was defined as twice the distance between the points. For three points the polygon is always a triangle. For more than three points (*e. g.* twelve, as in Figure 3) the polygon will have a point at each corner and a variable number of points scattered inside.

Upensky (1937:257-8) has calculated that the *mean* perimeter about pairs of points randomly located in a circle is about 1.82 radii, slightly less than my empirical estimate of the median. This difference is to be expected, because the distribution of distances from the center, which influences the distance between points, is skewed in the direction of larger values. The mean distance from the center (a problem formally similar to finding the center of gravity by integration, as treated in elementary texts) is two-thirds, slightly less than the median, $1/\sqrt{2}$.

Having obtained paired values, number of points and associated median perimeters (Table 1), one can fit the data to an equation that describes median perimeter as a dependent function of the number of points. Unfortunately, there is no set of principles that dictates the form that this equation should take, and there is no limit to the number of kinds of equations to which a given set of data can be fitted. Fortunately, there are some intuitive and aesthetic guides to the selection of an equation. A simple function is preferred to one that is more complex. A function that could conceivably hold over a wide range of the independent variable is preferable to one that would give patently erroneous or meaningless values outside the investigated range. For the present study, for instance, it would be undesirable to use an equation that could give negative perimeters or perimeters

TABLE 1

MEDIAN PERIMETERS OF CONVEX POLYGONS ABOUT POINTS SCATTERED
AT RANDOM WITHIN A CIRCLE

Number of Points	Polygons Measured	Medians	
		In Radii	Proportion of Maximum
2	300	1.93	0.480
3	100	2.84	0.547
5	70	3.93	0.688
8	40	4.51	0.737
12	32	4.80	0.783

that exceeded that of the enclosing circle. It is highly desirable that equations of the same form as that used for random distribution be applicable to observed fish distributions. Such equations would differ only in the values of constants that would furnish a basis of quantitative comparisons of each species with other species and with random distribution. One further consideration is that the number of fish (or points) be treated as a logarithmic rather than arithmetic function. We would expect the difference between two and three fish to be more closely comparable to that between twenty and thirty than to that between twenty and twenty-one. A logarithmic function emphasis proportionate, rather than absolute differences. An equation of the form (1) P =

$$(1) \qquad P = \int_{-\infty}^{m} G \ (a \log bm) \ dm$$

satisfies these conditions. P is the median perimeter, m the independent variable, and G the normal-curve ordinate of the parenthetical expression with constants a and b. The function G is, of course, mathematically quite complex, but the complexity has a ready resolution in standard statistical tables.

For the independent variable, it was convenient to use, not the number of points or fish, but the number minus one (*companion number*). This has an intuitive justification in that the smallest possible school is that formed by two individuals. Here the companion number is one, and this is a convenient origin on a logarithmic scale. The observed perimeter should be expressed as a proportion of the maximum possible perimeter that the group of points or fish could show (right-hand column in Table 1). With this selection of parameters the equation (1) plots as a straight line on log-probability paper. The fit is very good (see Figure 4) for random distribution, and certainly close enough for observed school perimeters. The function (Equation 1) also has the advantage of being readily visualized: As companion number increases, the observed perimeter approaches the maximum possible perimeter. The rates of approach for schooling fish and for randomly scattered points are, understandably, quite different, and the differences have a quantitative expression in the constants a and b.

Median perimeter as a proportion of the theoretical maximum is most easily visualized for two points. The maximum perimeter for these points (double the distance between) would be four radii and would occur when the points are at diametrically opposite positions on the edge of the circle. If observation showed that a group of two fish had a median perimeter of one radius, the plotted value would be 0.25.

OBSERVATIONS

Nine species were investigated in varying detail, and there was clearly evident schooling in all but two. The experimental animals formed a closely cohesive group most of the time and moved about the tank together. The schooling of the eel and the male guppy was weak and intermittent, and its demonstration requires a detailed analysis of distribution patterns. The male guppy proved too small to be clearly photographed at the 2.8-meter distance between camera and experimental tank. The eels, however, were clearly visible in photographs at this distance and therefore lent themselves to a photographic study that led to unforeseen complications. I will consider the eel first, therefore, to illustrate my statistical techniques for demonstrating and measuring schooling, and to indicate the intensity of this behavior at what is probably its lowest development among the species studied. The remaining species will be considered in a phylogenetic order.

Anguilla rostrata (LeSueur)

The eels were collected from seawater near Woods Hole, Massachusetts, by Dr. Carl J. George, now of the American University, Beirut. He netted them in August, 1958, as elvers, newly arrived from the sea. I conducted experiments with them in late 1958 and early 1959, when they averaged about 80 mm. in standard length. There was a steady mortality before, during, and after the period of experimentation, until the last of the original 33 died 15 months after collection. The behavior of this species may, therefore, have been influenced by the poor condition of some or all specimens.

In the experimental tank the eels moved listlessly near the bottom most of the time. They seldom remained completely still for more than a moment, but continuously made at least minor shifts of position. A contagious distribution was at least suggested most of the time, and was mediated by vision. When the lights were turned on after a few minutes of total darkness the fish appeared randomly scattered, but would usually come together in loosely cohesive groups in a minute or two. Sometimes, however, the fish would engage in rapid dashes about the tank, and such movements seemed independent. Groups of two to nine eels always lost cohesion when they started moving actively about. Their mutual attraction seemed weak at best.

TABLE 2

QUADRANT DISTRIBUTIONS FOR EELS (*Anguilla rostrata*)

Quadrant	Single Fish	Groups of Two and Three	Totals for 1, 2, and 3 Fish
I	14	165	179 (.30)
II	33	142	175 (.30)
III	14	107	121 (.20)
IV	11	105	116 (.20)

As indicated above, a single fish can be said to move at random if it shows about the same frequency of occurrence in each quadrant of the tank and a median distance from the center of about 0.707 of the tank radius. Apparently neither requirement was met by the animals tested. Consider first the addiction to Quadrant II (under "single fish" in Table 2). The deviations from the expected 18 occurrences in each quadrant are highly significant ($X^2 = 17$; n = 3; p « .01). The test, however, is based on the assumption of the independence of each count. The observations were clearly not independent events. In one experiment, for example, there were eleven consecutive observations in Quadrant II. The fish was sufficiently quiescent to stay in that quadrant for a long time. The degree of dependence of consecutive observations can be evaluated and suitable corrections made, but the problem can be more directly attacked in another way. The preference of single fish for a certain quadrant can cause a contagious distribution only if the majority in a group show the same quadrant preferences. To investigate this possibility I noted the quadrant distributions of each fish when in groups of two and three, the only group sizes for which I have photographic records (Table 2, last column). A preference for Quadrants I and II, the western half of the tank, is clearly apparent. The fish were in this half six-tenths of the time, and the 95 per cent confidence intervals for this proportion and sample size extends only from about 0.55 to 0.65 (Wilks, 1949:101).

Singly, the eels showed a median distance from the center of about 0.56 radii, about 0.79 of the expected random value of 0.71. About seven-tenths of the distances from the center were less than the random value. This fraction does not include 0.5 in its 95 per cent confidence interval. So individual eels not only deviated from random movement in their quadrant distribution, but also in their bias for the central and away from the peripheral parts of the tank. In groups this bias for the center was even more pronounced. In pairs, the median distance from the center was only 0.54 and in groups of three only 0.49 radii. The fish in groups spent about half the time in the central 25 per cent of the tank area. These observations can be interpreted as favorable to the schooling hypothesis by assuming that the presence of other eels near the center of the tank increases the attractiveness of that region of each eel. So their concentration in the center can be ascribed to the heterogeneous summation of schooling and aggregation.

Of the two kinds of departure from random distribution, the quadrant preference is the less serious. If each eel had a probability of 0.5 of occurring in each half of the tank, two eels would occur in the same half about half the time. If the probabilities were (as the data indicate) 0.4 and 0.6 for the two halves, the probability of two fish being in the same half or two different halves merely changes to 0.52 and 0.48, respectively. This

140

slightly greater likelihood of occurring in the same half could scarcely have a very important effect, by itself, on the median distances between individuals. If, however, as the data suggest, the median distance from the center is only about three-fourths of the expected random value, the median distance between two fish would also be reduced to about three-fourths of what it would be if they moved at random.

TABLE 3

MEDIAN PERIMETERS* OF EEL GROUPS IN PAIRS AND TRIOS

Date	Companion Number	Experiment	Median Perimeter Observed/Random
27 Jan. 1959	1	early	0.50/1.87 = 0.27
		late	1.16/1.87 = 0.62
6 April 1959	1	early	0.60/1.87 = 0.32
		late	0.43/1.87 = 0.23
30 May 1959	2	early	3.00/2.84 = 1.06
		late	1.28/2.84 = 0.45
		mean	0.49

*Expressed as a proportion of tank radius. For perimeter in mean fish lengths, multiply by 9.5.

Under these circumstances a demonstration of schooling would necessitate median perimeters well below three-fourths of the value expected from random movement. This requirement was certainly met in three of the six experiments (Table 3). In one experiment (on 30 May), the median perimeter was greater than the random value. This should not be regarded as over-dispersal caused by repulsion between individuals, because it resulted from the unusual behavior of one of the three fish, which spent nearly the whole time circling rapidly about the edge of the tank. Thus it showed, in contrast to the other specimens, a bias for the peripheral regions. The effect of this was to increase the perimeters resulting from independent movement. Six hours later this specimen had joined its fellows in the more central parts of the tank, and the resulting median perimeter was less than half the value calculated for independent motion.

For a group to show, in a series of 36 observations, such median perimeters as 0.23 and 0.27 of the calculated random value, it had to show a consistently contagious distribution, much stronger than could be attributed to slight tendencies to prefer certain regions of the tank. I conclude that the eels were schooling on these occasions. Obviously, however, they do not always school, because on at least two occasions they showed median perimeters not very different from what would have been expected on the basis of complete independence. The inconsistency is especially apparent from the observations made in June, when the same group that failed to school in the morning showed the most intensive schooling recorded for this species in the afternoon. The January observations indicate that the difference can not be resolved by taking account of the amount of time the fish spend in the experimental tank. I would conclude, therefore, that the species sometimes forms schools of low cohesion compared with other species that were investigated, that at other times it does not school, and that this inconsistency can not at present be explained.

Hyphessobrycon flammeus Myers

Twelve "flame tets," probably *Hyphessobrycon flammeus* (Family Characidae), were purchased from Pets Aquarium in Lansing, Michigan. I made no quantitative study of this species, but have no doubt that it schools. The twelve moved about the tank as a group that seldom presented more than what I would judge to be a third of the perimeter that random movement would have produced. A few usually dashed about in a frenzied manner when first introduced, but once they joined a school their behavior seemed normal. When the lights were switched on after a few minutes of darkness, the fish gave the appearance of random scattering, but soon started to school. The formation of a school, in this and other species of comparable size, seemed contingent on chance approaches of individuals to within about 20 cm of each other. Pairs formed of such encounters then swam about, in a seemingly undirected manner, until other individuals or small groups were closely approached and joined. Pairs always formed within a few seconds of sudden illumination, but sometimes a minute or two would pass before all of the small groups would coalesce into one school. Such a school of twelve would occasionally break up into two or three smaller groups, and sometimes a single individual would wander off for a brief period. Most of the time, however, the fish moved about in a single school.

The behavior of isolated individuals was quite different. One stayed immobile for 25 minutes where it had been placed in the tank. When eleven others were added, they formed a school that was joined by the first, although not for about four minutes. Another single individual showed marked disorientation when placed in the tank. It alternated between a head-up and head-down position for a few seconds and then sank, head downwards, to the bottom, where it rested immobile on its snout. I was convinced that I had done the fish some physical injury in transit from its aquarium. When I replaced it, however, it quickly entered a plant thicket and in a few minutes was swimming about normally. I saw similar but less extreme indications of disorientation in *Xiphophorus* on transfer to the experimental tank.

Hyphessobrycon, Xiphophorus, and also *Colisa,* indicated what seems to be distress caused by the all-white environment in another way. Some or all of them would rush to the net, instead of fleeing from it, when I undertook their removal at the end of an experiment. It would appear that the experimental fish were so cover-starved that a net, from which they would ordinarily flee as from a predator, was entered as if it were a haven.

Notropis atherinoides Rafinesque

Three emerald shiners, all about 55 mm long, were borrowed from exhibition tanks of the Department of Fisheries and Wildlife at Michigan State University and were returned a few days later. I made no quantitative study of this species, but observed them from time to time during their 19-hour stay in the experimental tank, and found that they school about as consistently and intensively as the following two native minnows.

Notropis stramineus (Cope)

Dennis W. Strawbridge (of Michigan State University) and I collected 23 specimens of sand shiner in October, 1956, from Little Long Lake, Kalamazoo County, Michigan. Most of them died within a month of collection, but five survived through the following October, when I initiated experiments on them. The experimental specimens ranged from 50 to 58 mm in standard length. Three were tested singly, two of them on two different occasions. Observations at 120-frame (about 7.5 second) intervals on motion-picture film gave occurrences in Quadrants I to IV of 23, 32, 35, and 34, respectively. Deviations from the expected 31 in each quadrant are not significant ($X^2 = 2.8$; $n = 3$; $p = .41$). The median distance from the center was 0.662 radii, which

143

is not sufficiently smaller than the 0.707 expected of random movement to cause appreciable aggregation. I assume that a single individual moves at random, although the number of observations is insufficient to demonstrate any but very marked deviations from randomness.

Groups of three individuals of 51-58 mm showed median perimeters of 0.47, 0.78, 1.20, and 1.30 radii in four experiments. The largest of these values is less than half of the 2.84 radii expected of independent movement. I conclude that this species schools consistently, although with variable intensity.

Pimephales notatus (Rafinesque)

Blunt-nosed minnows were collected at the same time and place as *Notropis stramineus* and had a similar history in captivity. There were fewer initially, but mortality was lower. One lived for more than three years in captivity. Four isolated individuals of this species were recorded 31, 41, 35, and 47 times in Quadrants I to IV, respectively ($X^2 = 3.84$; $p = 0.28$ with 3 degrees of freedom). So the deviations can reasonably be attributed to chance, especially since the extremes are in adjacent quadrants. The four showed median distances from the center of 0.60 to 0.86 radii, with a mean of 0.72, almost precisely the estimated random value. Individual movement, therefore, can be regarded as random.

Two groups of three fish (58 to 69 mm standard length) showed median perimeters of 0.41 and 0.86 radii, which average 0.64, a small proportion of the 2.84 radii expected from random movement, and a smaller value than the comparable measure for *Notropis stramineus,* although the difference is not significant. The blunt-nosed minnow seems to show schooling behavior to a marked degree.

Xiphophorus hybrid

I borrowed nine "black platies" from exhibition aquaria maintained by the Department of Natural Science, Michigan State University, and returned them shortly after the experiments. Their source and fate are unknown. The females were about 35 mm and the males about 30 mm in standard length.

I observed them in homo- and heterosexual groups of three and four individuals. One repeatedly jumped into the air immediately after introduction to the experimental tank, and two others jumped at least once after introduction. Of all the species tested, this seemed the most ill at ease. Even after several hours in the experimental tank, some individuals would hover immobile near the bottom, bent into a slight arc at the tail region. Fish more than a half meter apart would ignore each other, but at closer intervals, one would sometimes approach the other and station itself a few millimeters away. When one member of such a pair moved a short distance away, the other would follow. These were the main indications of schooling. The typical appearance of schooling behavior, that of two or more fish swimming about together, was seldom seen.

Poecilia reticulata Peters

The experimental guppies came from Pets Aquarium in Lansing, Michigan. There was a steady mortality, of perhaps 10 per cent per week, with no overt cause. Whatever was killing the fish might have influenced schooling behavior.

I observed males in a group of five on one occasion, long enough to convince me that schooling behavior was indicated. All five swam about as a group for a few seconds, and at other times two or three swam together for brief intervals. The schooling of the male guppies, however, was obviously intermittent and of low intensity. They were too small for a photographic study, but females were large enough to photograph clearly from the observation port. Unlike the males, they schooled most of the time. Two females tested singly gave 14, 18, 19, and 19 occurrences in Quadrants I to IV, respectively. Deviations from the expected 17.5 in each quadrant are not significant (p = .82 for X^2 = .972 with three degrees of freedom). They gave median distances from the center of 0.53 and 0.69 radius, which indicate that the species might have a tendency to aggregate in the center, but not enough to cause a markedly contagious distribution.

I tested female guppies in groups of two, three, five, and twelve (Table 4 and Figure 4). All median perimeters were markedly below the values expected from random movement. With respect to the general equation for the relationship of school perimeter to companion number, female guppies show a value of *b* of about 0.09 and a value of *a* of about 0.68. The low value of *b* (it is about 0.80 for random distribution) indicates strong social attraction among small numbers of individuals. The high value of *a* signifies a steep slope, much greater than with random movement (in which $a = 0.35$). The regression line for female guppies would, if continued, cross that of random distribution. Extrapolation on the graph gives a crossing at a companion number less than one hundred. The extrapolation is, of course, an uncertain basis for prediction, but it has a theoretical justification to be discussed later.

TABLE 4

MEDIAN PERIMETERS OF SCHOOLS OF FEMALE GUPPIES IN RADII*

Companion number	Random Perimeter	5 Feb. '60 Early	5 Feb. '60 Late	26 Feb. '60 Early	26 Feb. '60 Late	2 Apr. '60 Early	2 Apr. '60 Late	14 Apr. '60 Early	14 Apr. '60 Late	22 Apr. '60 Early	22 Apr. '60 Late	Means
1	1.93			0.10	0.30	0.15	0.17					0.18
2	2.84							0.21	0.29			0.25
4	3.93									0.50	1.72	1.11
11	4.80	3.13	3.78									3.46

*Multiply by 24 to obtain median perimeters in mean standard lengths.

Lepomis cyanellus Rafinesque

Two green sunfish of about 100 mm standard length were borrowed from exhibition tanks maintained by the Department of Fisheries and Wildlife at Michigan State University. Their origin and fate are unknown. They schooled intermittently during their twenty minutes in the experimental tank. I estimate the median perimeter to be about a third of that calculated for random distribution.

Colisa lalia (Hamilton-Buchanan)

The dwarf gouramis were purchased from Pets Aquarium in Lansing, Michigan. They always seemed normal and healthy, and mortality was negligible. The larger males frequently built and defended nests and courted the females. Spawning may have occurred, but reproduction could scarcely be successful with several pairs of fish in each aquarium and with frequent disturbance by nets.

This species was my primary experimental subject, chosen because it offered a number of advantages. I judged its size to be an optium compromise between the need for photographic clarity and for small size relative to the experimental tank. It also satisfied my desire for a species that would not ordinarily be thought of as a schooler. If such a fish schools in the experimental tank, it is obviously capable of schooling and inclined to do so in the special environment provided by the tank. Its disinclination to school in a planted aquarium must be attributed to an absence of the proper stimuli in such surroundings. This species also has the advantage of being readily obtainable and of having easily distinguished sexes. The males ranged from 35 to 42 mm in standard length, and the females from 33 to 38 mm.

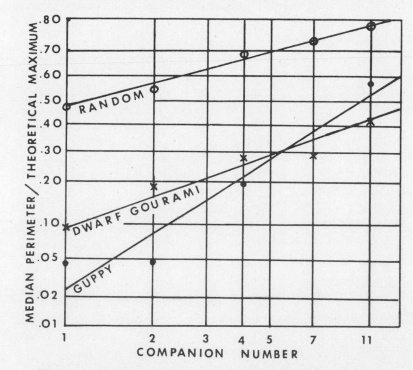

Figure 4. Median perimeters of schools of guppies and of dwarf gouramis compared with perimeters expected of random movement.

Two females tested singly each gave a median distance from the center of 0.86 radius in early observations. In late observations they gave 0.88 and 0.90. All these values are higher than the 0.71 calculated for random distribution. If they are representative, the independent movement of different individuals would result in an over-dispersed distribution (median perimeters larger than those calculated for randomness). The difference, however, hardly justifies consideration. The same females showed frequencies in Quadrants I to IV of 33, 29, 37, 32, respectively. Deviations from the expected 32.75 in each quadrant are not significant (p $= 0.80$ for $X^2 = 1.0$ with three degrees of freedom).

Two males similarly tested gave median distances from the center of 0.77 and 0.74 radius for early observations, and 0.64 and 0.63 for late. The mean of 0.695 is quite close to the random value. Occurrences in Quadrants I to IV were 30, 30, 29, and 21, respectively. Deviations from the expected 27.5 are not significant (p $= 0.63$ for $X^2 = 2.1$ with three degrees of freedom). I assume that both sexes move at random in isolation.

Both sexes showed pronounced schooling in all groups tested (Tables 5 to 8, Figure 4). It apparently makes little difference whether the school is composed of males or females or both. Hence the only variable considered in the graph is companion number. All experiments with the same companion number were averaged to obtain the plotted points. Such means of medians scarcely deviate from a straight line on the graph and thereby indicate the suitability of the graphic procedure. The schooling constants a and b are about 0.54 and 0.095, respectively, and indicate that social forces in this species are less than those of female guppies. Further consideration of these points will be undertaken in the discussion.

Gouramis and Guppies in Heterotypic Pairs

Median perimeters of one-companion schools of *Colisa* and *Poecilia* are about 0.39 and 0.18 radii, respectively. If schooling were a non-specific reaction to any object of appropriate size, one might expect heterotypic pairs to show intermediate perimeters. This, however, is decidedly not the case. In eight experiments with a female guppy and male or female gourami there was never a clear indication of schooling. Occasionally there were movements that could be interpreted as the guppy attempting momentarily to school with the gourami, but these events were infrequent and uncertain. The two fish usually gave the

TABLE 5
MEDIAN PERIMETERS IN RADII* OF ALL-MALE SCHOOLS OF *Colisa lalia*

Companion Number	Random Value	4 May '59		10 May '59		6 Dec. '59		15 Jan. '60		6 Feb. '60		7 Mar. '60		26 Mar. '60		8 Apr. '60		Means		
		Early	Late	Early	Late	Early	Late	Early	Late	Early	Late	Early	Late	Early	Late	Early	Late	Early	Late	Total
1	1.93	[blot]				0.48	0.38			0.30	0.75	0.46	0.27					0.41	0.47	0.44
2	2.84	0.46	0.80	0.73	0.66													0.63	0.73	0.68
4	3.93							0.87	1.29					2.18	1.64	1.22	0.80	1.43	1.45	1.44

*For perimeters in mean standard lengths of fish multiply by 19.

TABLE 6
MEDIAN PERIMETERS IN RADII* OF ALL-FEMALE SCHOOLS OF *Colisa lalia*

Companion Number	Random Value	16 May '59		13 Dec. '59		9 Jan. '60		13 Feb. '60		30 Mar. '60		7 Apr. '60		16 Apr. '60		Means		
		Early	Late	Early	Late	Early	Late	Early	Late	Early	Late	Early	Late	Early	Late	Early	Late	Total
1	1.93			0.58	0.37			0.44	0.17			0.29	0.22			0.44	0.25	0.34
2	2.84	0.67	1.10													1.09	1.28	1.19
4	3.93					1.47	1.38			1.51	1.46			2.51	1.83	1.99	1.60	1.80

*For perimeters in mean standard lengths of fish multiply by 19.

TABLE 7
PERIMETERS IN RADII* OF SCHOOLS OF EQUAL NUMBERS OF MALES AND FEMALES OF *Colisa lalia*

Companion Number	Random Values	13 Nov. '59		14 Nov. '59		15 Nov. '59		21 Nov. '59		22 Nov. '59		22 Jan. '60		22 Feb. '60		12 Mar. '60		Means		
		Early	Late	Early	Late	Early	Late	Early	Late	Early	Late	Early	Late	Early	Late	Early	Late	Early	Late	Total
1	1.93			0.26	0.29	0.35	0.40	0.21		0.83	0.32			0.34	0.40	0.46	0.30	0.41	0.34	0.38
7	4.50		1.77																1.77	1.77
11	4.80											2.59	2.58					2.59	2.58	2.59

*For perimeters in mean standard lengths of fish multiply by 19.

impression of complete independence. These intuitive impressions are verified by analysis of photographic records (Figure 5). Although the median perimeters of heterotypic pairs differ significantly from the calculated random value and from both homotypic combinations, they are closer to the random value.

I attempted to match all heterotypic pairs with respect to size. The two specimens were always within about two millimeters of the same standard length.

TABLE 8

MEANS OF MEDIAN PERIMETERS OF ALL SCHOOLS OF *Colisa lalia*

Companion Number	Random Median	Means of Medians			
		Early	Late	Total	Prop. of Max.
1	1.93	0.42	0.36	0.39	0.098
2	2.84	0.86	1.01	0.93	0.179
4	3.93	1.71	1.53	1.62	0.275
7	4.51	——	1.77	1.77	0.289
11	4.80	2.58	2.59	2.59	0.417

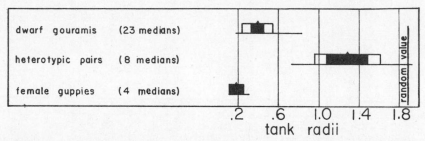

Figure 5. Median perimeters of homotypic and heterotypic pairs of dwarf gouramis and female guppies. A horizontal line shows the observed range, the open rectangle a standard deviation, and the blackened rectangle two standard errors on either side of the mean (central peak).

DISCUSSION

The proposals that have been made on the survival value of schooling can be put in three groups: (1) that schooling serves to "condition" the medium, (2) that it relates to reproduction, and (3) that it affords protection against predators.

It is well established that various organisms can modify their environments to their own benefit, but can succeed in this only when their population densities are high. The removal of toxic substances from the water is of demonstrated effectiveness in aquaria (Allee *et al.*, 1949:360), where concentrations of toxins may be high while their absolute quantity is low. In nature, such effects may be of some importance in areas of industrial pollution, but they could not have been important in the evolution of schooling in the oceans and other large bodies of water where this behavior is most in evidence.

It can hardly be doubted that some schools play an important role in reproduction. The young of many species move in a school of their sibs under parental supervision. Likewise, adults of some species form spawning schools, although most of the apparent examples may really be aggregations on restricted spawning areas. In the present discussion I am not concerned with these specialized examples, but with the schools formed both in and out of the breeding season by individuals of about the same size regardless of sex.

Breder (1959:470) pointed out that schooling keeps individuals of the same species together and must therefore assure, at least occasionally, that each fish can find a mate when the breeding season arrives. This is undoubtedly true, but it can hardly be an important factor in the evolution of schooling. If it were we would expect a certain efficiency in the realization of this effect. We would expect that schooling would generally be initiated or intensified as the breeding season approaches, and we would expect it to be most characteristic of species of low population densities in which the finding of mates would be a serious problem. Neither of these expectations is realized.

A number of people have suggested that schools afford protection against predation. Allee (1951:100, Allee *et al.*, 1949:399) maintained that schooling reduces the total area exposed to attack, or produces a "confusion effect" on the predators and thereby reduces the loss by predation. Brock and Riffenburgh (1960) proposed that predators quickly feed to satiation when they find a large school of their prey, but that such encounters are reduced in number as the schools become fewer and larger. Since any predator has a limited capacity at any one time, the reduction in the number of encounters may reduce the total predation. Knipper (1955) interpreted some very dense schools as "collective mimicry." A school of a large number of small fish presumably resembles a single large fish or other invulnerable object well enough to deceive its usual predators. These effects are probably real in some instances, but there are known examples of predation being facilitated by the schooling of the prey. An obvious example is predation by man. To be commercially exploitable a fish must be individually so large that it is economical for fishermen to pursue a single individual, or it must form schools, so that fishermen may pursue a large number in one operation. No small fish that does not school can support an important fishery. It is likely that schooling plays a similar role in the activities of predators other than man. The long tail of the thresher shark is reputed to be used to herd fish into localized schools so as to facilitate attack (Bigelow and Schroeder, 1953). The sword of a swordfish may be a weapon specialized for attacking schools. Rich (1947) described swordfish moving through large schools and swinging their two-edged weapons so as to incapacitate large numbers of smaller fish, some of which are then recovered and consumed. Enough are missed, however, to allow fishermen to glean "bushels" of dead or maimed individuals. The wastefulness of attacks on schools by swordfish and other species seriously impugns the concept of protection by the rapid satiation of the predator. The saw of the sawfish may likewise be a specialized device for feeding on schools (references cited by Breder, 1952). The spear of the marlins apparently has no such significance (Wisner, 1958). Breder (1959: 414), Bullis (1961) and Fink (1959) gave other probable examples of the facilitation of predation by the schooling of the prey.

The normal effect of schooling on predation, either positive or negative, can not be decided by the citing of examples, real or hypothetical. Extensive and unbiased evidence is required for such an approach and is not likely to be forthcoming. The matter must be decided on the less direct evidence of relating the properties of schools to the presumed protective function. It is incumbent on those who believe that schools are protective adaptations to show that the structure and workings of a school are such as to achieve mass protection in a functionally efficient manner.

I see no evidence that schools do have such properties. On the contrary, a school typically shows no trace of any functional organization. It has no leader; it shows no division of labor; it undergoes no ordered sequence of endogenous changes. As Breder (1959:459) asserted in cybernetic terms, a school is highly redundant and has a low information content. In these respects, a school shows a sharp contrast with such groups as ant colonies, in which all important features have an inescapable functional basis.

This lack of any apparent functional organization is an eloquent argument for the conclusion that the properties of schools have not been established by natural selection on the basis of their survival values. I would therefore propose that a school is not an adaptive mechanism itself, but rather an incidental consequence of adaptive individual behavior. The adaptation is the reaction of each individual to the school. The fact that the reaction contributes to the properties of the stimulus need not complicate the basic issue. A similar, but less extreme position has been suggested by Breder (1959 and earlier).

I suggest that schooling could be expected to arise in any species subject to aggregation. Even pelagic fishes would aggregate where their food is concentrated, as at zooplankton concentrations caused by convergence of surface waters. Under such conditions, the first fish to be sighted and pursued by a predator would be one on the periphery of the aggregation. Any individual that had an urge to place itself among others of its own kind would often be putting them between itself and sources of danger. Such an individual would be found within the group more often and on the periphery less often than its fellows, and would therefore be favorably selected for survival in the presence of vision-dependent predation. Genetic tendencies in this direction would accumulate until the advantages of increased gregariousness would be balanced by some disadvantage, such as the depletion of food in the center of a school.

I believe that this is an adaquate explanation of the historical initiation of schooling behavior. Once started, however, other factors would become important. It would be an advantage to any fish, normally found among others of its own kind, to be able to recognize fright or distress in its companions. Such a fish would not only be using them to lure predators away from itself, but also as a distant early warning system for the detection of dangers beyond its own sensory range. Adaptations to exploit this possibility could be expected to arise.

The seeking of conspecific individuals or groups is only one way of achieving the protection of their proximity. Another way is to attract other individuals by making species-recognition cues as effective as possible. Both species specificity and size homogeneity would be important for making a fish inconspicuous in a school. The schooling phenomenon would be based on four factors: (1) The recognition of conspecific individuals, (2) the desire to approach them, (3) the giving of species-recognition cues, and (4) the initiation of defensive reactions when such reactions are detected in other members of the school. A school and all of its properties can be explained as the statistical consequence of these individual adaptations. Nothing is implied here of any functional organization of the school as such. Such organization would be implied by the existence of any specialized warning signals, any sounds or displays that function as warnings and can not be explained as incidental to the flight reactions themselves. Flocks of birds, which give alarm notes and display conspicuous tail feathers when they take flight, do appear to be functionally organized in this respect, although these mechanisms may relate more to the survival of dependent offspring than of the flock. Evidence for such mechanisms in fish schools would invalidate my position on their lack of a functional organization.

Such a mechanism has been recognized in the *Schreckstoff* effect. Skinner, Mathews, and Parkhurst (1962) state that "As a means of warning other members of a school, alarmed fish communicate fright by releasing a chemical substance into the water." Investigations of the *Schreckstoff* effect were reviewed by Pfeiffer (1962). It has been established for many of the Cypriniformes that an injured individual gives off an olfactory stimulus that induces defensive reactions in other individuals. These reactions are obviously adaptive. The release of the stimulus, however, is explainable as an incidental effect of damage, and gives no indication of being "a means of warning other members of the school." My reasons for this conclusion are: (1) Fishes provide a wealth of examples of communication by visual and auditory social releasers, which are transmitted instantaneously between individuals, and schooling itself is a reaction to visual stimuli. Why then, for communicating a message for which speed of reception would be especially important, would fishes rely on the slow process of chemical diffusion. (2) *Schreckstoff* is not actively secreted into the water. It escapes from the skin as a result of mechanical injury. (3) The phenomenon is almost entirely confined to a freshwater order, and within this group it can be as easily demonstrated in

species that seldom school as in those that do so regularly.
(4) *Schreckstoff* has a low specificity, and a damaged fish may
stimulate alarm reactions in other genera and families. I suspect
that *Schreckstoff* may be functionally a repellent, which would
stimulate distaste and rapid release from the mouths of some
predators. A number of observations reported by Pfeiffer suggest this interpretation.

I have assumed that schooling is a form of cover-seeking,
and should be most in evidence in cover-deficient habitats. This
is certainly in accordance with general observations, which indicate that schooling is a conspicuous phenomenon in pelagic
oceanic waters, scarcely detectable in weedy or rocky shallows,
and of intermediate development in intermediate habitats. No
doubt there are species with strong innate tendencies to form
schools, and other species in which such tendencies are never
more than weakly developed. My experiments, however, raise
the possibility that much of the observed variation is due directly
to the habitats in which different species are found. It was with
this possibility in mind that I designed the experimental environment to be as deficient in cover as possible. A fish in such an
environment would have no possible outlet for its cover-seeking
reactions other than whatever outlet could be provided by other
fish. Every species observed in this tank schooled in it, at least
part of the time. These species included such forms as eels and
dwarf gouramies, which would not ordinarily be expected to
school. The fact that they did so indicated the pertinence of
schooling to cover-seeking behavior. There are many other
observations that indicate that schooling is initiated or intensified when escape or cover-seeking would be appropriate (Breder,
1959:412).

Yet the great contrast in the intensity of schooling between
homotypic and heterotypic pairs of the dwarf gourami and the
guppy indicates that these fish show a specialized reaction to
conspecific individuals. Even in these species, schooling has
evolved to the point at which specific recognition cues play an
important role. In nature, the schooling reaction would probably
be weakly shown, because these species normally inhabit cover-
rich, heterogeneous environments. Their schooling, however,
would probably reinforce any aggregating tendencies, and cause
a more contagious distribution than would otherwise prevail.
I do not wish to imply that the reinforcement of schooling by

species recognition cues can be demonstrated in all fishes nor that I would expect them in all of the species treated in this report. I would seriously doubt their existence in the juvenile eels. Their schooling is probably in its most primitive state, a mere reaction to other objects that afford a slight amount of visual concealment.

One objective of the study was to attempt to measure schooling intensity by parameters that would be independent of the number of individuals present. The constants a and b in the schooling equation are uniformly descriptive over the investigated range of two to twelve individuals. This gives a minor advantage over comparing deviations from the binomial distribution of fish numbers in different halves of an aquarium (Breder and Halpern, 1946). These deviations provide meaningful data only with respect to specific numbers of individuals. No one has yet used a measure of schooling that would not be influenced by the size of the experimental container.

I have quantitative data on groups of widely different companion number in two species, and in at least one the regression of school perimeter as a function of companion number was steeper than the corresponding random function. If these slopes showed even an approximate constancy for one more logarithmic cycle, the regressions for the fishes would certainly cross that of random distribution, probably with companion numbers in the range of fifty to a hundred. This crossing is to be expected on the basis of the theory of equilibrium distance, of which Breder (1954) provides a mathematical derivation based on the observed behavior of fish as they approach each other and form a school. Similar Soviet studies are summarized by Nikolsky (1963). If schooling results from the establishment of preferred average distances between individuals it would, if distances were short, explain the small perimeters observed in my experiments. With increasing numbers in any finite container, however, it would sooner or later be impossible for the fish to avoid being forced closer together than the equilibrium distance. As this condition is approached, the fish would pass from a contagious to an over-dispersed distribution (crossing of the regressions). A hundred guppies or dwarf gouramis in the experimental tank would be over-dispersed if they maintained distances from each other equal to the median observed for pairs.

SUMMARY

Schooling (a contagious distribution due to social attraction) was apparent in all nine species investigated, including such normally territorial or solitary forms as the dwarf gourami and the eel. This behavior was attributable to the special conditions of the circular experimental tank, which provided a topographically featureless and visually uniform white background. Dwarf gouramis, female guppies, and groups of points scattered at random within a circle showed median perimeters (of the smallest circumscribed convex polygons) closely approximated by a mathematical function of fish number that plots as a straight line on a log-probability graph. Such lines and two constants are useful for comparing species with each other and with the calculated random distribution. The schooling behavior seen in homotypic pairs of either dwarf gouramis or female guppies was virtually absent for heterotypic pairs of these species.

Schooling can be expected to evolve in any population subject to aggregation (contagious distribution due to non-social forces), because peripheral or isolated individuals would be especially liable to predation. Genetic tendencies to avoid such positions by associating with other individuals would accumulate. Social releasers effective in the attraction of other individuals would evolve for the same reason. Another expected adaptation would be the ability to react appropriately to the appearance of flight or distress in other members of a school. In this way a school can be explained as the statistical consequence of the adaptations of its members. The regularities of the properties of schools are those of statistics, not of a functional organization of the whole.

LITERATURE CITED

ALLEE, W. C.
    1951. Cooperation among animals. Schuman, New York, 233 pp.

ALLEE, W. C., ALFRED E. EMERSON, ORLANDO PARK, THOMAS PARK, AND
KARL P. SCHMIDT
    1949. Principles of animal ecology. W. B. Saunders Co., Philad., xii +
        837 pp.

ATZ, JAMES W.
    1953. Orientation in schooling fishes. Proc. of a conf. on orientation
        in animals (Naval Res. Lab. Publ.) :103-14.

BIGELOW, HENRY B., AND W. C. SCHROEDER
    1953. Fishes of the Gulf of Maine. Fish. Bull. U. S., 53:1-577.

BREDER, CHARLES M., JR.
    1929. Certain effects in the habits of schooling fishes, as based on the
        habits of Jenkinsia. Amer. Mus. Novit., 382:1-5.
    1952. Studies on the structure of the fish school. Bull. Amer. Mus. Nat.
        Hist., 96:1-28.
    1954. Equations descriptive of fish schools and other animal aggrega-
        tions. Ecology, 35:361-70.
    1959. Studies on social groupings in fishes. Bull. Amer. Mus. Nat.
        Hist., 117:393-481, pl. 70-80.

BREDER, CHARLES M., JR. AND FLORENCE HALPERN
    1946. Innate and acquired behavior affecting the aggregation of fishes.
        Physiol. Zool., 19:154-90.

BROCK, VERNON E. AND ROBERT H. RIFFENBURGH
    1960. Fish schooling: a possible factor in reducing predation. J. Cons.
        Int. Explor. Mer, 25:307-17.

BULLIS, HARVEY R.
    1961. Observations on the feeding behavior of white-tip sharks on
        schooling fishes. Ecology, 42:194-5.

CLARK, P. J. AND F. C. EVANS
    1955. On some aspects of spatial pattern in biological populations.
        Science, 121:397-8.

FINK, BERNARD D.
    1959. Observation of porpoise predation on a school of Pacific sardines.
        Calif. Fish Game, 45:216-7.

KNIPPER, HELMUT
    1955. Ein Fall von "Killectiver Mimikry." Umschau, 13:398-400.

NIKOLSKY, G. V.
    1963. The ecology of fishes. Academic Press, Inc., London and New York, xv + 352 pp.

PARR, A. E.
    1927. A contribution to the theoretical analysis of the schooling behavior of fishes. Occ. Pap. Bingham oceanogr. Coll., 1:1-32.

PFEIFFER, WOLFGANG
    1962. The fright reaction of fish. Biol. Rev., 37:495-511.

RICH, WALTER H.
    1947. The swordfish and swordfishery of New England. Proc. Portland Soc. Nat. Hist., 4 (2):5-102.

SKINNER, W. A., R. D. MATHEWS, AND R. M. PARKHURST
    1962. Alarm reaction of the top smelt, *Atherinops affinis* (Ayres). Science, 138:681-2.

SPOONER, G. M.
    1931. Some observations on schooling in fish. J. Mar. Biol. Ass. U. K., (N.S.) 17:421-48.

TINBERGEN, N.
    1951. The study of instinct. Oxford Univ. Press, 228 pp.

UPENSKY, J. V.
    1937. Introduction to mathematical probability. McGraw-Hill, New York, ix + 411 pp.

WILKS, S. S.
    1949. Elementary statistical methods. Princeton Univ. Press, xi + 284 pp.

WISNER, ROBERT L.
    1958. Is the spear of istiophorid fishes used in feeding? Pac. Sci., 12:60-70.

YULE, GEORGE U. AND M. G. KENDALL
    1950. An introduction to the theory of statistics. C. Griffin and Co., London, xxiv + 701 pp.

8

Reprinted from *Zoologica*, 52:25-40 (1967)

On the Survival Value of Fish Schools

C. M. BREDER, JR.

The American Museum of Natural History

and

Cape Haze Marine Laboratory

INTRODUCTION

THE QUESTION of whether the typical schools or other groupings of fishes have survival value has often been raised, but few investigators have gone into the matter in any depth. One approach, which might be called the anecdotal or the naturalist's approach, is usually given as a general verbal interpretation, based on simple observation. Another, which might be called the mathematician's approach, is typically given as a rigorous analysis of a schematic abstraction of a fish school, usually as an oversimplification, in which prey and predators are considered as making more or fewer encounters, based primarily on random movements. The first may be exemplified by Breder and Halpern (1946), Hiatt and Brock (1948), Sette (1950), Springer (1957), Milanovskii and Rekubratskii (1960) and von Wahlert (1963). The second may be illustrated by Brock and Riffenburgh (1960) and Olson (1964). There is, of course, merit in both these approaches, but neither, by itself, would seem to be adequate to develop a full understanding of the phenomenon. A third approach would, of course, be the experimental one, but there have been only two reports directed toward the possible significance of schools to survival (Williams, 1964; John, 1964). The recent great activity in the study of schooling, on aspects other than possible survival value, has nonetheless useful data to contribute to this subject.

The primary purpose of the present paper is to indicate clearly that all fish schools are not necessarily similar structures, nor that they could be encompassed in a single formulation. A considerable amount of material has been examined and various theoretical considerations have been drawn into the present study. This treatment makes it possible to show, at least at minimum, some of the complications necessarily involved in any attempt to assign a specific survival value to a given fish school under definite conditions of existence.

Valuable assistance has been freely given by Dr. Eugenie Clark on matters concerning visibility and certain aspects of assemblage and by Dr. William N. Tavolga on features of underwater sound and its consequences. The complete manuscript has been critically examined by Dr. James W. Atz. Drs. Donn E. Rosen and W. N. Tavolga examined those sections in detail pertinent to their interests. To these people the author is grateful for the help rendered.

DEFINITIONS

As in most fields that are undergoing rapid growth there is considerable variety in the usage of words and terminology. This is a normal symptom of an active and changing field. It is brought about primarily by differences in the interests and purposes of the earlier writers on the subject. Evidently there are still too many new facts and ideas developing to expect an early stabilization or general agreement on usage. Thus it behooves all workers in the area to indicate scrupulously just how they are using any terms that could possibly lead to confusion and misunderstanding. Also readers should use great care to be sure that they understand an author's precise meaning. In addition to definitions in this section, differences in point of view and usage are indicated wherever clarification would seem needed.

The word "school" has a long history of common usage in connection with fishes and many dictionaries give as a definition, "a large number of fish swimming together," or some equivalent. The connotation would ordinarily be that if they were swimming together they would be going in the same direction, as opposed to churning about or simply resting. Parr (1927), Atz (1953) and Breder (1929 through 1965) have used the word essentially in the sense of the ordinary dictionary definition. Spooner (1931) attempted to restrict the use of "school" to cover only social groups, as opposed to groups drawn to one place by non-social influences. Certainly there is no objection to redefining a word for technical purposes when such a modification of usage is justified. However, it is seldom possible to determine what motivations are effective in the formation of a school. In most cases it is difficult or impossible to define what determines the formation of any type of fish group. There is always at least a residue of both a social and non-social influence present for the fishes must at least tolerate each other and must be located where they are because of non-social influences, such as temperature, nearness to surface or bottom, light, *et cetera*.

It is partly for the above reasons that the more nearly objective and always recognizable measure, not concerned with what drives the fish may or may not have, has been used here. Williams (1964) objects to this usage stating, "It may be of some value to distinguish these two phases of activity,[1] but the difference between social and non-social groupings is in greater need of terminological distinction." This, of course, is a measure of the difference between two approaches, needs and purposes. It very nicely illustrates the point made in the preceding comments. The field is still in such an uncongealed state that it is possible for two very thoughtful papers (Brock and Riffenburgh, 1960; Williams, 1964) to express essentially opposite points of view.

Milanovskii and Rekubratskii (1960) who consider ". . . schooling behavior as one of the adaptive features of a population of a single species . . ." use the word "school" in an even broader way than does Williams and indicate that the usages of Parr, Keenleyside (1955) and others are "one-sided" and "contain mechanistic elements." At least there is agreement in the present paper with their comment, "At any stage of elaboration of the problem we find it necessary to have a working hypothesis—tentative definitions of school and schooling behavior, which should be based on the present level of our knowledge."

Williams (1964) develops the idea that schooling and aggregating are basically rooted in a tendency to hide behind something, as a response to "fright." He carried out experiments bearing on this idea, with a number of species of essentially aggregating types of fishes, as did John (1964) on *Astyanax*. Both found that in a "blank" environment their fishes tended to stay together and formed aggregations or "fright schools." With these experiments there is no disagreement, but an examination of them may help to further clarify the different usages and attitudes toward the word "school." If the correspondingly opposite experiments be made of placing permanently-schooling fishes in an environment of abundant and varied cover these fishes will not hide behind anything, even if completely isolated from others of their kind. They merely go into a period of fast and erratic swimming, evidently in search of companions —behavior that looks surprisingly like "panic." It is not uncommon for them to exhaust themselves, collapse and promptly expire. For this reason the term "obligate schoolers" would seem to be appropriate in contrast to fish that may be called "facultative schoolers." Under such conditions of isolation, obligate schoolers will attempt to school with practically any fish, solitary or not, that may be presented. These may be very unlike, for example, *Mugil* "schooling" with *Canthigaster* (Breder, 1949). Evidently it is the motion of another swimming fish that induces the otherwise isolated obligate schooler to react, while they do not respond at all to inert objects. Formal experiments hardly seem necessary in this connection, as the action seems to be entirely evident. The attempt to experiment with these extreme types is, in any case, difficult. They are notoriously difficult to even establish and keep in aquaria. This is the principal reason why scombriform, carangiform or clupeiform species are seldom seen on display in public aquaria.

Williams performed his experiments on *Anguilla rostrata* (LeSueur), *Hyphesobrycon flammeus* Myers, *Notropis antherinoides* Rafinesque, *N. stramineus* (Cope), *Pimephales notatus* (Rafinesque), *Xiphophorus* hybrids, *Poecilia reticulata* Peters, *Lepomis cyanellus* Rafinesque and *Colisa lalia* (Hamilton-Buchanan).[2] Not

---

[1] That is, "schooling" and "aggregating" in the sense used here. This footnote mine.

[2] Breder and Halpern (1946) and Breder and Roemhild (1947) performed somewhat related experiments in which they analyzed the statistical deployment of a number of similar species of fishes, none of which were obligate schoolers. All such work on aggregating forms, while useful, is not adequate to determine the behavior of obligate schoolers.

one of these is an obligate schooler. They may form aggregations in the non-polarized sense, fright schools, schools in rapidly-flowing water or other facultative assemblages. If this was all there was to the matter no one would have thought to differentiate chronic schoolers from the others. It is here that the confusion about this term and its usage arises. In an attempt to clarify the present point of view, the following details are brought together to enable a perhaps clearer separation of the obligate from the facultative.

To be considered obligate, schoolers must be coherently polarized; can only be forced to stop schooling momentarily, and then only by means of considerable violence; and will not maintain a state of random orientation. The group is permanent, excepting only when physical conditions in the environment suppress the functioning of some essential system, usually the optical, as on an extraordinarily dark night. Isolated members display erratic locomotion and commonly cannot exist for long in the solitary state. The drive to associate with others in a body of great unanimity of orientation is clearly a positive matter of great strength, quite unlike the fragile schools of fright, or other temporary mutual orientations seen in fishes which otherwise are found more commonly in non-polarized aggregations or as solitary individuals. For fully evident mechanical reasons only schooling fishes are able to form fish mills, a type of circular swimming which occurs regularly in obligate schools. The non-polarized aggregations are fully unable to form the mill structure. See Breder (1965) for an extended discussion of this phenomenon.

In the terminology proposed by Williams, "school" refers to any group of fishes ". . . that owes its persistence to social (but not sexual) forces" and "aggregation" refers to ". . . groups that arise by individuals independently seeking the same localized conditions", which is in agreement with Spooner (1931). Probably all groups contain an element of both "school" and "aggregation" in the above usage, except the obligate schoolers, as here used. That is, the obligate schooler is so locked to its fellows that it ignores other things in its environment to a remarkable extent while the facultative schooler clearly is more actively involved with other environmental details, often ignoring his fellows to the point of losing its group altogether. Of this Williams was aware when he wrote that ". . . when one observes a dense con-

centration of the same species moving about in a pelagic or other uniform habitat . . . he is probably safe in calling it a school . . . ," but that "schools and aggregations cannot be as confidently distinguished in heterogeneous environments, and it must often happen that groups are formed that owe their cohesiveness to both schooling and aggregation in mutual reinforcement (heterogeneous summation of Tinbergen, 1951)."

Thus it appears that what would seem to be two very different positions are not as far apart as might be thought for, in many cases, if not all, by designating species by either system a very similar listing would develop. That is to say, what are designated as schools in the present view are assembled on a great preponderance of social tendencies, while aggregations are assembled with a far greater content of general environmental influence. This is precisely the view propounded by Williams, and in an area where so little is yet known, may be very useful as a first approximation on what holds the group together, that is, primarily social influences or non-social influences.

ANALYSIS OF PERTINENT DETAILS

The treatment of the available data in this section has been broken down into several subsections, bringing together the controlling influences of the environment and their effects under various conditions of predation.

THE INFLUENCE OF ENVIRONMENT

A suitable point of departure is a consideration of the sensory modalities that are dominant in schooling fishes and the effects of various environmental influences on their functioning.

Visibility and transparency of water

It has been abundantly shown that vision is necessary for the formation and maintenance of fish schools, see for instance, Parr (1927), Atz (1953), Breder (1959), and Blaxter and Parrish (1965). Also, blind fish and fish in total darkness are unable to maintain this highly polarized arrangement. Obviously the transparency of the water is of great importance to any behavior so largely dependent on vision. This was clearly recognized by Brock and Riffenburgh (1960), in connection with vision's role in school maintenance, when they wrote, "A consideration of the optical peculiarities of water is pertinent in this connection. The distance an object of given size can be seen depends upon two factors: the intercept angle at the eye and the contrast difference between the object and the background. Due to backscatter and light absorption an object of high contrast will fade from sight regardless of size at a relatively small distance, say 200 feet or less, even in the clearest water. This means that for objects above a fairly moderate size, large enough to give an intercept angle adequate for effective vision at the distance where light absorption and backscatter reduce contrast difference to a point of invisibility, taken at 2 per cent, for man (Duntley, 1952), any increase in size of the object will not effectively increase the distance at which it may be seen. The critical intercept angle for the human eye is taken to be one minute which would occur for an object 0.72 inches in diameter at 200 feet." This obviously gives a measure of the extreme visibility range under water, which in most places is not even closely approached. It may be that this estimate, although based on Duntley's paper, is too high, for he, in another place, wrote, "It is expected that water having hydrological range[3] as great as 130 feet will be found in the Sargasso Sea and in the Mediterranean." In a personal communication, Dr. Eugenie Clark estimated that horizontal visibility as great as 180 feet occurred off the Caribbean coast of Yucatan.

Before distances as great as those mentioned above are brought into the discussion, there are considerations involving the geometry of fish schools operating completely within the area of full visibility, which can properly be discussed at this point. Since fish often tend to accumulate into "balls," see for instance Breder (1959), they thereby also tend to occupy the minimum space and show the least surface area. This is also done by a droplet of fluid for purely mechanical and geometrical reasons. The result is to incidentally produce a figure of least conspicuousness and therefore to possess some presumed selective value. This we might call "primary selective value," in which the *direct* response to a stimulus, which may be a simple physical condition, produces a result of definite selective value. Viewed this way, it follows that departure from the spherical form may be taken as a measure of the extent to which other influences make the fishes independent of this or

other similar constraints. In its place come other constraints, which may be thought of as "secondary survival values." These are, of course, the types of selective activity ordinarily referred to as simply "selective values," by evolutionists. When the primary and secondary selective processes both press in the same direction, it is often difficult, if not impossible, to clearly separate them, but it is here that one would expect the development of great stability of behavior or structure or whatever the selective processes have been directing. Anyone familiar with schooling fishes can attest to the strength and rigidity of the habit. Departures from it are clearly associated with special circumstances. Some of these may be considered merely various deformations of a primary tendency toward a globular school, or even a non-polarized aggregation, for in this feature both schools and aggregations show similar tendencies.

Deformations may be related to groups forming close to the water surface and spreading out like a rising globule of very viscous oil. A similar case, on the bottom, would be like a globule of heavy oil spreading out. In very shallow water both surface and bottom would exert deformative influences. Also elongate schools are normally associated with fish migrating or under other kinds of highly directional travel.

As a more generalized concept of the geometry of schools, their size and the restrictions of lateral visibility in water, the following situation may be postulated. Given a case where a single individual, prey or predator has a useful visual range of, say, 30 feet, each solitary individual fish may be considered at the center of a sphere with a 30-foot radius. This is too much of an oversimplification, however, for the restrictions on vision from above and below are somewhat less than in any horizontal direction. In the case of looking down, an object below is more fully illuminated than any side view of one at the same depth. In the case of looking up, the fish is silhouetted against the illumination from above. The resultant increase of visibility, both up and down, increases the visual range vertically to an extent determined by turbidity, light angle, *et cetera*, except for the following facts. These differences in visibility, owing to direction, are the precise ones that are minimized by countershading. In most clear open waters countershading is notably efficient. Consequently it is more nearly correct to think of an individual fish as at the center of a geometric figure approximating a very slightly prolate ellipsoid with its long axis vertical and the horizontal axis, coinciding with that of the fish, longer than the transverse axis.

---

[3] Duntley (1952) defined "hydrological range" as follows. "The clarity of water can usefully be specified in terms of *hydrological range* (v). This is the distance measured along the path of sight, at which the apparent contrast of any object seen against a deep water background is reduced to two per cent of its inherent value. Along a horizontal path of sight hydrological range (v_o) is related to the transmittance (T) of the water (as measured by a hydrophotometer) by the equation

$$T = e\text{-}3.912x/v_o \qquad (4.1)$$

where x is the distance from the object to the observer." For the derivation of this expression see the original.

As a matter of simple geometry, several propositions follow. One fish or a "school" of two has practically the same lateral range of vision and there is little increase in the ability of two fish, forming a "school," to detect a predator, over the ability of a single fish. This is because the "inner side" of each is either blocked by its companion or, if not, their fields of view are almost completely duplicative. As a school increases in the number of fishes the range of vision increases proportionately to the area of the side presented. As both prey and predators wander about, there is thus twice the chance of an encounter with two single fish, not encroaching on each other's field of vision, as with two fish together in a school.

The above is precisely calculable and is independent of concerns of Duntley (1952) so long as the fishes do not wander beyond their mutual visibility ranges. Examples cover only certain fishes indigenous to very shallow water, or living near the surface where illumination is not notably attenuated. Here small fishes such as *Jenkinsia* or *Sardinella* are often preyed upon by immature carangids and *Sphyraena* or mature *Strongylura*. The schools may be large, up to over a thousand or more individuals, with the predators cruising about with the prey in full view. The predators may be typically solitary (*Sphyraena*) or in small bands themselves (*Strongylura* and *Caranx*). Any of them may strike into a school and pick off their prey at will, either alone or as two, or rarely a few, actively-feeding predators. In the above named fishes, multiple attacks are most common in the carangids. Presumably the predators under such situations are generally filled to satiation. Field observations have shown that individuals coming in from some distant point beyond the range of visibility, and new to the school of prey, usually pick off a few fish and then rest idly nearby. From then on it is only occasionally that one will dash in to take a single fish, with extended idle intervals between. The length of these intervals is presumably a measure of the degree of digestive satiation which an individual predator has reached. The situation above described is one that can be generally found in regions where such fishes abound and is apparently the normal circumstances under which they usually exist. This could be conceived of as the degenerate limit of the situations involving no limitation on visibility, as earlier discussed. Here the predators are never under prolonged hunger and escape of a school unscathed never occurs. Also here the maintenance of a population of prey species must depend more on reproductive potential or continued recruit-

ment from "safer" environments, with little or no dependence on locomotor activity for escape in flight. However, even within the limits defined, the least healthy, alert or most awkward, would on the average, be systematically eliminated. From the standpoint of selection theory, this in itself could be valuable to the long-time survival of the population.

Sound production and its prevention

The problem of sound production by the swimming efforts of schooling fishes or their predators is presently unclear for several reasons, see for instance Winn (1964), Wodinsky and Tavolga (1964) and van Bergeijk (1964). Ordinarily most fishes make no appreciable sound incident to their locomotor activity but may do so on sharp turns, see Moulton (1960). His observations check well with our own in this respect, considering that different fishes were under sonic observation. Very little on swimming sounds has been reported by acoustical students on either individual fishes or fish schools. This is most notable in observations made in light. It is possible that there has been selection tending to reduce activities and structures responsible for the production of sounds. If this is the case, then schooling fishes that are reported to produce sounds in the dark, see for instance Takarev (1958), Shishkova (1958), Moulton (1960) and Marshall (1962), could represent an overriding nocturnal specialization toward the prevention of too-wide dispersal under lightless conditions. To predators with sonar echo mechanisms, such as porpoises, fish sounds or their absence would apparently make little difference, if any. These forms are able to feed by locating fishes by means of their echo-ranging mechanisms alone, the data on which is summarized by Norris (1964).

The above should not be interpreted to mean that a complete silence is present in a school of fishes, but only that its magnitude is too small to be effective at distances under which predators have to operate. The sounds noted by Moulton (1960) when sharp turns are made by fish schools are evidently only produced under some fright-inducing stimulus. This means only that fish already sensing the near presence of a predator in their locomotor escape efforts, exceed some physical limit above which higher sound levels are reached. This occurs at a time when quietude is evidently no longer as important as flight.

Tavolga, in a personal communication, wrote as follows about the quality of sounds produced by a "smoothly" moving school, "The quality of this noise is interesting in that it would tend to be random since all the fish-tail movements are not perfectly in phase. Such a noise might tend to be masked by ambient noise. Therefore, even if a predator might be in the range of this school noise, he might perceive it as only a slight increase in ambient noise level, as might be produced by wave action or some other physical phenomenon." These sounds would, of course, be quite different from the various nighttime sounds described by authors, often as clicks or taps, and which are clearly not sounds made incidental to locomotion.

A point to be considered about the above is related to the information provided by the sonar instruments such as those used by anglers to locate fishes. These devices, because of the Doppler effect, provide not only an indication of the presence and species but also an estimate of the size of the fish or fishes and the numbers present. This information is based on the pulsations provided by the motion of swimming fishes, which are characteristic for most species and sizes. Of course, the reflected high frequencies used by these instruments, brought down to the audible range by electronic means, are not identical with the low frequency, faint sounds produced by the fishes themselves. However, if these are audible at all, they must have a beat basically similar to that of the ultra-sonic reflected frequencies. It is certainly true that many schools are so lacking in swimming synchronization that only a broad band of low frequency noise could be expected. However, schools vary from those in which the individuals are completely out of phase to those that have well over 50 per cent of the members in good swimming synchronization. Occasionally small schools, usually of not more than a dozen individuals as seen in various species of *Mugil, Caranx* and a variety of scombrids, are clearly in near perfect phase. Schools, other than the ones lacking any substantial synchronization, would introduce a type of "noise" containing a beat, more or less masked, but which should be able to convey information to a predator, including estimates of species, size, number and direction of travel. These thoughts introduce an unexplored area, including the extent of synchronization in fish schools, the reasons for its presence or absence and a study of its sonic product, including volume and characteristic beat. All this should be amenable to an instrumental approach. Indeed the schools without individuals in phase may be an adaptation to the need for the suppression of telltale sounds rather than the other way around.

Bearing on this is the question of the ability of fishes to detect the direction from which a

sound emanates. It has been argued by Harris and van Bergeijk (1962), van Bergeijk (1964) and Harris (1964) that far-field effects are virtually non-directional for fishes, while near-field effects are highly directional. Thus, a school out of visual range and beyond the near-field might not give a predator sonic cues as to its location, but nonetheless, the sounds might stimulate intensified ranging activities on the part of the predator that could lead the latter to its target on a basis of increasing intensity of sound as it approached the school during random searching. This is a matter distinctly different from following up a sound gradient, the phenomenon whose existence has been questioned by van Bergeijk.

In order to present some idea of the areas and limits of the near-field and far-field effects and their somewhat complicated relationships, the following comments and calculations are given.

How far near-field directional cues extend from a sound producing source will, to a considerable extent, determine their utility to the listener. This distance varies with the frequency, being greatest at low frequencies and least at high, and with the amount of the energy output of the source. For instance, holding the energy output constant, through the temperature range at which *Galeichthys* emits its characteristic "percolator"[4] sound, approximately between 20 and 30°C, a frequency of 1000 Hz has the calculated limit of its near field between 9½ and 9¾ inches from the origin, respectively. Other values in feet follow:

Temp.		Frequencies in Hz			
°C	25	100	200	300	800
20	31′+	7′+	3′+	3′−	1′+
30	32′+	8′+	3′+	3′−	1′+

These relationships were calculated from the given temperatures and frequency by means of the empirical equation of Albers (1960).

$c = 141,000 + 421t - 3.7t^2 + 110s + 0.018d$, where c = velocity in cm/sec, t = temp. in °C, s = salinity in ppt and d = depth below surface in cm. Using s = 34.8 ptt and d = 150 cm, values of c were calculated for various values of t. Changes in s and d were negligible for present purposes and were held at the values given, reducing the equation to

$$c = 145,829.7 + 421t - 3.7t^2.$$

The values of the wavelengths were obtained from the relationship

$$\lambda = c/f$$

where λ = wavelength and f = frequency in Hz. From van Bergeijk (1964) the point of equal

---

[4] So designated by Kellogg (1953).

amplitude of the pressure waves and the displacement waves, from a pulsating bubble, which he indicates as a convenient measure of the range of the usefulness of near-field effects, were calculated from the expression

$$n = \lambda/2$$

where n = the distance of the point of equal amplitude from the point of origin.[5] The values obtained are, of course, rather rough approximations, but are fully adequate for the present discussion. The data on the temperature range at which *Galeichthys* is sonic are original, having been established for a certain locality in connection with another project, only vaguely related to studies on schooling. Differences in the attenuation of the various wavelengths concerned are not significant within the spread of frequencies here discussed (Albers, 1960). At much higher frequencies, that is, within the k Hz. range, there is some differential absorption, but this is far removed from the sounds fish usually produce. It should be emphasized, however, that these calculations do not include the influence of the absolute energy of the original signal which, of course, can be of great importance.

Since the range of hearing in fishes has been calculated in general terms to run from about 100 to 3,000 Hz and the range important to the lateral-line organs from about 20 to 500 Hz (Harris, 1964), it follows that the statements made here all fall within the accepted range of fish auditory powers. Also, that when the producer is separated from the receiver, ". . . both near-field and far-field effects must be considered for the organs of hearing as well as the organs of the lateral line." At a frequency of 25 Hz, the wavelength is about 200 feet and at 1000 Hz it is about five feet.

From the preceding it should be possible to estimate at about what distance a fish would lose the directionality of, say, the percolator sound by knowing the frequencies and temperatures involved. Tavolga (1960) stated that there was a predominance of frequencies around 300 Hz in these sounds, and his sonogram indicated that they ranged to below 100 and above 800 Hz. If a fish loses its sense of directionality at about the distance calculated, then if a fish was receiving cues from a *Galeichthys* producing the "percolator" sound at a frequency of 300 Hz or higher, it would not be useful beyond something less than three feet. However, in the spectrum of this sound there are abundant frequencies of 200 and some of less than 100 Hz. Presumably

these would be considerably more attenuated at their respective ranges which are about three and one-half and eight feet. At a frequency as low as 25 Hz the range reaches some 30 feet, and one may assume that there are some effective frequencies between these two extremes, at perhaps ten to 20 feet from the sound source. At this distance the ability to receive directional cues, especially at night, could be of great value, as will be developed, especially since there is some observed behavior of fishes that may be accounted for by a range similar to the one given above. In a personal communication, Dr. Tavolga indicated that he has also observed differences in the behavior of both "lost" schooling fishes and predators that could perhaps represent a passing out of or into the limits of the near-field.

Other influences

Other sensory modalities, such as olfaction or taste, would not seem to be importantly involved in the interactions of schooling fishes and their predators, if at all, or at least there is no clear evidence or theory which would indicate such involvement. Brock and Riffenburgh (1960) considered olfaction a possibility, writing, ". . . predators may attempt to remain with a school of prey even though satiated, and it is not unlikely that a large school of prey may leave an easily detectable trail of odor for a predator to follow," but present no data to support this opinion. Skinner, Mathews and Parkhurst (1962) concluded that the *Schreckstoff* effect served to warn other members of a school, because ". . . alarmed fish communicate fright by releasing a chemical substance into the water." This statement was questioned by Williams (1964) as follows, "Why then for communcating a message for which speed of reaction would be especially important, would fishes rely on the slow process of chemical diffusion?" With apparently a single exception, the *Schreckstoff* reaction is confined to the Cypriniformes, an almost entirely freshwater order. This group does exhibit some schooling, usually in a facultative form. Fishes of this group are not to be considered as obligate schoolers. Strangely, in this connection, Thines and Vandenbussche (1966) indicate that in *Rasbora* the alarm substance is more effective in the daytime, even in a dark room. Pfeiffer (1962) has reviewed the entire subject of the "fright reaction" and his analysis indicates it to be rather remote from the present problem.

Breaks in ontogeny, or more properly, points at which step functions occur, such as in the case of certain fishes, pelagic from hatching,

---

[5] n is expressed in the same units used to measure wavelength.

when they reach a sufficiently advanced but still transparent post-larval stage, and encounter shallow water, will permanently change their attitudes, develop pigment and settle close to the bottom. These, at this time, usually break up their schools into single individuals or small parties, as the life history unfolds. This type of ontogenetic change seems to be present in a life history where one stage is required to vanish abruptly, so that the species concerned either becomes a permanent schooler or abandons the habit entirely.

THE STRUCTURE AND SIZE OF SCHOOLS

Brief reference has already been made to the range of visibility under water and the relation of the conspicuousness of fishes to its degree of transparency. Here a return is made to that subject and its more immediate implications. Because of the considerable mathematical difficulty of dealing with three-dimensional structures of complex outline, see Cullen, Shaw and Baldwin (1965), the case of a simple surface type school, which is often not more than one or two fish deep, will be discussed for illustrative purposes.

It is not merely accidental that most fusiform fishes, not in a school, usually face toward any disturbance less than one that instigates immediate flight. Aside from visual demands in an animal that cannot turn its head alone, there is an immediate reduction of conspicuousness, as the frontal view is much less conspicuous than the corresponding lateral aspect. Anyone who has operated under water is well aware of the phenomenon of having a fish effectively disappear before one's eyes merely because it had turned so as to point at the observer. Such turning to face a disturbance is much less likely in the case of a chunky fish such as an ostracid or diodontid in which such a maneuver would do little to alter its aspect. These, moreover, are distinctly non-schooling types.[6]

The shape of schools

Since circles and spheres enclose the maximum amount of area or volume respectively for a given perimeter or surface, it follows that these or other shapes have a distinct bearing on the conspicuousness of fish schools and aggregations. For these reasons it could be argued that the commonness of such approximations as are found in real schools is a result of selection. As has, however, been indicated in other connections, it happens that many non-living

[6] All these comments are related to the less specifically expressed view of Allee *et al.* (1949) and Allee (1951) on the reduction of total area exposed by fishes in a school.

systems show the same kind of behavior which depends only on their innate cohesiveness. That is, a drop of suitable oil in water of the same specific gravity will be found to be spherical or a drop of mercury on a flat surface will be found to be a badly deformed sphere, flattened on one side and of other curvature on the top side. In other words, departures from the form showing minimum surface may be considered as a measure of some special influence. In this sense the spherical schools discussed by Breder (1959) and the flowing schools of Breder (1951) all could be following simple physical influences, with the first presenting the least conspicuous form possible and the second exposing a much greater area. The latter are usually seen in very shallow water, commonly shallow enough to eliminate the species' predators. Also with the bottom and water surface so close together only globular groups of small size could occur, as for instance the globular pods of *Plotosus* reported by Knipper (1953 and 1955) and observed and discussed by Clark, in a personal communication. However, large sheet-like schools can naturally "fit" most easily into such vertically limited environments. Where this dimension is greater, schools tend to deepen, culminating in approximate spheres of some bulk. Here also larger predators may swim and view such gatherings from greater distances, up to the point where visibility ceases and the schools have protection not so much based on their own geometry as on the peculiarities of underwater vision. Springer (1957) considered huge schools of small fishes, whose bulk at a little distance could resemble some single large creature, to have a discouraging influence on possible predators. This would represent a case where visibility instead of invisibility became of positive advantage to the schoolers.

The problem of enormous schools

Data on details relevant to the present studies are not yet available on the truly huge schools, often involving many thousands of fishes, as exemplified by the great assemblages which are frequently formed by *Clupea* and *Scomber*. Suggestive information, however, would seem to indicate that they are not as uniform in their size composition as smaller schools are usually seen to be. It is conceivable that such lack of uniformity may be based on the manner in which they develop. If so, it may be that they represent an agglomeration of all the smaller schools in a given area. If, say, several hundred schools, each normally uniform in size range within itself, merged with others acceptably similar, it could cause the assembled mass to show a larger variation, from place to place within the whole group. If the combined schools mixed sufficiently, large fish encountering much smaller ones, a disruptive influence could develop, or at least induce an internal realignment so that the large fish were somewhat restricted to one part of the group and the small to another part, with intermediate fishes bridging between them. Then more or less temporary gradients in respect to size, or other characteristics, could develop and stream about within the group, establishing a continual movement driven by the realignment activities of all individuals. This sort of continual adjustment, with respect to locomotor facility is actually to be seen, on a much smaller scale, in smaller schools, and Breder (1965) thought that it formed the basis of the continual small adjustments found in most ordinary schools. This could easily lead to a shearing action breaking up the different size-groups into smaller, but still large schools. Such effects may in fact be responsible for the eventual disintegration of gigantic schools.[7] Also, it has been shown by Hunter (1966) that angular divergencies between school members are greater between individuals of greater variation in size.

Milanovskii and Rekubratskii (1960) performed some experiments with *hoxinus* that have an indirect bearing on the preceding comments and on the amalgamation and disruption of groups composed of merely facultative schoolers, as follows.

"We noted that under natural conditions, several schools of minnows which fed in the same place, and which appeared from the outside to be one unit, reacted differently to changes of the surrounding environment. In the beginning of our observations, a school of small minnows was feeding; then a school of larger minnows approached cautiously, followed by the school of largest minnows, even more cautious and rapid than the fish of the first two schools. All the fishes, small, medium and large, mingled together and had we not seen them approaching gradually we might have considered them to be a single school. However, after some time, the large minnows hid behind the nearest stone, which they found somewhat downstream. From their hiding place, they swam to the food, grabbed it, and swam back. Such a phenomenon of utmost cautiousness in the search for food we designated by the term "withdrawal." At the slightest movement of the observer, the large

minnows swam away, while the small and medium-sized ones continued to feed undisturbed. When the experimenter stretched his hand over the feeding spot, the school of medium-sized minnows fled while the smallest remained, fleeing only after the hand was immersed in the water. Thus, fishes of three different schools reacted in different ways to changes in the environment, while fishes belonging to each of the three schools reacted as one whole. The natural movements of fishes, obtaining food, fleeing in the face of danger, etc., have definite signal values (of different orders of importance) for the remaining fishes of the school. Among these movements one can distinguish between searching movements, alimentary movements and movements of fear." Also they wrote, again of fishes in a stream, "The strongest biological signal is the natural movement of fear. If, being frightened by something, one or several fishes move aside, the whole school follows them. We tried to give the fishes food in such small quantities that only one or two fishes could obtain it. Once satiated, these specimens became more fearful and went to shelter; they were followed by all the other, still hungry, fishes."[8]

The bearing that the various preceding notes have on ideas concerning the survival value of schooling is, among others, as before intimated, that such massive groups may have a deterrent influence over approaching predators.[9] However, it is also reasonable that such an influence would wane in a short time, to be replaced by an opposite one based primarily on habituation of nearby predators to such tremendous schools. The slow drawing in of predators from perhaps a considerable distance would be expected to follow, because individuals of the prey species concentrated in one place in an enormous mass would proportionally restrict their numbers elsewhere. Thus, a situation of positive survival value could transform to a negative one, and possibly also could become a force for the disintegration of the huge group.[10]

---

[7] The finding of Allee and Dickinson (1954) that when a *Mustelus* was as little as 6.7 per cent smaller than another, the lesser dogfish would avoid the greater. This does not imply aggression on the part of the large fishes. This kind of avoidance is basic to the matters discussed above.

[8] These observations are also related to those of Breder (1965) on the feeding of schools of very small *Mugil*. The avoidance reactions these workers described is, no doubt, caused at least partly by the general refusal of fishes of slightly different sizes to mix.

[9] Such a situation is probably related to or identical with the "confusion effects" of Allee *et al.* (1949) and Allee (1951). Also related to this is evidence that fishes eat more when in groups than when alone (Allee, 1938).

[10] According to the English translation of Nikolsky (1963), the Russian usage is to apply "shoal" to such large groups as those here under discussion and to limit "school" to groups so small that presumably all members could have visual or other contact with every other member. In English and American usage "shoal" has apparently always been used as a synonym of "school."

The maximum advantage, then, is enjoyed by relatively small groups; that is, with additions of a few fish to a small group, the conspicuousness of the assemblage increases at a much smaller rate than does the number of its members. This advantage is lost, however, when the number becomes so vast that the volume occupied by the group, although remaining proportional to the number of individuals, becomes a conspicuous mass in terms of absolute size.

These various factors are necessarily influential in limiting the sizes of fish schools. Field observation demonstrates that in a wide variety of species this vague but very real "limit" is not very large, at least under normal circumstances. Although Breder (1965) could find no theoretical upper limit to the size that a fish school might attain on a hydrodynamic basis, such limitation may well be rooted in the aspect here under consideration.

Williams suggests that the tendency for schools to increase in size without limit until ". . . the advantages of increased gregariousness would be balanced by some disadvantage, such as depletion of food in the center of a school." This is something that under ordinary conditions would call for an extremely large school because of the internal churning of schools, exposing first one and then another of its members to the periphery as well as the general conditions of having the school move about or holding a position in a flow of water through it.

MacFarland and Moss (1967) were able to measure dissolved oxygen within and outside of large schools of *Mugil cephalus* Linnaeus. They report that there was a reduction of the oxygen concentration within the schools. Also that there were areas of disruptive activity in the locations showing the lowest oxygen readings. These areas sometimes broke up into several smaller schools. They refer such intra-school activity to oxygen depletion, carbon dioxide increase and pH reduction. As they indicate, this could account, at least in part, for such behavior and may be a factor in limiting school size on a basis of respiratory need.

Here the problem of mill formation originally analyzed by Parr (1927) and extended by Breder (1965) is pertinent. Does mill formation actually have deleterious[11] effects on the fish in a school or is it an occasional occurrence of it without significant effect on them, making an interest in mills merely a matter of the mechanics of its origin and eventual destruction? This

will have to remain an unanswered question, as so far there appear to be no facts or ideas that could begin a structure of theory building.

THE RELATIVE SIZE OF PREY AND PREDATOR

The manner of feeding of predators on schools would seem to have a distinct bearing on the success of the school as a survival device. Commonly predator fishes may be seen to dash into a school and pick off an individual member and immediately retreat, usually swallowing the fish whole. The predator seldom takes more than one fish at a time, but returns again and again, apparently until satiated. Typical examples of this type of predator are *Caranx, Tylosurus* and *Sphyraena*. This type of feeding is probably the least disruptive and the most conservative of the predators' food supply.

Other manners of feeding on schools, as that shown by *Pomatomus,* is destructive of much more of the food supply than that described above. Commonly an individual *Pomatomus* or small group of them will race through a school of smaller fishes, snapping right and left while they go, leaving a trail of half-fish behind. Usually it is the anterior end that is left, and this probably means that less than half of each fish destroyed becomes food for the predator.[12] Similar modes of "wasteful" feeding on schooling fishes have been described by Rich (1947) for *Xiphias,* and Breder (1952) for *Pristis.* Wisner (1958), however, exonerates *Makaira* from such destructive activity, as flailing about with its elongated rostral process in a school of much smaller fishes.

In fishes the ratio of the size of prey to predator may vary widely, ranging from extreme cases where the predator may be more than 20 million times the weight of its normal prey's weight, as for instance *Manta* preying on near microscopic plankton.[13] From this extreme the ratio ranges to unity or even to cases in which the prey may be larger than the predator, as in *Histrio* and the extreme example of *Chiasmodon.* This range of differences in size has a bearing on the nature of the utility of schooling. The phenomenon of herding, for instance,

---

[11] These could be extrinsic, possibly leading to greater predation for instance, or intrinsic, holding the fish uselessly or dangerously in a place of poor feeding or other disadvantage.

---

[12] These mutilated fish-remains usually become food of other types of fishes or invertebrates which otherwise would be scavenging for other organic matter. Occasionally some of them survive but are no longer members of the schooling population. See Breder (1934) and Gunter and Ward (1961) for records of this sort.

[13] Based on a *Manta* of 3,000 lbs. compared to a plankter of 0.1 oz., which is probably much too heavy for the average plankton organism. The value given for the difference in size is certainly minimal, possibly even 3 to 5 times too small.

can only take place within certain relative size ranges between prey and predator, for if the two be of approximately the same size, the predator's approach becomes one of stalking, and if the prey is vastly smaller, as above noted for *Manta,* it becomes a matter of ranging about in search of streaks of plankton where neither stealth nor herding is involved.[14]

SCHOOLS, THEIR MODELS AND DISCUSSION

The only serious mathematical treatment of the possible protective value of schooling has been presented by Brock and Riffenburgh (1960). See also Brock (1962) for supplementary data. This was followed by a note from Olson (1964) who called attention to the work of Koopman (1956a and b and 1957). The latter, which is concerned with the development of "the theory of search" from the mathematical approach, discusses cases involving situations where both target and searcher are moving, as in naval battles. Olson recognized the identity of this with the situation of prey and predator, especially among oceanic fishes. The contributions of both Brock and Riffenburgh, and Koopman are given in convincing mathematical terms.

The usage of the word "school" by Brock and Riffenburgh and by Olson is different from the usage here employed, both in implication and in context. In their usage, a school of fish covers both schools and aggregations as here used, irrespective of the individual orientations or the distances between individuals, up to the limit of the range of visibility and without reference to the drives and circumstances that created the group.

As the equations of Brock and Riffenburgh do not take the orientation of individuals into account, they apply equally well to either polarized or non-polarized assemblages. One of the marked characteristics of schools, in the present sense, is that they consist of individuals spaced a "standard" distance apart. Thus equation (28) of Brock and Riffenburgh is applicable to schools only when c, the distance between individuals in the group, is very small, for if it becomes large, the polarization loosens and the group can no longer be recognized as a closely ordered array of fishes, all swimming side by side in a common direction. This distance, (axis to axis between adjacent fishes) is usually from

[14] See Bigelow and Schroeder (1948) for a discussion of herding in *Alopias* and Hiatt and Brock (1948) for a discussion of it in *Euthynnus.* More complex prey-predator relationships are described by Springer (1957) for *Rhincodon* and others, by Fink (1959) for Porpoises and *Sardinops* and by Bullis (1961) for *Carcharinus longimanus.*

one-half to three-quarters the length of the individuals (Breder, 1954, 1965).

The whole possible confusion is further complicated by the fact that Brock and Riffenburgh, although dealing with ". . . assumptions . . . and conclusions . . . not related to the observed behavior pattern of any particular species of fish . . . ," obviously are concerned primarily with scombriform fishes, a group with which the senior author of that paper has had wide experience. These fishes form excellent material for such studies, being one of the notable schooling groups. It so happens, however, that as many of these species age they tend to lose their strong propensity to school. Consequently, at least in the larger species such as *Thunnus,* the giant-sized individuals occur as solitary fishes or at least do not form the tightly organized schools of their youth. Large fishes in general tend less toward schooling than do small ones. This may be associated with the fact that the larger the fish, the less likely it is to fall prey to some predator of still larger size. Certainly if schooling serves a protective function, the above should naturally follow.

Lest any of the above comments be thought a criticism of a very thoughtful piece of work, this is to emphasize that these remarks are given here only as a warning to the reader to beware of possible misunderstanding because of differences in the usage of terms.

Koopman (1956a and b, 1957) divides his work into three parts, which he describes as follows. "I. The kinematic bases, involving the positions, geometrical configurations, and motions in the searchers and targets, with particular reference to the statistics of their contacts and the probabilities of their reaching various specified positions. II. The probabilistic behavior of the instrument (eye, radar, sonar, etc.) when making a given passage relative to the target. III. The over-all result—the probability of contact under general stated conditions, along with the possibility of optimizing the results by improving the methods of directing the search." Koopman considers much of his theory concerned with the probability of situation to be a special case of the theory of stocastic processes. Obviously much of this has direct bearing on predator and prey relationships, especially as displayed by open water fishes.

Both Brock and Riffenburgh, and Olson express regret for the small amount of field data available to compare with mathematical models. The former wrote, "The general lack of field data concerning the behavior pattern for a prey species and its predator renders either the confirmation or refutation of conclusions reached in this paper by the elaboration of some scheme of predator strategy rather futile." The latter wrote, referring to the Koopman equations, "These are two basic equations, but to put reasonable numbers in them is another matter." For similar reasons, no attempt will be made here to apply any of these equations. Our intent is to bring together the mathematical and observational aspects of work on fish schools, to present some field observations hitherto unpublished, and to give some general considerations on the whole matter.

Although there is a large literature on prey and predator relationships, almost none of it is concerned with features that would seem to have bearing on the problems of fish schools. The work on bird flocks, such as those formed by starlings, indicates that these are evidently operating in a similar manner about as closely as could be expected, considering the large basic differences between birds and fishes, see for instance Horstmann (1950).

In discussing the possible evolutionary course of the schooling habit Williams wrote that ". . . the lack of any apparent functional organization is an eloquent argument for the conclusion that the properties of schools have not been established by natural selection on a basis of survival values." By "functional organization" Williams means any or all specializations such as "alarm notes," markings displayed in flight, *et cetera*. His detailed comments on the above are followed by, "Evidence for such mechanisms in fish schools would invalidate my position on their lack of functional organization." This extreme position is here considered, at least, premature, as there are a number of valid instances when just such mechanisms seem to be indicated. Considering the difficulties in obtaining adequate data and in interpreting their significance, the slow progress in this direction is not surprising. Relevant evidence suggestive of just such "functional organization" is to be found in practically all the current work on sound production among aggregating and schooling fishes, such as seen in Fish (1954), Kellogg (1953), Moulton (1956, 1958, and 1960), Tavolga (1958a, b, c and 1960), Marshall (1962), Stout (1963a and b), and Winn (1964). The consensus of these workers is in general that there are two primary functions provided by the sounds produced by fishes, evidently being either of sexual or social significance. The evidence that sound production is relevant to organization is indicated by various schooling fish that become sonic only at night, when the visual system is inoperable or only feebly so (Takarev, 1958; Shishkova, 1958; Moulton, 1960; and Marshall, 1962).

Bearing on the question of functional organization of fish schools are recent, more refined measurements of the spacing of individuals in a school that have shown that both extrinsic and intrinsic influences can vary these distances. John (1966), working on *Tilapia nilotica* (Linnaeus) and *Notemigonus crysoleucas* (Mitchill), for instance, showed that at very low light levels, below 10^{-3} f. c., schools tended to break up and that individuals served as, ". . . mutual distractions for one another and also as sources of fright." Previously, it had been thought that schools in little light broke up merely because of visual difficulties. This indicates that there is a positive repelling factor involved, that appears as light fades.

Hunter (1966), by means of computer techniques, showed that schools of *Trachurus symmetricus* (Ayres) deprived of food swam at greater distances from each other than did the same fishes after feeding. Although schools and aggregations appear to be leaderless, there are some special cases, such as a white *Carassius* being the focal point for aggregating by yellow companions (Breder, 1959).

There is no disagreement with the William's view of how schools may have arisen, namely ". . . that schooling could be expected to arise in any species subject to aggregation." In accordance with Williams' definition of schooling, this means, in effect, that fishes drawn to a given area by some non-social influence may then in some cases become social. He also wrote, ". . . that a school is not an adaptive mechanism itself, but rather an incidental consequence of adaptive individual behavior. The adaptation is the reaction of each individual to the school."

It seems most likely that schooling in fishes arose from a wide variety of causes, including some that are purely mechanical (Breder, 1965). Further speculations on this matter would seem hardly to be worthwhile, until some time when data and theory have reached a higher level of development.

Evidently the schooling habit becomes established because of purely mechanical or biological reasons, but it would certainly be expected that gene flow could re-enforce the habit, if it proved to be advantageous to the group.

Levins (1962, 1963, 1964) expresses the idea that the adaptive significance of gene flow is that it permits appropriate response to long-term general fluctuations of environment, while ". . . damping the responses to local ephemeral oscillations." This undoubtedly has bearing on the distribution of fish assemblages of all kinds. Levins indicates that migration tends to increase the above condition. It is noteworthy in this connection that obligate schooling forms generally have a large geographic range, produce large numbers of young and commonly show migratory movements. The population density is, of course, extremely high within the limits of the close confines their schools delimit. It is, however, extremely thin if their numbers are considered in reference to the huge areas the schools pass over, even more so if extensive migrations are involved. The fact of schooling precludes nest building or other protective reproductive modes that presumably permit the production of fewer young. The formation of great schools, and their subsequent dissolution, as discussed herein, may well exercise a regulatory role in the gene flow of the species involved.

All that precedes in this paper could be used to support the view that the functioning of prey-fish schools, as well as of unpolarized aggregations, represents just another method of attaining a manner of behavioral homeostasis. This implies that schooling is effective against excessive predation through a wide range of activities, but fails when various limits are exceeded. Backed up by adjustment of reproductive potential, all under the control of selective processes, including those of both predator and prey, as parts of a dynamic system, it is evidently sufficient to produce a situation of considerable stability in the observed populations. While these systems are probably not as closely controlled as, for instance, the hydra populations of Slobodkin (1964), it would be extremely difficult to attempt such analysis and experimental procedures on schooling fishes as he gives his material. Nevertheless, it would seem that the basic activity is similar. This view accepts schooling as a biologically useful activity seen against the appropriate ecological background.

SUMMARY

1. The range of sight, limited as it is by transparency of the water and the amount of light present, governs the effectiveness of schooling as a form of predation control, which varies widely with environmental features.

2. The geometry of the school shape and its motion affects the conspicuousness of schools where water transparency permits good visibility.

3. In situations where visibility is not a limiting factor, the system presents the degenerate limit where schooling fails to protect effectively.

4. The general quietness of fish schools, except under special conditions where some sound may be inevitable or others in which it may be desirable, suggests that there may have been suppression of sound in schooling fishes, probably by way of selection.

5. The physical form and attitudes of the constituent fishes bear on the effectiveness of schools as a protective device, as do the shape and motion of them, schooling being associated chiefly with streamlined fishes, less often with chunky or odd-shaped fishes.

6. Sufficiently large schools may act as a repellent to predators because of their size and shape.

7. School size is related to the availability of fishes of sufficient similarity of size to compose a coherent group, as well as the mechanics of flow within the group, and to this extent becomes amenable to treatment by hydrodynamic means.

8. The size of the predators relative to the size of the prey leads to "stalking" if the sizes are about equal and to planktonic sifting if the prey is extremely small compared with the predator.

9. The whole matter of schooling and aggregating is looked upon as a mechanism of behavioral homeostasis and as such is subject to the influences of selective processes.

BIBLIOGRAPHY

ALBERS, V. M.
   1960. Underwater acoustics handbook. Penn. State Univ. Press: i-xiii, 1-290.

ALLEE, W. C.
   1938. The social life of animals. New York, W. W. Norton Co.: 1-293.
   1951. Cooperation among animals. New York, Schuman: 1-233.

ALLEE, W. C. and J. C. DICKINSON, JR.
   1954. Dominance and subordination in the smooth dogfish, Mustelus canis (Mitchill). Physiol. Zool., 27(4): 356-364.

ALLEE, W. C., A. E. EMERSON, O. PARK, T. PARK, K. P. SCHMIDT
   1949. Principles of animal ecology. Phila., W. B. Saunders: i-xii, 1-837.

ATZ, J. W.
   1953. Orientation in schooling fishes. In Proceedings of a Conference on Orientation in Animals, Feb. 6 and 7, 1953, Washington, D.C., Office of the Naval Research, Department of the Navy, pp. 1-242.

BERGEIJK, W. A. VAN
   1964. Directional and non-directional hearing in fish. In Marine Bio-acoustics, W. N. Tavolga (ed.), Proc. Symp. Lerner Marine Lab., Bimini, Bahamas, pp. 281-298.

BIGELOW, H. B. and W. C. SCHROEDER
   1948. Sharks. Mem. Sears Foundation Marine Res., no. 1, pt. 1, pp. 59-546.

BLAXTER, J. H. S. and B. B. PARRISH
1965. The importance of light in shoaling, avoidance of nets and vertical migration by herring. Journ. Cons. perm. int. Explor. Mer., 30(1): 40-57.

BREDER, C. M., JR.
1929. Certain effects in the habits of schooling fishes, as based on the habits of *Jenkinsia.* Amer. Mus. Novitates, 382: 1-5.
1934. The ultimate in tailless fish. Bull. N. Y. Zool. Soc., 37(5): 141-145.
1949. On the relationship of social behavior to pigmentation in tropical shore fishes. Bull. Amer. Mus. Nat. Hist., 94, art. 2: 83-106.
1951. Studies on the structure of the fish school. Bull. Amer. Mus. Nat. Hist., 98: 1-28.
1952. On the utility of the saw of the sawfish. Copeia, no. 2: 90-91.
1954. Equations descriptive of fish schools and other animal aggregations. Ecology, 35 (3): 361-370.
1959. Studies on social groupings in fishes. Bull. Amer. Mus. Nat. Hist., 117: 393-482.
1965. Vortices and fish schools. Zoologica, 50: 97-114.

BREDER, C. M., JR. and F. HALPERN
1946. Innate and acquired behavior affecting aggregations of fishes. Physiol. Zool., 19: 154-190.

BREDER, C. M., JR. and J. ROEMHILD
1947. Comparative behavior of various fishes under differing conditions of aggregation. Copeia, no. 1: 29-40.

BROCK, V. E.
1962. On the nature of the selective fishing action of longline gear. Pacific Science, 16(1): 3-14.

BROCK, V. E. and R. H. RIFFENBURGH
1960. Fish schooling: a possible factor in reducing predation. Journ. du Conseil, 25 (3): 307-317.

BULLIS, H. R.
1961. Observations on the feeding behavior of white-tip sharks on schooling fishes. Ecology, 42: 194-195.

DUNTLEY, S. Q.
1952. The visibility of submerged objects. Final Report, Visibility Lab., Mass. Inst. Tech., pp. 1-74. Reissued Visibility Lab., Scripps Inst. Oceanography (Chap. 1 through 4), pp. 1-36.

FINK, B. D.
1959. Observations of porpoise predation on a school of Pacific sardines. Calif. Fish and Game, 45: 216-217.

FISH, M. P.
1954. The character and significance of sound production among fishes of the Western North Atlantic. Bull. Bingham Ocean. Coll. 14: 1-109.

GUNTER, G. and J. W. WARD
1961. Some fishes that survive extreme injuries and some aspects of tenacity of life. Copeia, no. 4: 456-462.

HIATT, R. W. and V. E. BROCK
1948. On the herding of prey and the schooling of *Euthynnus yaito* Kishinouye. Pacific Science, 2(4): 297-298.

HARRIS, G. G.
1964. Considerations on the physics of sound production by fishes. *In* Marine Bio-acoustics, W. N. Tavolga (ed.), Proc. Sympos. Lerner Marine Lab., Bimini, Bahamas, pp. 233-247.

HARRIS, G. G. and W. A. VAN BERGEIJK
1962. Evidence that the lateral-line organ responds to near-field displacements of sound sources in water. Journ. Acoustical Soc. Amer., 34: 1831-1841.

HORSTMANN, E.
1950. Schwarm und Phalanx als überindividuelle Lebensformen. Forsch. Spiekeroog Univ. Hamburg, Stuttgart, no. 1: 7-25.

HUNTER, J. R.
1966. Procedure for analysis of schooling behavior. Journ. Fish. Res. Bd. Canada, 23(4): 457-562.

JOHN, K. R.
1964. Illumination, vision and schooling of *Astyanax mexicanus* (Filippi). Jour. Fish Res. Bd. Canada, 21(6): 1453-1473.
1966. Schooling of fresh-water fishes under varying illumination. Abstract, Bull. Eco. Soc. Amer., Autumn Issue, Sept., p. 145.

KEENLEYSIDE, M. H. A.
1955. Some aspects of the schooling behavior of fish. Behavior, 8(2, 3): 183-248.

KELLOGG, W. N.
1953. Bibliography of the noises made by marine organisms. Amer. Mus. Novitates, 1611: 1-15.

KNIPPER, H.
1953. Beobachtungen an jungen *Plotosus anguillaris* (Bloch). Veröffentl. Überseemus. Bremen, 2(3): 141-148.
1955. Ein fall von "Killectiver Mimikry." Umshau, 13: 398-400.

KOOPMAN, B. O.
1956a. The theory of search. I. Kinematic bases. Operations Research, 4: 324-346.

1956b. The theory of search. II. Target detection. Ibid.: 503-531.

1957. The theory of search. III. The optimum distribution of searching effort. Ibid.: 613-626.

LEVINS, R.
1962. The theory of fitness in a heterogenous environment. I. The fitness set and adaptive function. Am. Nat. 96: 361-373.

1963. The theory of fitness in a heterogenous environment. II. Developmental flexibility and niche selection. Ibid.: 97: 75-90.

1964. The theory of fitness in a heterogenous environment. III. The adaptive significance of gene flow. Evolution, 18: 635-638.

MacFarland, W. N. and S. A. Moss
1967. Internal behavior in fish schools. Science, 156(3772): 260-262.

MARSHALL, N. B.
1962. The biology of sound producing fishes. *In* Biological Acoustics, Haskell, P. T. and F. C. Fraser (eds.). Symposia Zool. Soc. London, no. 7: 45-60.

MILANOVSKII, YU. E. and V. A. REKUBRATSKII
1960. Methods of studying the schooling behavior of fishes. Nauchnye Dok. Vysshe. Shkoly, Biol. Nauki. No. 4: 77-81. (English translation from U. S. Off. Tech. Serv., Dept. Comm. OTS 63-11116.)

MOULTON, J. M.
1956. Influencing the calling of sea robins (*Prionotus* spp.) with sound. Bio. Bull., 111: 393-398.

1958. The acoustical behavior of some fishes in the Bimini area. Biol. Bull., 114: 357-374.

1960. Swimming sounds and the schooling of fishes. Biol. Bull. Woods Hole, 114: 357-374.

NIKOLSKY, G. V.
1963. The ecology of fishes. London and New York, Academic Press, pp. i-xv, 325.

NORRIS, K. S.
1964. Some problems of echo location in cetaceans. *In* Marine Bio-acoustics, W. N. Tavolga (ed.), Proc. Sympos. Lerner Marine Lab., Bimini, Bahamas, pp. 317-336.

OLSON, F. C. W.
1964. The survival value of fish schooling. Journ. du Conseil., 29(1): 115-116.

PARR, A. E.
1927. A contribution to the theoretical analysis of the schooling behavior of fishes. Occas. Papers Bingham Oceanogr. Coll., no I: 1-32.

PFEIFFER, W.
1962. The fright reaction of·fish. Biol. Rev., 37: 495-511.

RICH, H. W.
1947. The swordfish and swordfishery of New England. Proc. Portland Soc. Nat. Hist., 4(2): 5-102.

SETTE, O.
1950. Biology of the Atlantic mackerel (*Scomber scombrus*) of North America. Part II. Migrations and habits. Fish. Bull., Fish and Wildlife Serv., 51(49): 251-358.

SHISHKOVA, E. V.
1958. Recording and study of sounds made by fish. Trud. VNIRO, 36: 280.

SKINNER, W. A., R. D. MATHEWS and R. M. PARKHURST
1962. Alarm reaction of the top smelt, *Atherinops affinis* (Ayer). Science, 138: 681-682.

SLOBODKIN, L. B.
1964. The strategy of evolution. Amer. Sci., 52(3): 342-357.

SPOONER, G. M.
1931. Some observations on schooling in fish. Jour. Marine Biol. Assoc. United Kingdom, 17: 421-448.

SPRINGER, S.
1957. Some observations on the behavior of schools of fishes in the Gulf of Mexico and adjacent waters. Ecology, 38(1): 166-171.

STOUT, J. F.
1963a. Sound communication during the reproductive behavior of *Notropis analostanus* (Pisces: Cyprinidae). Ph.D. thesis, Univ. Maryland.

1963b. The significance of sound production during the reproductive behavior of *Notropis analostanus* (family Cyprinidae). Animal Behavior, 11(1): 83-92.

TAKAREV, A. K.
1958. On biological and hydrodynamic sounds produced by fish. Trud. VNIRO, 36: 272.

TAVOLGA, W. N.
1958a. The significance of underwater sounds produced by males of the gobiid fish, *Bathygobius soporator*. Physio. Zool., 31: 259-251.

1958b. Underwater sounds produced by males of the blenniid fish, *Chasmodes bosquianus*. Ecology, 39: 759-760.

1958c. Underwater sounds produced by two species of toadfish, *Opsanus tau* and *Opsanus beta*. Bull. Mar. Sci. Gulf and Carribean, 8: 278-284.

1960. Sound production and underwater communication in fishes. *In* Animal Sounds and Communications, Pub. No. 7, AIBS, Lanyon, W. E. and W. N. Tavolga (eds.): 93-136.

THINES, G. and E. VANDENBUSSCHE

1966. The effects of alarm substance on the schooling of *Rasbora heteromorpha* Duncker in day and night conditions. Animal Behaviour, vol. 14, no. 2-3, pp. 296-302.

TINBERGEN, N.

1951. The study of instinct. Oxford Univ. Press: 1-228.

WAHLERT, G. and H. VON

1963. Beobachtungen an Fischschwärmen. Veröffentlichungen Instit. für Meeresforschungin Bremerhaven, 8(2): 151-162.

WILLIAMS, G. C.

1964. Measurements of consociation among fishes. Publ. Mus. Mich. State Univ., Biol. Ser., 2(7): 349-384.

WINN, H. E.

1964. The biological significance of fish sounds. *In* Marine Bio-acoustics, W. N. Tavolga (ed.) Proc. Sympos. Lerner Marine Lab., Bimini, Bahamas, pp. 213-231.

WISNER, R. L.

1958. Is the spear of istiophorid fishes used in feeding? Pac. Sci. 12: 60-70.

WODINSKY, J. and W. N. TAVOLGA

1964. Sound detection in teleost fishes. *In* Marine Bio-acoustics, W. N. Tavolga (ed.) Proc. Sympos. Lerner Marine Laboratory, Bimini, Bahamas, pp. 269-280.

9

Reprinted from *J. Theor. Biol.*, 31:295–311 (1971)

Geometry for the Selfish Herd

W. D. HAMILTON

Department of Zoology,
Imperial College, London, S.W.7, England

(*Received* 28 *September* 1970)

This paper presents an antithesis to the view that gregarious behaviour is evolved through benefits to the population or species. Following Galton (1871) and Williams (1964) gregarious behaviour is considered as a form of cover-seeking in which each animal tries to reduce its chance of being caught by a predator.

It is easy to see how pruning of marginal individuals can maintain centripetal instincts in already gregarious species; some evidence that marginal pruning actually occurs is summarized. Besides this, simply defined models are used to show that even in non-gregarious species selection is likely to favour individuals who stay close to others.

Although not universal or unipotent, cover-seeking is a widespread and important element in animal aggregation, as the literature shows. Neglect of the idea has probably followed from a general disbelief that evolution can be dysgenic for a species. Nevertheless, selection theory provides no support for such disbelief in the case of species with outbreeding or unsubdivided populations.

The model for two dimensions involves a complex problem in geometrical probability which has relevance also in metallurgy and communication science. Some empirical data on this, gathered from random number plots, is presented as of possible heuristic value.

1. A Model of Predation in One Dimension

Imagine a circular lily pond. Imagine that the pond shelters a colony of frogs and a water-snake. The snake preys on the frogs but only does so at a certain time of day—up to this time it sleeps on the bottom of the pond. Shortly before the snake is due to wake up all the frogs climb out onto the rim of the pond. This is because the snake prefers to catch frogs in the water. If it can't find any, however, it rears its head out of the water and surveys the disconsolate line sitting on the rim—it is supposed that fear of terrestial predators prevents the frogs from going back from the rim—the snake surveys this line and snatches *the nearest one*.

Now suppose that the frogs are given opportunity to move about on the rim before the snake appears, and suppose that initially they are dispersed in some rather random way. Knowing that the snake is about to appear, will all the frogs be content with their initial positions? No; each will have a better chance of not being nearest to the snake if he is situated in a narrow gap between two others. One can imagine that a frog that happens to have climbed out into a wide open space will want to improve his position. The part of the pond's perimeter on which the snake could appear and find a certain frog to be nearest to him may be termed that frog's "domain of danger": its length is half that of the gap between the neighbours on either side. The diagram below shows the best move for one particular frog and how his domain of danger is diminished by it:

But usually neighbours will be moving as well and one can imagine a confused toing-and-froing in which the desirable narrow gaps are as elusive as the croquet hoops in Alice's game in Wonderland. From the positions of the above diagram, assuming the outside frogs to be in gaps larger than any others shown, the following moves may be expected:

What will be the result of this communal exercise? Devious and unfair as usual, natural justice does not, in general, equalize the risks of these selfish frogs by spacing them out. On the contrary, with any reasonable assumptions about the exact jumping behaviour, they quickly collect in heaps. Except in the case of three frogs who start spaced out in an acute-angled triangle I know of no rule of jumping that can prevent them aggregating. Some occupy protected central positions from the start; some are protected only initially in groups destined to dissolve; some, on the margins of groups, commute wildly from one heap to another and yet continue to bear most of the risk. Figure 1 shows the result of a computer

10° segments of pool margin (degrees)

Position number	0				90						180							270						360		
1	2 3 3 3		6 1 3 3		4 1 3 5		1 3 4 5 2 9 4		5 6 3 5 4		1 2 2		4 3													
2	5 2 2	8	3 2	6 1 1 7		2 4 7 11 2		7 5 2 5 5		3 2		4 4														
3	6 1 2	8	3 1	8	9		4 9 11		9 4 1 7 5		2 2		4 4													
4	6 1	9	3	9	8		4 11 10		10 3	8 6		1 2		4 5												
5	7	9	2	9	8		5 12 8		12 1	9 6		3		4 5												
6	7	9		9	8		5 14 7		13	9 5		5		3 6												
7	8	7		9	8		5 16 5		15	9 3		6		3 6												
8	9	5		9	9		4 18 3		17	9 1		6		5 5												
9	10	4		8	10		3 20 1		19	8		6		7 4												
10	12	3		7	11		2 22		20	6		6		9 2												
11	13	2		6	12		1 22		22	4		7		9 2												
12	14			6	13		22		24	3		7		9 2												
13	14			5	13		22		26	2		8		8 2												
14	15			3	13		22		28	1		9		7 2												
15	16			1	13		22		30			10		6 2												
16	17				12		22		32			9		6 2												
17	17				11		22		33			8		7 2												
18	18				9		22		35			6		8 2												
19	19				7		22		37			5		9 1												

FIG. 1. Gregarious behaviour of 100 frogs is shown in terms of the numbers found successively within 10° segments on the margin of the pool. The initial scatter (position 1) is random. Frogs jump simultaneously giving the series of positions shown. They pass neighbours' positions by one-third of the width of the gap. For further explanation, see text

simulation experiment in which 100 frogs are initially spaced randomly round the pool. In each "round" of jumping a frog stays put only if the "gap" it occupies is smaller than both neighbouring gaps; otherwise it jumps into the smaller of these gaps, passing the neighbour's position by one-third of the gap-length. Note that at the termination of the experiment only the largest group is growing rapidly.

The idea of this round pond and its circular rim is to study cover-seeking behaviour in an edgeless universe. No apology, therefore, need be made even for the rather ridiculous behaviour that tends to arise in the later stages of the model process, in which frogs supposedly fly right round the circular rim to "jump into" a gap on the other side of the aggregation. The model gives the hint which I wish to develop: that even when one starts with an edgeless group of animals, randomly or evenly spaced, the *selfish avoidance of a predator can lead to aggregation.*

2. Aggregations and Predators

It may seem a far cry from such a phantasy to the realities of natural selection. Nevertheless I think there can be little doubt that behaviour which is similar in biological intention to that of the hypothetical frogs is an important factor in the gregarious tendencies of a very wide variety of animals. Most of the herds and flocks with which one is familiar show a visible closing-in of the aggregation in the presence of their common predators. Starlings do this in the presence of a sparrowhawk (Baerends & Baerends-van Roon, 1950; Hostman, 1952; Lorenz, 1966; Tinbergen, 1951); sheep in the presence of a dog, or, indeed, any frightening stimulus (Scott, 1945). Parallel observations are available for the vast flocks of the quelea (Crook, 1960), and for deer (Darling, 1937). No doubt a thorough search of the literature would reveal many other examples. The phenomenon in fish must be familiar to anyone who has tried to catch minnows or sand eels with a net in British waters. Almost any sudden stimulus causes schooling fish to cluster more tightly (Breder, 1959), and fish have been described as packing, in the presence of predators, into balls so tight that they cannot swim and such that some on top are thrust above the surface of the water (Springer, 1957). A shark has been described as biting mouthfuls from a school of fish "much in the manner of a person eating an apple" (Bullis, 1960).

G. C. Williams, originator of the theory of fish schooling that I am here supporting (Williams, 1964, 1966), points out that schooling is particularly evident in the fish that inhabit open waters. This fits with the view that schooling is similar to cover-seeking in its motivation. His experiments showed that fish species whose normal environment afforded cover in the form of weeds and rocks had generally less marked schooling tendencies. Among mammals, similarly, the most gregarious species are inhabitants of open grassy plains rather than of forest (Hesse, Allee & Schmidt, 1937). With fish schools observers have noted the apparent uneasiness of the outside fish and their eagerness for an opportunity to bury themselves in the throng (Springer, 1957) and a parallel to this is commonly seen in the behaviour of the hindmost sheep that a sheepdog has driven into an enclosure: such sheep try to butt or to jump their way into the close packed ranks in front. Behaviour of this kind certainly cannot be regarded as showing an unselfish concern for the welfare of the whole group.

With ungulate herds (Galton, 1871; Sdobnikov, 1935) with bird flocks (Tinbergen, 1951; Wynne-Edwards, 1962) and with the dense and sudden-emerging columns of bats that have been described issuing at dusk from great bat caves (Moore, 1948; Pryer, 1884) observations that predators do often take isolated and marginal individuals have frequently been recorded. Nor are such observations confined to vertebrate or to mobile aggregations. Similar observations have been recorded for locusts (Hudleston, 1958), for gregarious caterpillars (Tinbergen, 1953a) and, as various entomologists have told me, for aphids.

For the aphids some of the agents concerned are not predators in a strict sense but fatal parasites. Insect parasites of vertebrates are seldom directly fatal but, through transmitted diseases or the weakening caused by the activities of endoparasitic larvae, must often cause death indirectly nevertheless. Thus escape from insect attack is another possible reward for gregarious instincts. From observations in Russia, V. M. Sdobnikov (1935) has stated that when reindeer are standing in dense herds only the outermost animals are much attacked by insects. Among the species which he observed attacking the reindeer, those which produced the most serious affliction were nose flies and warble flies, larvae of which are endoparasites of the nasal passages and the skin respectively. Recently Espmark (1968) has verified and extended most of Sdobnikov's information. His work reinforces the view that such oestrid flies are important and ancient enemies of their various ungulate hosts, as is suggested by the fact that their presence induces a seemingly *instinctive* terror. Reindeer seem to be almost as terrified of them as they are of wolves, and cattle react as though they feared the certainly painless egg-laying of warble flies far more than they fear the painful bites of large blood-suckers (Austen, 1939).

The occurrence of marginal predation has also been recorded for some of the aggregations formed by otherwise not very gregarious animals for the purpose of breeding. The best data known to me concerns nesting Black-headed gulls. The work of Kruuk (1964) has been reinforced by further studies, summarized by Lack (1968). The latter seem to have shown that *all* marginal nests failed to rear young, mainly due to predation. Perhaps, nevertheless, the gulls that could not get places in the centre of the colony were right in nesting on the edge rather than in isolation where, for a conspicuous bird like a gull, the chances would have been even worse.

It is perhaps worth digressing here to mention the *temporal* aspect of marginal predation. The "aggregation" in timing already alluded to for bats issuing from bat caves parallels the marked synchrony in breeding activities which is seen in most aggregations and which has been called the "Fraser Darling effect". In explanation of this, Lack (1968) points out that late and early breeders in terns and the black-headed gull do worse in terms of young raised than those best-synchronized with the mass, and he implies that this is mainly due to predation. Individuals coming into breeding condition late or early may also have a problem in sexual selection—that of finding a mate. This point will be touched on later. There are similar influences of temporal selection for flowering plants.

The securing of a nest site in the middle of a colony area is certainly likely to be an achievement of protected position in a sense related to that explained in the story of the frogs, but in the relative immobility of such positions the case diverges somewhat from the initial theme. In all the foregoing examples, except perhaps that of the insect parasites of vertebrates, close analogy to that theme has also been lost through the assumption that the predator is likely to approach from outside the group. This is difficult to avoid.†

When a predator habitually approaches from outside it is comparatively obvious how marginal pruning will at least maintain the centripetal instincts of the prey species. Whether predation could also initiate gregariousness in an originally non-gregarious species is another matter. As mentioned before, this will begin to seem likely if it can be shown that when predators tend to appear within a non-aggregated field of prey geometrical principles of self-protection still orientate towards gregarious behaviour. So far I have only shown this for one highly artificial case: that of jumping organisms in a one-dimensional universe. For the case in two dimensions a more realistic story can be given. The most realistic would, perhaps, take reindeer and warble flies as its subject animals, but in order to follow an interesting historical precedent, which has now to be mentioned, the animals chosen will be cattle and lions.

In 1871 Francis Galton published in *Macmillan's Magazine* an article entitled "Gregariousness in Cattle and in Men". In it he outlined a theory of the evolution of gregarious behaviour based on his own observations of the behaviour of the half wild herds of cattle owned by the Damaras in South Africa. In spite of the characteristically forceful and persuasive style of his writing, Galton's argument is not entirely clear and consistent. Some specific criticisms will be mentioned shortly. Nevertheless it does contain in embryo the idea of marginal predation as a force of natural selection leading to the evolution of gregarious behaviour. The main predators of the Damaraland cattle, according to Galton, were lions, and he states clearly that these did prefer to take the isolated and marginal beasts. The following passage shows sufficiently well his line of thought. After stating that the cattle are unamiable to one another and do not seem to have come together due to any "ordinary social desires", he writes:

"Yet although the ox has so little affection for, or individual interest in, his fellows, he cannot endure even a momentary severance from his herd. If he be separated from it by strategem or force, he exhibits every sign of mental agony; he strives with all his might to get back again and when he succeeds, he plunges into its middle, to bathe his whole body with the comfort of closest companionship."

---

† Cannibalism in gregarious species raises different problems and may help to explain why the nests of gulls do not become very closely aggregated [see "the gull problem" in Tinbergen (1953b)]. Insects have another kind of "wolf in sheep's clothing": the predator mimicking its prey (Wickler, 1968).

3. A Model of Predation in Two Dimensions

Although as Galton implies, lions, like most other predators, usually attack from outside the herd, it is possible to imagine that in some circumstances a lion may remain hidden until the cattle are feeding on all sides of it. Consider therefore a herd grazing on a plain and suppose that its deep grass may conceal—anywhere—a lion. The cattle are unaware of danger until suddenly the lion is heard to roar. By reason of some peculiar imaginary quality the sound gives no hint of the whereabouts of the lion but it informs the cattle of danger. At any moment, at any point in the terrain the cattle are traversing, the lion may suddenly appear and attack the nearest cow.

As in the case of the "frog" model, the rule that the predator attacks the nearest prey specifies a "domain of danger" for each individual. Each domain contains all points nearer to the owner of the domain than to any other individual. In the present case such domains are polygonal (Fig. 2). Each polygon is bounded by lines which bisect at right angles the lines which join the owner to certain neighbours; boundaries meet three at a point and an irregular tesselation of polygons covers the whole plane. On hearing the lion roar each beast will want to move in a way that will cause its polygonal domain to decrease. Not all domains can decrease at once of course: as in the case of the frogs, if some decrease others must grow larger; nevertheless, if one cow moves while others remain stationary the one moving can very definitely improve its position. Hence it can be assumed that inclinations to attempt some adaptive change of position will be established by natural selection. The optimal strategy of movement for any situation is far from obvious, and before discussing even certain better-than-nothing principles that are easily seen it will be a cautionary digression to consider what is already known about a particular and important case of such a tesselation of polygons, that in which the "centres" of polygons are scattered at random.

Patterns of a more concrete nature which are closely analogous to the tesselation defined certainly exist in nature. In two dimensions they may be seen in the patterns formed by encrusting lichens on rocks, and in the cross-sectional patterns of cracks in columnar basalt. The corresponding pattern in three dimensions, consisting of polyhedra, is closely imitated in the crystal grains of some metals and other materials formed by solidification of liquids. The problem of the statistical description of the pattern in the case where centres are distributed at random was first attacked with reference to the grain structure of metals (Evans, 1945). More recently the N-dimensional analogue has been studied by E. N. Gilbert (1962) on the incentive of a problem arising in communication science. Yet in spite of great expertise many simple facts about even the two-dimensional case remain

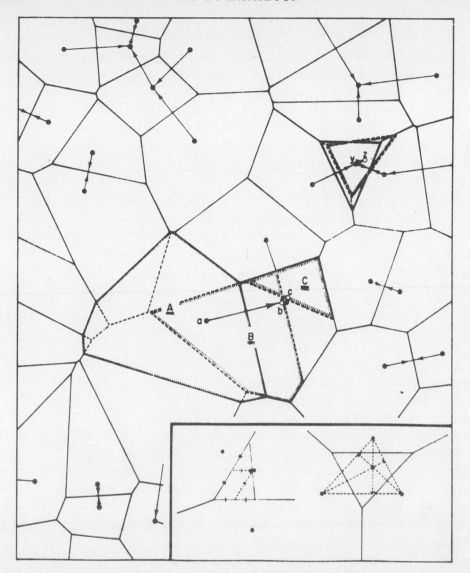

FIG. 2. Domains of danger for a randomly dispersed prey population when a hidden predator will attack the nearest animal. Thin arrowed lines indicate the nearest neighbour of each prey. Thicker arrowed lines show supposed movements of particular prey. Position of prey is given a lower case letter and the domain obtained from each position is given the corresponding upper case letter. The pattern underlining the domain letter corresponds to the pattern used to indicate the boundary of the domain in question. Dashed lines are used for the boundaries of domains that come into existence after the first movement.

Approach to a nearest neighbour usually diminishes the domain of danger (as a → b), but not always (as y → z). On reaching a neighbour the domain may, in theory, be minimized by moving round to a particular side (as b → c). The inset diagram shows geometrical algorithms for minimum domains attainable by a (on reaching c), and by y. In the case of y the minimum domain is equal to the triangle of neighbours and is obtained from the orthocentre of that triangle.

unknown. The distribution of the angles of the polygons is known (J. L. Meijering, personal communication); something is known of the lengths of their sides, and Gilbert has found the variance of their areas.† Nothing, apart from the fairly obvious value of six for the mean, is known of the numbers of sides. Since a biologist might be interested to know whether, in the ideal territory system of these polygons, six neighbours is the most likely number, or whether five is more likely than seven, some data gathered from plots of random numbers is given in Table 1.

TABLE 1

Distribution of the number of sides of 376 polygons constructed from random number plots

No. of sides		3	4	5	6	7	8	9	10	11
No. observed		3	38	97	123	70	29	11	4	1
Total no. observed	376									

Data for 100 of these polygons were received from E. N. Gilbert. The rarity of triangles indicates that the sample area chosen for Fig. 2 is not quite typical.

After glancing at Gilbert's paper a non-mathematical biologist may well despair of finding a theory of *ideal* self-protective movement for the random cattle. Nevertheless experimentation with ruler and compasses and one small evasion quickly reveal a plausible working principle. As a preliminary it may be pointed out that the problem differs radically from the linear problem in that domains of danger change continuously in size with movements of their owner. (In the linear case no change takes place until an individual actually passes his neighbour; this was the reason for choosing organisms that jump.) Now suppose that one cow alone has sensed the presence of the lion and is hastily moving through an otherwise motionless herd. In most circumstances such a cow will diminish its domain of anger by approaching its nearest neighbour. A direct approach can seldom be the path it would choose if it requires that every step should bring the maximum decrease of danger, and when the nearest neighbour is a rather isolated individual it may even happen that a domain increases during approach to a nearest neighbour (Fig. 2, y → z). Such increases will tend to be associated with polygons with very low side number. When an individual is enclosed in a ring of others and consequently owns a many-sided domain, decrease in area is almost inevitable (Fig. 2, a → b). Since the average number of sides is six and triangles are rare (Table 1), it must be a generally useful rule for a cow to approach its nearest

† Recently R. E. Miles (1970) has greatly extended analysis of this case (*Math. Biosci.* **6**, 85) (footnote added in proof).

neighbour. This is a rule for which natural selection could easily build the necessary instincts. Behaviour in accord with it has been reported in sheep (Scott, 1945). The evasion, then, is to imagine that these imaginary cattle are too slow-witted to do anything better. Most readers will agree, I hope, that there is no need to call them stupid on this account.

If the rest of the herd remains stationary and the alarmed cow reaches its nearest neighbour it will usually further decrease its domain of danger by moving round to a side that gives it a minimal "corner" of the neighbour's now enlarged domain (Fig. 2, b → c; left inset diagram shows the principle for finding such a minimal "corner" domain. In this position the cow of the story may be supposed to reach equilibrium; it is some sort of equivalent, albeit not a close one, of the stable non-jump positions that sometimes occur with the frogs. Even more than with the stable positions of the frogs, however, the stable positions of this model must be almost impossible to attain. As in the case of the frogs this is seen as soon as we imagine that all the other cows are moving and have similar aims in view.

Nearest-neighbour relationships connect up points into groups (Fig. 2). Every group has somewhere in it a "reflexive pair"; that is, a pair for which each is the nearest neighbour of the other. A formula for the frequency of membership of reflexive pairs when the points are randomly dispersed has been derived by Clark & Evans (1955). It is $6\pi/(8\pi + \sqrt{27})$. Halving and inverting this expression gives us the average size of groups as 3·218. No further facts are known mathematically about the relative frequencies of different sizes of groups, but a summary of the forms and sizes found in the random number plots is given in Table 2.

Clearly, if randomly dispersed cattle condense according to the statistical pattern indicated in Table 2, the groups so formed can hardly be described as herds. Nevertheless there is undoubtedly a gregarious tendency of a kind, and once such primary groups have formed the cattle in each group will see a common advantage in moving through the field of groups using the same principles as before. Thus an indefinite series of such condensations will take place and eventually large herds will appear. Condensation is not prevented by having the points initially more evenly spaced than in the random dispersion. For example, when the points are initially in some lattice formation the inherent instability of the system is actually easier to see than it is for the random dispersion: every point, if moving alone, can definitely diminish its domain of danger by approaching any one of its equidistant neighbours. Thus even for the most unpromising initial conditions it remains evident that predation should lead to the evolution of gregarious behaviour.

No story even as poorly realistic as the foregoing can be given for the

TABLE 2

No. in group	2	3	4	5	6	7
No. of cases found	46	37	24	8	5	1
Configurations found						

(The configurations consist of small point-and-line diagrams with associated frequencies. For group size 2: 46. For group size 3: 37. For group size 4: frequencies 9, 4, 2, 1. For group size 5: frequencies 5, 4, 1, 1, 1. For group size 6: frequencies 1, 1, 1, 1. For group size 7: frequency 1.)

$$\text{Mean } N \text{ in group} \begin{cases} \text{observed: } 3\cdot107 \\ \text{expected: } 3\cdot218 \end{cases}$$

The nearest neighbour relation allocates points of a random dispersion to groups (see arrowed lines on Fig. 2). Here the configurations found in 121 groups formed by 376 points are recorded together with their frequencies of occurrence. Double connecting lines indicate the reflexive nearest neighbour relationship.

case of predation in three dimensions. In the air and in water there is still less cover in which the predator can hide. If there were a hiding place, however, for what it is worth the argument can be repeated almost exactly and we arrive at $3\frac{3}{8}$ for the mean size of primary groups. For celestial cattle in a space of infinitely many dimensions the mean group size attains only to 4.†

Whether or not marginal predation is, as I believe it to be, a common primary cause of the evolution of gregariousness, it is surprising that the idea of such a cause has received so little attention from biologists. The reason for the neglect of Galton's views cannot lie in the non-scientific nature of the journal in which it was originally published. Galton repeated them in his book *Inquiries into Human Faculty* (1883), and this is one of

† The general formula, where n is the number of dimensions, is

$$\frac{4n}{n+1} + \frac{2}{(n-1)\pi^{\frac{1}{2}}} \left(\frac{3^{\frac{1}{2}}}{2}\right)^{n-1} \frac{\Gamma\{n/2\}}{\Gamma\{(n-1)/2\}}.$$

his best known works. Yet both the hints of the present hypothesis in writings on the schooling of fishes and the admirable study and discussion of this hypothesis by G. C. Williams (1964) were completely independent of Galton's publications. Galton himself is certainly largely to blame for this. With one exception he did not relate his idea to species other than cattle and sheep, and the one exception was man. Something of the tenor of his views on the manifestations of gregariousness in man may be gathered from the fact that the relevant passage in the book mentioned above is headed: "On gregarious and slavish instincts". Needless to say his analogy between human and bovine behaviour now seems somewhat naive. His views remain, as always, interesting and evocative, but their dogmatic and moralizing tone and their obvious connection with their author's widely distrusted line on eugenics can be imagined to have scared off many potentially interested zoologists. Another probable reason for the neglect of Galton's idea is that he himself presented it mixed up with another really quite separate idea which he treated as if it were simply another aspect of the same thing. This was that every cow, whether marginal or interior, benefited by being part of a herd, and that therefore herding was beneficial to the species. His supporting points are undoubtedly forcible; he mentions mutual warning and the idea that by forming at bay in outward-facing bands the cattle can present a really formidable defence against the lions. However, whether or not the cattle actually do, in the last resort, overcome their centripetal inclination and turn to face the predator (as smaller bands of "musk ox" certainly do), these points raise a different issue, as do the mass attacks on predators by gregarious nesting birds. My models certainly give no indication that such mutualistic defence is a necessary part of gregarious behaviour. Moreover, mutual defence and warning can hardly be described as "slavish" behaviour and so seem not to be covered by Galton's heading. There is no doubt, of course, that mutual warning and occasional unselfish defence of others are sometimes shown by gregarious mammals and birds, but where such actions occur they are probably connected on the one hand with the smallness of the risk taken and, on the other, with the closeness of the genetical relationship of the animals benefited (Hamilton, 1963). With the musk ox, for example, the bands are small and clearly based on close family relationships. Sheep and cattle also take risks in defending their young (as Galton pointed out), but this is based on recognition of their own offspring. Apart from the forced circumstances, and the unnatural dispositions engendered by domestication, females usually do not associate with, still less defend, young which are not their own (Williams, 1966). The ability to recognize particular individuals of the same species is highly developed in mammals and birds and there is no difficulty either practical or theoretical in supposing that the mutualistic behaviour of the adult musk oxen, for instance, is an evolutionary development from the altruism involved in parental care. Positive social relations between members of family groups probably exist submerged in most manifestations of mass gregariousness in animals higher than fish. For example the flock of sheep driven by a shepherd owes its compactness and apparent homogeneity to the presence of the sheep dogs: when it is left undisturbed on the mountain it arranges itself into a loosely clustered and loosely territorial system. The clusters are usually based on kinship and the sheep of a cluster are antagonistic to any strange sheep that attempts to feed on their ground (Hunter, 1964). It accords well with the present theory that the breeds of sheep that are most readily driven into large flocks are those which derive from the Merino breed of Spain (Darling, 1937), which is an area where, until very recently, predation by wolves was common. Galton likewise gave the relaxation of predation as the reason for European cattle being very much less strongly gregarious than the cattle he observed in Africa. He noted that the centripetal inclinations evolved through predation would be opposed continually by the need to find ungrazed pasture, so that when predation is relaxed gregarious instincts would be selected against. This point has also been made by Williams.

Most writers on the subject of animal aggregation seem to have believed that the evolution of gregarious behaviour must be based on some advantage to the aggregation as a whole. Many well known biologists have subscribed, outspokenly or by implication, to this view. At the same time some, for example Hesse *et al.* (1937), Fisher (1953) and Lorenz (1966), have admitted that the nature of the group advantage remains obscure in many cases. On the hypothesis of gregarious behaviour presented here the apparent absence of a group benefit is no cause for surprise. On the contrary, the hypothesis suggests that the evolution of the gregarious tendency may go on even though the result is a considerable lowering of the overall mean fitness. At the end of our one-dimensional fairy tale, it will be remembered, only the snake lives happily, taking his meal at leisure from the scrambling heaps of frogs which the mere thought of his existence has brought into being. The cases of predators feeding on apparently helpless balls of fish seem parallel to this phantasy: here gregariousness seems much more to the advantage of the species of the predator than to that of the prey. Certain predators of fish appear to have evolved adaptations for exploiting the schooling tendencies of their prey. These cases are not only unfavourable to the hypothesis of a group advantage in schooling but also somewhat unfavourable to the present hypothesis, since it begins to seem that there must be an advantage to fish which do not join the school. The thresher shark is said to use its much elongated tail to round up fish into easily eaten schools (Williams, 1964).

Even worse, the swordfish is recorded as feeding on schools by first im-mobilizing large numbers with blows from its sharp-edged sword. This may be its usual method of feeding, and the sawfish may use its weapon similarly. Such predators presumably do best by striking through the middle of a school; lone fish might fail to attract them. It is obvious, however, that such cases should not be assessed on the assumption that a particular predator is the only one (Bullis, 1960). At the time of attack by a swordfish there may be other important predators round about that are still concentrating on the isolated and peripheral fish. In the case of locusts the occurrence of such contrary influences from predation is recorded (Hudleston, 1957): large bird predators (ravens and hornbills) attacking bands of hoppers of the desert locust did tend to disperse the bands, and to decimate their central members, by settling in their midst, but at the same time smaller birds (chats and warblers) captured only marginal hoppers and stragglers.

Adaptations of the predator to exploit the gregariousness of its prey certainly suggests the possibility of a changeover to a disaggregating phase of selection. In terms of the earlier story, if the snake evolves such a lazy preference for taking its prey from a heap that it comes to overlook a nearer lone frog, then it may come to be the mutant gregariphobe which survives best to propagate its kind. There are in fact many examples of species which are gregarious in only part of their range or their life cycle and possibly this may reflect the differing influences of different predators.

A surprising number of discussions of animal aggregation have mentioned the occurrence of marginal predation and yet shown apparently no apprecia-tion of its possible evolutionary significance. Whether marginal predation or similarly orientated pressure of selection on individuals is sufficiently powerful to account for the gregariousness observed in any particular case cannot be decided by any already existing body of data. Nevertheless it can be claimed that for none of the cases so far discussed does the data exclude the possibility. It can also be claimed that all the rival theories based on the idea of "group selection" are theoretically insecure. Consider for example the theory of Wynne-Edwards (1962) which, from this basis, attempts to bring many facts of animal aggregation under a common explanation. This theory suggests that aggregations serve to make individuals aware of the current level of population density and that this awareness reacts on repro-ductive performance in a way that holds off the possibility of a disastrous crash due to over-exploitation of the food supply or other limited resources. Except in cases where groups are exclusive and mutually competitive for territory it is very difficult to see how there can be positive selection of any tendency by an individual to reproduce less effectively than it is able (Hamilton, 1971). Wynne-Edwards's cases of massive aggregations, even those having a kin-based substructure as in the gregarious ungulates, are unlikely candidates for the class of exceptions. The alternative theory supported here has no parallel difficulty as to the underlying processes of selection since it can rest firmly on the theory of genetical natural selection.

Certain other versions of the idea of a group benefit, unlike the Wynne-Edwards theory, do not require the concept of self-disciplined restraint by individuals. With such versions, statements that aggregation has evolved because it aids the survival of the species may be treated as merely errors of expression, in which a possibly genuine effect is given the status of a cause. The factor of communal alertness, for instance, may really make life more difficult for a predator and safer for a gregarious prey, especially if the predator is one which relies on stealth rather than speed. In Galton's perhaps over-persuasive words,

> "To live gregariously is to become a fibre in a vast sentient web over-spreading many acres; it is to become the possessor of faculties always awake, of eyes that see in all directions, of ears and nostrils that explore a broad belt of air; it is to become the occupier of every bit of vantage ground whence the approach of a lurking enemy might be overlooked."

But there is of course nothing in the least altruistic in keeping alert for signs of nervousness in companions as well as for signs of the predator itself, and there is, correspondingly, no difficulty in explaining how gregariousness on this basis could be evolved.

Returning to the more interesting and controversial postulate of a population regulatory function for mass aggregation, consider, as a point of detailed criticism, the problem presented by aggregations in which several species are mixed. All the main classes of aggregation that have so far been mentioned —fish schools, bird flocks (both mobile and nesting) and grazing herds— provide numerous examples of species mixture. The species involved may be as widely related as ostriches and antelopes (Hesse *et al.*, 1937). If it is difficult to see how the supposed group selection can work within a species it is certainly even more difficult when the groups are mixed. On the other hand the theory that gregariousness is essentially due to the need for cover finds no difficulty over this point provided that the species which mix have at least one important predator in common. None of the vertebrate predators that have been mentioned as possible agents for the moulding of gregarious inclinations in their prey seem so specialized in their hunting as to make it improbable that individuals of one prey species could expect to gain some protection by immersing themselves in an aggregation of another species.

Perhaps most of the examples of mixed aggregation lie outside the class for which Wynne-Edwards and his supporters would claim a population regulatory function. The examples which have been cited in support of the theory are, in general, more directly concerned with reproduction. Apart from the case of nest aggregations, which has been discussed, there remain Wynne-Edwards's numerous citations of nuptial gatherings. With these the case for the effectiveness of marginal predation is admittedly weaker. Marginal predation seems somewhat less likely on general grounds and there is little evidence. However, there are other ways in which selection is likely to favour individuals which are in the nuptial gathering (or at least on its margin) over those that are isolated. Such selection may be sexual; it may work through differences in the chances of obtaining a mate. As one example, consider the swarms of midges that are so common in damp, still, vegetated places in summer. Many species of nematocerous flies have the habit of forming such swarms. Each swarm usually consists of males of a single species and tends to hover in a fixed spot, often near to some conspicuous object. Females come to the swarm and on arrival each is seized by a male. Passing over possible stages in the initiation of such habits, it is at least clear that as soon as proximity to the swarm itself becomes a key to the female's further co-operation in copulation there is likely to be little chance of mating for the male which does not join the swarm. In such a case the optimal position for a male is probably, not, as it is under predation, at the centre of the throng, and considering how males might endeavour to spend the maximum time in relatively favoured positions, downwind or upwind, above or below, possible explanations for the dancelike motion of such swarms become apparent.

I wish to thank Dr I. Vine for the stimulus to publish this paper: his paper (*J. theor. Biol.* **30**, 405), treating the influence of predation from a different point of view and showing one way in which gregariousness may be beneficial to the group as a whole, was sent to me in manuscript.

REFERENCES

AUSTIN, E. E. (1939). In *British Bloodsucking Flies*. (F. W. Edwards, H. Olroyd & J. Smart, eds.), pp. 149–152. London: British Museum.
BAERENDS, G. P. & BAERENDS-VAN ROON, J. M. (1950). *Behaviour* Suppl. **1**, 1.
BREDER, G. M. (1959). *Bull. Am. Mus. nat. Hist.* **117**, 395.
BULLIS, H. R., JR. (1960). *Ecology* **42**, 194.
CLARK, P. J. & EVANS, F. C. (1955). *Science, N.Y.* **121**, 397.
CROOK, J. H. (1960). *Behaviour* **16**, 1.
DARLING, F. F. (1937). *A Herd of Red Deer*. Oxford: Oxford University Press.
ESPMARK, Y. (1968). *Zool. Beitr.* **14**, 155.
EVANS, U. R. (1945). *Trans. Faraday Soc.* **41**, 365.
FISHER, J. (1953). In *Evolution as a Process*. (J. Huxley *et al.*, eds.), pp. 71–82. London: Allen and Unwin.
GALTON, F. (1871). *Macmillan's Mag., Lond.* **23**, 353.
GALTON, F. (1883). *Inquiries into Human Faculty and its Development*. London: Dent; (1928) New York: Dutton.
GILBERT, E. N. (1962). *Ann. math. Statist.* **33**, 958.
HAMILTON, W. D. (1963). *Am. Nat.* **97**, 354.

HAMILTON, W. D. (1971). *Proc. 3rd int. Symp. Smithson Instn* (in the press).

HESSE, R., ALLEE, W. C. & SCHMIDT, K. P. (1937). *Ecological Animal Geography*. New York and London: Chapman, Wiley and Hall.

HORSTMANN, E. (1952). *Zool. Anz.* Suppl. 17, 153.

HUDLESTON, J. A. (1958). *Entomologist's mon. Mag.* 94, 210.

HUNTER, R. F. (1964). *Advmt. Sci., Lond.* 21 (90) 29.

KRUUK, H. (1964). *Behaviour* Suppl. 11, 1.

LACK, D. (1968). *Ecological Adaptations for Breeding in Birds*. London: Methuen.

LORENZ, K. (1966). *On Aggression*. London: Methuen.

MOORE, W. G. (1948). *Turtox News* 26, 262.

PRYER, H. (1884). *Proc. zool. Soc. Lond.* 1884, 532.

SCOTT, J. P. (1945). *Comp. Psychol. Monogr.* 18 (96), 1–29.

SDOBNIKOV, V. M. (1935). *Trudy arkt. nauchno-issled Inst.* 24, 353.

SPRINGER, S. (1957). *Ecology* 38, 166.

TINBERGEN, N. (1951). *The Study of Instinct*. Oxford: Oxford University Press.

TINBERGEN, N. (1953a). *Social Behaviour in Animals*. London: Methuen.

TINBERGEN, N. (1953b). *The Herring Gull's World*. London: Collins.

WICKLER, W. (1968). *Mimicry in Plants and Animals*. London: Weidenfeld & Nicolson.

WILLIAMS, G. C. (1964). *Mich. St. Univ. Mus. Publ. Biol. Sev.* 2, 351.

WILLIAMS, G. C. (1966). *Adaptation and Natural Selection*. Princeton: Princeton University Press.

WYNNE-EDWARDS, V. C. (1962). *Animal Dispersion in Relation to Social Behaviour*. Edinburgh: Oliver & Boyd.

10

Reprinted from *Auk*, 69:160–170 (Apr. 1952)

FLOCKING BEHAVIOR IN BIRDS

BY JOHN T. EMLEN, JR.

MODERN studies of social behavior in birds have concentrated on two problems: 1) the mechanisms of integration of the organized social unit, and 2) the effects of the social environment on the activity, fecundity, and survival of the individual. The first of these has been brought to your attention by Dr. Collias in his discussion of the development of social behavior (Auk, 69: 127–159, 1952). The second will occupy the attention of the two remaining papers in this symposium, those by Dr. Davis and Dr. Darling (Auk, 69: 171–191, 1952). In the interim I would like to bring up for your consideration a third aspect of social behavior—that of flocking responses, gregariousness, and the various factors which determine the size and density characteristics of bird flocks.

The actuality of flocking behavior in birds does not need to be proved. It is everywhere in evidence, indeed it is difficult to find situations and species which do not show at least a trace of it. Large and dense bird flocks are familiar to the most casual observer. Chattering hordes of migrating blackbirds, swirling clouds of swallows in a pre-roosting flight, jostling crowds of sea-birds on a rocky islet; such scenes provide some of the most thrilling spectacles to be seen in bird life.

The term flock, while commonly associated with such spectacular phenomena, will be applied in this discussion to any aggregation of homogeneous individuals, regardless of size or density. The word homogeneous as used here is not to be interpreted in too strict a manner, but is employed in order to exclude the special heterogeneous groupings of sex and age categories occurring in the breeding pair and the parent-young family group. A flock in this broad sense might result simply from a convergence of independent individuals at a common, localized source of attraction such as a patch of shade or a feeding station. It might, on the other hand, arise as a result of a mutual attraction between individuals. In many bird flocks it is probable that both of these factors operate, the relative rôles of each varying with the species and with the circumstances.

SOCIAL FORCES IN BALANCE

The convergence of birds in response to external physical factors presents, in itself, no great problems to the student of bird behavior. Social responses, on the other hand, are highly complex and fraught with challenging problems.

193

The tendency of birds to respond positively to the presence of others of their kind, commonly referred to as gregariousness, is little understood despite its conspicuousness and widespread occurrence. Various writers have compared it with hunger, a craving or sensation of discomfort which arises in the absence of a physical requirement. Trotter (1916:30) described gregariousness as an impulse in individuals to be in and remain with the flock and to resist anything which tended to separate them from it. Craig (1918) classified it as an appetite, which he defined as "a state of agitation which continues so long as a certain stimulus—is absent," and which is resolved as soon as the appeted stimulus is received. Wheeler (1928:11) compared it with the appetites of hunger and sex and noted its persistent nature and its striking effects on segregated individuals. Various psychologists have regarded it as a condition of responsiveness to social stimuli which, if blocked, leads to frustration activities.

Illustrations of gregarious behavior are not hard to find. Nearly everyone has watched stragglers from a flock of Starlings hurry to join their confreres, or has seen passing Crows respond to a flock of their kind on the ground. Duck and goose hunters are thoroughly familiar with the effect that a group of decoys has on their quarry. Alverdes (1927:108) noted how the artificial isolation of a social animal such as a dog produces numerous signs of discomfort while the presence of a companion, even one belonging to another species, will quiet these "social cravings." Stresemann (1917) discussed the fascination which a flock of birds holds for a segregated individual.

While few would deny this positive social reaction among the members of a flock, Allee (1931) and others have pointed out that there is another factor operating in the formation and regulation of aggregations, the factor of tolerance of social encroachment.

Social tolerance may be considered as promoting flocking behavior by permitting the members of a population to converge in response to either environmental or internal (gregarious) factors. For our purposes, however, it is convenient to consider tolerance in its negative aspect as intolerance, an expression of independence or self-assertion acting in opposition to forces which tend to bring birds together into flocks. Social intolerance is functionally the antithesis of gregariousness. If we follow Craig in calling the craving for companionship an appetite, this second, negative factor is an "aversion," a state of agitation which continues so long as a certain stimulus is present but which ceases when that stimulus is withdrawn (Craig, 1918).

This negative reaction of individuals to crowding is of widespread occurrence in animal populations of all kinds and needs little elaboration. It is particularly conspicuous in birds and is nowhere better illustrated than in the territorial behavior of both colonial and non-colonial species. It underlies most of the situations which lead to conflict between individuals in nature and is considered basic in our modern concepts of the mechanisms of population regulation. Problems of social tolerance and the rôle of aggression in the integration and regulation of bird flocks have recently been reviewed by Collias (1944).

We thus have two opposing forces, a positive force of mutual attraction and a negative force of mutual repulsion, interacting in the formation of bird flocks. The positive force initiates the process and acts centripetally in drawing membership; the negative force serves a regulatory rôle, limiting the size of the flock and preventing close crowding through its centrifugal action. Such a concept may be criticized by those who, encountering difficulties in elucidating emotions in subhuman subjects, object to the word "force" as applied to bird behavior. In the present case, however, I am not referring to any stored or penned-up energy but simply to the cause of the centripetal or centifugal movements observed, whatever that might be (see Webster's unabridged dictionary, 2nd edition, definition No. 15). The responses might be regarded as essentially tropistic and comparable to the prototaxes proposed by Wallin (1927).

No matter what terminology we choose, there seems little doubt that positive and negative social responses occur and interact. Craig observed this interaction in the behavior of his caged Ring Doves as they settled on their roosts for the night. Each bird, he says sought a perch close to friendly companions, *but not too close*, and the difficulties involved in satisfying both the appetite for companionship and the aversion for crowding often kept the birds busy for more than an hour. I have observed similar performances in roosting crows and in loafing flocks of starlings and swallows. By way of illustration, I would like to relate some observations which I made on Cliff Swallows, *Petrochelidon pyrrhonota*, during the past summer.

Cliff Swallows near a group of nesting colonies at Moran, Wyoming, spent much of their time loafing on spans of telephone wires. Positive social forces were immediately apparent in this behavior for the distribution of the birds over the available perches on the roosts was far from random. Of the thousands of linear feet of wire to be found in the area only a few relatively small sections were used at any one time. One or two birds, alighting apparently at random, typically served as the nucleus for a potential gathering. Other birds followed until an aggregation of 100 or more had accumulated, all within a space of 100 to 150 feet.

Negative social forces were also apparent, for in spite of the overall compactness of the group no bird ever held a perch closer than about four inches from its nearest neighbor. Birds were constantly arriving and leaving, and an unstable situation occasionally arose when a bird attempted to secure a perch too close to another, already-settled bird.

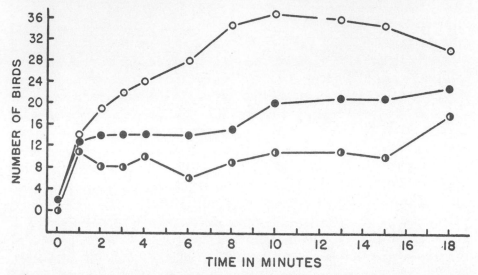

FIGURE 1. Sample curve showing increase in number of Cliff Swallows perching on the central third (open circles), east third (solid circles), and west third (half circles) of a section of telephone wires near Moran, Wyoming, August 4, 1950.

Shuffling and reshuffling inevitably followed until the proper spacing had been reestablished all down the line. Ten-inch gaps were filled centrally and without trouble, a six-inch space on the other hand was usually avoided and, if chosen, was invaded only with the accompaniment of aggressive displays and a local reshuffling of perches. Swallows move their feet but little when once settled and this tolerance limit of four inches may be related to the maximum reach of a bird from a fixed perch.

In an accumulating aggregation of this sort the central portion of the perching area tended to fill more rapidly than did the peripheral portions, and shifts from peripheral to central positions occurred more frequently than did shifts in the other direction. Thus when the perching area was optically divided by the observer into three comparable sections and the number of birds in each of the sections counted at one minute intervals, the central section was seen to grow most rapidly. If the process continued for some time, however, and the central section became filled to capacity, a change in the growth picture ensued and the central section stabilized or even declined while the peripheral sections advanced (Fig. 1).

Observations of territorial defense at the nests indicate that the same narrow but definite tolerance limitations that were seen in the resting flocks on telephone wires applied.   The spatial arrangements of nests in the colonies also indicated that the proximity of nest openings was limited by the reach of a perching swallow.   Details of these observations will be published shortly in another paper (Emlen, *in press*, Condor, 1952).

As I have already suggested, the size and density characteristics of a bird flock are determined by the balance of centripetal and centrifugal forces acting on an innate pattern of behavioral response. Without such forces we may assume that the dispersal of the members of a population over its range would be random.   A positive social force acting alone on such a randomly distributed population would tend to produce clusters and might, if unchecked, lead to the complete aggregation of all individuals into one great and compact flock. Negative forces would presumably intervene, however, to limit and regulate this process of aggregation and, in balance with the positive forces, produce a pattern of small dispersed aggregations each in dynamic balance within itself and with neighboring aggregations.

Variations in the size and density characteristics of flocks may be interpreted as resulting from different balances of these positive and negative social forces.   A few species under certain circumstances may exhibit the positive social attraction with very little of the negative element of intolerance.   This apparently is the case in the sleeping clusters of tree swifts (Hemiprocnidae), wood swallows (Artamidae), and colies (Coliidae) (Allen, 1925:270; Pycraft, 1910:138).   Roosting Passenger Pigeons were said by Kalm (1759) to pile up in heaps on the branches and by Audubon (1831:324) to form solid masses as large as hogsheads.   Other examples of clustering occur, particularly among species which roost in cavities, but a far more common situation is to find the birds spaced and jealously defending their immediate surroundings even in large and dense flocks.   Thus, swallows space themselves along telephone wires and rarely if ever tolerate physical contact with neighbors.   Murres nest in dense colonies yet vigorously defend their immediate surroundings against others of their kind (Howard, 1920:143).   Roosting crows, despite early accounts of clustering (Godman, 1842; Wright, 1897:178), characteristically adjust and re-adjust their perches on the outer twigs of the roosting trees until a suitable interval of three or four inches is achieved (unpublished personal observations made in California and New York).

Small flocks and flocks of low density may be regarded as representing a further shift in the balance through a reduced gregariousness, an increased intolerance, or a combination of both.   Even dispersed social groupings and territorial societies may be included in this concept as cases in which negative forces are strongly developed.   Thus territorial Red-wings, *Agelaius phoeniceus*, for all their pugnacity, show many evidences of a colonial social bond (Robert Nero, unpubl. notes).   Indeed, as Mr. Darling shows in the concluding paper of this symposium, social attraction may be an essential element of territorial display in relatively dispersed populations of non-colonial species.

The effects of non-social, physical factors of the environment such as a localized center of attraction or a localized source of repulsion may be superimposed on the pattern of dispersal set by the balance of social forces.   When such factors are present aggregations may develop which violate the limits of social intolerance and precipitate fighting within the flock.   Fighting, for instance, is common at winter feeding stations among species which rarely quarrel back in the brush where food is more generally dispersed.

Artificial confinement or other restrictions to free movement may have a similar effect.   Fighting is frequent when birds are placed together in cages, and this is particularly noticeable among non-flocking species (Tompkins, 1933).   It is also prevalent in very dense breeding populations of territorial birds where crowding creates the equivalent of spatial restriction (Palmer, 1941:100; Kendeigh, 1941: 42).   The increase in aggressiveness which accompanies the spring recrudescence of sexual activity in flocks of California Quail (Sumner, 1935:214) may perhaps reflect a temporary violation of social tolerance resulting from the inertia of the population in adjusting to internal changes in sociality.

With a fluctuating environment such as is encountered in northern and temperate latitudes and a fluctuating physiology such as is characteristic of nearly all birds, the balance of factors determining flocking behavior is obviously far from stable.

THE PHYSIOLOGY OF FLOCKING

Having noted the existence of centripetal forces related to social attraction and of centrifugal forces related to social repulsion operating in dynamic balance in bird flocks, it might be profitable to give brief attention to the physiological basis of flocking responses.

Since several forms of social behavior, such as those exhibited in the pairing relationship or in the parent-offspring relationship, are demonstrably stimulated and regulated by specific hormones, one is tempted to search for a hormonal basis for the definite and emotionally expressed social responses.   No hormone for gregariousness has been demonstrated, however. The emotional aspects of social behavior are probably related to nervous tensions arising as a result of frustrated attempts to follow a stereotyped neural pattern of social responsiveness.

Social intolerance, the disruptive element in flocking behavior, is, by contrast, often clearly related to the activity of specific hormones. Male sex hormones have been repeatedly shown to induce birds to fight with their flock associates, apparently as a part of the adjustment for sexual activity. Injections of a male hormone into free living California Quail in winter produced aggressive displays and, within a few days, withdrawal from the flock (Emlen and Lorenz, 1942). Castration, conversely, suppressed aggressiveness and fighting in male pigeons (Carpenter, 1932:522). Such responses in experimental birds suggest that the natural increase in sex hormone secretion by the gonads in spring is directly related to the disintegration of wintering flocks in many species.

Other hormones may influence general aggressiveness under special conditions. Prolactin induces maternal reactions (Riddle, 1935), a form of behavior which involves intolerance of flock associates. Thyroxin has little effect on domestic hens except at high levels when it decreases aggressiveness (Allee, Collias, and Beeman, 1940). Estradiol produces similar effects in hens at high levels (Allee and Collias, 1940).

It thus appears that flocking responses have their physiological basis in stereotyped neural patterns and are influenced by hormonal factors only as these incite disruptive responses associated with sexual or parental activity. The aggregated pattern of distribution, reflecting the unrestricted action of positive social responses of gregariousness may thus be regarded as the neutral or "resting" state, and any deviation from it toward a dispersed pattern, a state of tension effected by the introduction of negative social elements.

ENVIRONMENTAL FACTORS

The fluctuations in sociality which are characteristic of most if not all species of birds correlate with two great rhythms of the environment, the seasonal and the diurnal. In both of these the factors which are associated with increased flocking are those that may be considered unfavorable. This correlation of aggregative tendencies with unfavorable conditions has been noted and emphasized by Alverdes (1927) and various other writers concerned with a wide variety of organisms, both vertebrate and invertebrate. Such generalizations should be extended with caution, however, for they may tend to hide the true relationships.

Seasonal factors fluctuate at a slower tempo than diurnal factors, and their effects may thus become integrated with the slower forms of response mechanisms in the bird's physiology, such as those which involve conspicuous morphological changes of the primary, secondary, and accessory sexual structures. Diurnal factors, on the other hand, fluctuate at a tempo which precludes major morphological adjustments, and behavioral fluctuations associated with them are more superficial.

Two factors associated with the seasonal cycle which commonly promote flocking are low temperature and low precipitation. These correlations may be seen in the normal seasonal activity cycles of many birds but are best illustrated, freed from the possible effects of innate rhythms, in the behavioral responses of local populations to irregular fluctuations of weather.

Although the spring recrudescence of sexual activity with its corollary aggressiveness is basically a physiological response to increased day-length, cold temperatures have a profound modifying influence. Red-winged Blackbirds, for instance, respond to a cold spell during early stages of the nesting cycle by abandoning their aggressively defended territories and returning to a winter flocking behavior (Beer and Tibbitts, 1950:63). Many other species behave similarly. Flocking should not be regarded as a general response to cold, however, for in those species which habitually breed in flocks, the effect of cold may be quite different. Thus Cliff Swallows in several colonies near Moran, Wyoming, in 1950 responded to an unseasonable cold spell early in the nesting season by temporarily abandoning their half-built nests and shifting for two days from one type of flocking behavior at the nesting site to another, slightly more dispersed type on foraging grounds several miles away (Emlen, *in press*, Condor, 1952).

Drought is another environmental factor which, occurring abnormally, may promote flocking responses during the non-flocking season. This has been noted in Gambel Quail, *Lophortyx gambelii*, in arid portions of southern California and Arizona where a failure of the usual spring rains inhibits the normal spring dispersal of the winter coveys (Leopold, 1936:28). Endocrine disturbances resulting from nutritional deficiencies are thought to be responsible (MacGregor and Inlay, 1951).

It thus appears that the behavioral responses of birds to unseasonable cold or drought constitute essentially a return to the non-breeding pattern of the species. This implies a suppression of sexual activity and suggests a reduction either of hormone output or of responsiveness to persisting levels of hormone in the blood. Regardless of the mechanisms, however, we may conclude that unfavorable weather promotes flocking behavior by suppressing the disruptive element of social intolerance. There is no evidence that it directly modifies gregariousness itself.

Light intensity is the principal environmental variable of the diurnal cycle, and it is quite evident through observations of roosting behavior that darkness is associated with increased flocking in a great many species. One has only to recall the great roosting assemblages of such diverse birds as herons, gulls, vultures, pheasants, pigeons, swifts, crows, swallows, blackbirds, and robins to capture a realization of this response to the light cycle.

Such sudden fluctuations in sociality are probably not associated with hormonal changes such as those involved in the seasonal cycle; at least no such relationships have been demonstrated. Perhaps they can best be explained by relating them to the schedule of general activity imposed on the birds by alternating periods of light and dark. Night is a period of enforced inactivity for most birds, while the hours of daylight provide the only time during which foraging and other essential activities of self-maintenance can be performed. Self-maintenance calls for independent action which, while not necessarily involving social intolerance, entails a certain amount of freedom from interference. Thus, the members of a covey of quail disperse slightly from their compact roosting aggregation during the morning foraging period, may reunite to loaf during the noon hours, and then fan out again for a second feeding period before finally congregating for the night. The spectacular flights of blackbirds to and from their huge roosting assemblages reflect the same alternation of periods of activity and rest in a pattern and on a scale compatible with their special feeding habits and greater mobility. The limited acreage of a blackbird roosting-site could not conceivably support, even briefly, the hundreds of thousands of birds which congregate on it nightly.

Thus, after foraging in a relatively dispersed pattern, and in a relatively independent manner during the day, the members of a population are temporarily released from the demands of self-maintenance and are permitted to respond freely to their basic gregarious appetites.

RESUMÉ AND CONCLUSION

Symposia such as the present one provide perhaps a legitimate excuse for a little speculation. I hope that I have not abused this privilege and stepped too far beyond the firm shore of established facts.

My objective has been to develop a theoretical basis for interpreting flocking behavior and the various factors, both internal and external, which affect it. For this I have proposed that the form and density characteristics of bird flocks are determined by the interplay of positive and negative forces associated with gregariousness on the one hand and intolerance and independence on the other. Gregariousness has its basis in stereotyped neural patterns, and there is no evidence at present that it is affected directly by hormonal or environmental influences. Social intolerance and independence, on the other hand, are highly variable, and in their variations regulate and determine the dispersal or flocking pattern of the population. Flocking reaches its highest development when gregariousness is given free rein, unrestricted by conflicting demands of reproduction and self-maintenance.

LITERATURE CITED

ALLEE, W. C. 1931. Animal aggregations, a study in general sociology. (Univ. Chicago Press, Chicago), pp. ix + 431.

ALLEE, W. C., AND N. COLLIAS. 1940. The influence of estradiol on the social organization of flocks of hens. Endocrinology, 27:87–94.

ALLEE, W. C., N. E. COLLIAS, AND E. BEEMAN. 1940. The effect of thyroxin on the social order in flocks of hens. Endocrinology, 27:827–835.

ALLEN, G. M. 1925. Birds and their attributes. (Marshall Jones Co., Boston), pp. xiii + 338.

ALVERDES, F. 1927. Social life in the animal world. (Transl. by K. C. Creasy). (Harcourt, Brace, New York), pp. ix + 216.

AUDUBON, J. J. 1831. Ornithological biography. (Dobson & Porter, Philadelphia), pp. xxiv + 512.

BEER, J., AND D. TIBBITTS. 1950. Nesting behavior of the Red-wing Blackbird. Flicker, 22:61–77.

CARPENTER, C. R. 1932. Relation of the male avian gonad to responses pertinent to reproductive phenomena. Physiol. Bull., 29:509–527.

COLLIAS, N. 1944. Aggressive behavior among vertebrate animals. Physiol. Zool., 17:83–123.

CRAIG, W. 1918. Appetites and aversions as constituents of instincts. Biol. Bull., 34:91–107.

EMLEN, J. T., AND F. W. LORENZ. 1942. Pairing responses of free-living Valley Quail to sex-hormone pellet implants. Auk, 59:369–378.

GODMAN, J. D. 1842. American natural history, Vol. II. (H. Carey & I. Lea, Phila.), pp. 337.

HOWARD, H. E. 1920. Territory in bird life. (John Murray, London), pp. xiii + 308.

KALM, P. 1759. Beskrifning på de vilda Dufvor, som somliga år i så otrolig stor myckenhet komma til de södra Engelska nybyggen i Norra America. Kongl. Vetenskaps-Akad. Handl. för ar 1759, 20: 275–295 (transl. Gronberger, Auk, 28: 53, 1911).

KENDEIGH, S. C. 1941. Territorial and mating behavior of the house wren. Ill. Biol. Monog., 18: 1–120.

LEOPOLD, A. 1936. Game management. (Scribner's, New York), pp. xxi + 481.

MACGREGOR, W., AND M. INLAY. 1951. Observations on failure of Gambel Quail to breed. Calif. Fish and Game, 37: 218–219.

PALMER, R. S. 1941. A behavior study of the common tern. Proc. Bost. Soc. Nat. Hist., 42: 1–119.

PYCRAFT, W. P. 1910. A history of birds. (Methuen & Co., London), pp. xxxi + 458.

RIDDLE, O. 1935. Aspects and implications of the hormonal control of the maternal instinct. Proc. Amer. Philos. Soc., 75(6): 521–525.

STRESEMANN, E. 1917. Über gemischte Vogelschwärme. Verh. der Ornith. Ges. in Bayern, 13(2): 127–151.

SUMNER, E. L., JR. 1935. A life history study of the California Quail with recommendations for its conservation and management. Calif. Fish and Game, 21: 167–256.

TOMPKINS, G. 1933. Individuality and territoriality as displayed in winter by three passerine species. Condor, 35: 98–106.

TROTTER, WM. 1916. Instincts of the herd in peace and war. (T. Fisher, Unwin. Ltd., London), pp. 213.

WALLIN, I. E. 1927. Symbionticism and the origin of species. (Williams and Wilkins, Baltimore), pp. xi + 171.

WHEELER, W. M. 1928. The social insects, their origin and evolution. (Harcourt Brace, New York), pp. xviii + 378.

WRIGHT, J. S. 1897. Notes on crow roosts of western Indiana and eastern Illinois. Proc. Ind. Acad. Sci., 1897: 178–180.

Department of Zoology, University of Wisconsin, Madison, Wisconsin, November 28, 1951.

11

ON AGGRESSION IN BIRDS

Peter Marler

The Rockefeller University and The New York Zoological Society

An organism that is to survive and reproduce must have access to an adequate supply of the necessary facilities and raw materials. If it reproduces sexually, it needs a mate. There are many circumstances in which the supply of the resources required will be limited, forcing the animal into competition for them with its fellows. A lack of success in such competition is likely to hinder the prospects of effective reproduction. Natural selection has inevitably favored changes in behavior that permit individuals or groups to compete more effectively. A great deal of the social behavior of animals is concerned with the disposition of resources among the members of the community.

One way to accomplish this function is by the use of social interactions that ensure the spatial distribution of individuals and groups in patterns in some sense appropriate to the distribution of the vital resources. It is a primary function of aggression in animals to facilitate the accomplishment of such patterns of animal dispersion.

The circumstances surrounding reproduction are vitally important in the evolution of a species. The behavior associated with it is often conspicuous and intricate, thus becoming a natural focus for the attention of ethologists interested in evolutionary biology. In the reproductive phase of the life cycle, aggression is manifest in many animals as a form of territoriality, in which animals actively exclude others from a certain fixed space. Most ethological discussions of animal aggression are concerned with this type of fighting.

Aggression is in fact equally common in the nonreproductive phase of the lives of many animals. This is true for example of many birds, which are thought of as providing

classical illustrations of territoriality. Such is the case with the European chaffinch, *Fringilla coelebs*, and the studies I shall describe were motivated by the conviction that careful analysis of aggression among birds in nonreproductive condition might throw some light on the function and significance of animal aggression in general.

The experiments were conducted at the University of Cambridge in England, while I was working at the Madingley Ornithological Field Station, as it was then called. The subjects were captured locally and studied in captivity, in large aviaries. Under these conditions, as in the field, chaffinches display a variety of behavior patterns that are included here under the rubric of fighting, ranging from an extreme form of aggressive posturing—the head-forward display—to the mildest sleeking of the feathers or turning toward another bird that can also be an effective prelude to withdrawal of the opponent. The aim of these experiments was to define the conditions under which such aggressive behavior appears.

BEHAVIOR IN RELATION TO THE SOCIAL HIERARCHY

In the first of four studies, two flocks of captive chaffinches were observed during successive winters, in a large aviary. There was one flock of four birds, another of eight; all birds were marked with colored leg rings, and the sexes were equally represented in each flock. Observations were made from a blind at one end of the aviary, from the end of September to December, when chaffinches are in nonreproductive condition.

Social Hierarchy at the Feeding Station

Two seed dishes were placed in the center of the aviary, 3 ft above the ground. The fighting that occurred at this

Revised and adapted for BIO from material previously published in P. Marler, Studies of fighting in chaffinches. *Brit. J. Animal Behaviour* 3: 111–117, 3: 137–146, 4: 23–30, 5: 29–37 (1955–1957).

feeding station during normal feeding behavior was observed to be organized in a peck-right hierarchy of the simplest straight-line type; the despot, or alpha bird, won all fights, the omega bird none.[1] Males were dominant over females, as in the wild state. When the seed dishes were moved to other locations in the aviary, the hierarchy was unchanged. Thus the outcome of fighting between individual chaffinches over food was independent of where feeding occurred in the cage. It was also found to be unchanged in time; the hierarchy, once established, remained more or less stable throughout the winter, although occasional changes in rank did occur.

Records were kept of the time spent at the food dish by a group of four birds. During 30 hours of observation spread evenly throughout each day for several weeks, the average number of visits per hour by the four birds was, in order of their dominance rank, 2.3, 4.4, 5.3, 5.1. Thus the low-ranking birds visited the food dish more often than those of high rank. Since subordinates are continually displaced from food by dominants, this is to be expected. The average duration of visits to the food decreased down the hierarchy; again in order of dominance, 3.2 minutes, 2.01 minutes, 1.64 minutes, 1.29 minutes. Thus subordinates made a greater number of visits, of shorter duration. Combining the figures for number of visits and duration, we derive the time per hour spent feeding: 7.36 minutes, 8.85 minutes, 8.69 minutes, 6.58 minutes. Although, as expected, the omega bird is able to spend the least time feeding, it is remarkable that the despot, which is free to feed at any time it wishes, seems to feed less than the two middle birds. This may be the result of their having learned, as a result of previous deprivation by the despot, to feed as much as they can. In retrospect, it is regrettable that the birds were not weighed at the time.

The aggressive displays observed were most intense and prolonged between combatants close in rank. The fiercest fighting of all accompanied inversions of dominance; high-ranking birds would make direct supplanting attacks on low-ranking birds, without any preliminary display. Male attacks on females were often made with greater confidence than on other males. Some birds showed submissive behavior, and by this means often prevented others from attacking them. This behavior was more common in females and was especially typical of the omega female while close to the despot.

The general deportment of the birds while not fighting also often reflected their rank. The despot could usually be identified by his alert behavior, with free movement and visual examination of the environment. Movements of pivoting and tail flipping were freely used, in contrast with the other birds. Although they too showed the alert posture, it was used tensely, and movement was notably restrained when in close proximity to the despot.

Effects of Increased Competition for Food

In our study, because of the small size of the seed dishes provided, generally only one bird fed at a time, and the birds would usually avoid approaching a dish already occupied by a dominant bird. If they did approach, they were repelled, unless they managed to inhibit the domi-

nant's attack by submissive behavior. Members of an established flock know that they can attack subordinates freely, but they will generally flee promptly from dominants.

It might be expected that increased competition for food would lessen the preponderance of approaches by dominant birds to a dish occupied by a subordinate as compared with the approaches of subordinates to a feeding dominant. In comparing these ratios in small and large aviaries having different numbers of birds, this was found to be generally true. Further, when the results for each flock were broken down according to sex—that is, approaches of males to males, females to females, and females to males—it was found that females tended to approach feeding dominants of either sex more often than males approached dominants of their own sex.

Effects of Starvation

Another method of increasing competition is to starve the birds and then provide them with a limited food supply. Several such experiments were conducted with a single seed dish and flocks of eight birds previously deprived of food for various periods up to 2 hours, with results as shown in Table 1. The effect on a male flock was to increase the number of approaches by subordinates to feeding dominants until they roughly equaled the number of approaches by subordinates (i.e., a dominant/subordinate approach ratio of about 1.0); with a flock of females the effect was similar. In a mixed flock, however, the ratio was never above 0.8 and fell as low as 0.3. Here there were sometimes more than three times as many subordinate as dominant approaches. Thus, passing from male, to female, to mixed flocks, the fear that subordinates had of dominants, under the effects of starvation, was found to be progressively less.

TABLE 1. EFFECTS OF SHORT PERIODS OF STARVATION ON RATIO OF DOMINANT APPROACHES TO OCCUPIED FOOD; SUBORDINATE APPROACHES (d/s) IN THREE FLOCKS

	Starvation, min						
	0	15	30	45	60	90	120
d/s: eight males	10.4	2.1	5.5	0.73	1.6	0.8	1.3
d/s: eight females	3.9	2.3	1.4	0.55	0.80	0.75	0.97
d/s: four males and four females	0.79	0.74	0.63	0.49	0.66	0.29	

The meaning of these differences may be more clearly understood if we consider how the increasing numbers of subordinate approaches were received by the dominants. Under normal conditions a subordinate approaching a feeding dominant may either be repelled from the dish or may be tolerated, especially if it shows submissive behavior. The proportion of subordinate approaches to feeding dominants that resulted in toleration after various periods of starvation is shown in Table 2. In the males, subordinate approaches were greeted usually with aggression and hardly ever with toleration, just as under normal conditions. In the female flock, on the other hand, the majority were greeted with toleration. Thus under starvation females tolerated each other. It might be inferred

---

[1] A peck-right hierarchy is one in which it is sufficient to identify the individuals concerned in a fight to predict its outcome, regardless of circumstances or location. If it is a straight-line hierarchy, the top-ranking individual dominates all others in the group, the second in rank dominates all except the top individual, and so on down to the one at the bottom of the hierarchy, which is subordinate to all other members of the group.

TABLE 2. PROPORTION OF SUBORDINATE APPROACHES (s) TO FEEDING DOMINANTS THAT RESULT IN TOLERATION (t) BY DOMINANTS, EXPRESSED AS THE RATIO t/s*

	Starvation, min						
	0	15	30	45	60	90	120
t/s: eight males	0.43	0.57	0.05	0.06	0.09	0	0.12
t/s: eight females	0.54	0.66	0.77	0.90	0.84	0.82	0.79
t/s: four males and four females	0.90	0.89	0.82	0.58	0.65	0.64	—

*A ratio of 1.0 implies complete toleration; a ratio of 0, no toleration at all.

from the results shown in Table 2 that dominant males actually became more aggressive after starvation. But this is only an apparent effect resulting from the fact that under normal conditions subordinate males would ordinarily only approach dominants when they were sure of being tolerated, whereas after starvation they would approach dominants indiscriminately. This point will be pursued again later.

Since aggressiveness in itself is not affected by starvation, the results suggest that dominant females become tolerant of subordinates when their fear has been reduced, in this case by starvation. In other words, dominant females are more ready to attack a subordinate who shows a tendency to avoid them. This phenomenon is known in the normal fighting behavior of many species, domestic hens for example. A strongly aggressive hen will attack others no matter how they behave, while a less aggressive one waits until its opponent shows signs of fleeing before she will press the attack. We may conclude that male chaffinches come into the first category and females into the second, but that the balance in females is sufficiently delicate that even if a dominant has learned to attack subordinates freely under normal conditions, a reduction of the subordinates' fear will cause marked reduction in the tendency to attack.

The effects of starvation on the mixed flock were intermediate between those on the male and female flocks (Table 2): there was more toleration of subordinates than in the male, less than in the female. This was partly the result of the averaging out of male aggressiveness with female toleration in the mixed flock, but this factor alone would have led to a much lower proportion of tolerations than appeared. There was a counterbalance from the males, who had been more tolerant of females than of their own sex under normal conditions and continued to tolerate them after starvation.

Avoidance responses other than those in the feeding context seemed also to be reduced as a result of food deprivation. Starved birds showed less agitation than under normal conditions on the approach of a human observer. This was most apparent in low-ranking birds, the ones most likely to be short of food. Even under the usual aviary conditions, some subordinates developed the habit of taking the chance to feed during disturbances while the rest of the flock showed escape behavior. Starvation appears to reduce further the response to escape stimuli in general, not merely those emanating from a food source. This has been clearly demonstrated in mammals. C. S. Hall, in his studies of emotional behavior in rats, showed hungry rats to be less timid in strange surroundings than satiated

individuals, and he concluded that "needs other than the need to escape inhibit the display of emotional behavior by distracting the animal from the fear-provoking aspects of the situation."

The fact that males continued to attack subordinate males whose fear reaction was reduced by starvation whereas they would tolerate subordinate females under the same conditions suggests a different response to the morphological characteristics of males and females, and this interesting point was explored in subsequent experiments.

EFFECTS ON DOMINANCE RELATIONS OF DISGUISING FEMALES AS MALES

The female chaffinch is predominantly grayish brown; the male differs in its blue crown and nape, its chestnut mantle, and its reddish-orange underparts. The most conspicuous difference between the sexes is the contrasting color of the breast feathers. With this difference as the basis for the study, we took females at random from captive flocks of nonreproductive birds, moistened their underparts with soap, and dyed them with a mixture of red, orange, and burnt sienna inks in imitation of the male. The disguised females were placed in open-air aviaries with normal males and females that they had not previously encountered, and we observed the effects of the red coloration on aggressive behavior in a variety of pairings.

Aggressive encounters occurred in several different ways: most frequently, one bird would evict another from the food or water dish, or from a perch; but occasionally a bird approaching a food dish would be repelled by another already in occupation. Only those fights in which one individual was clearly victorious are included in the results.

Matching of "Red" and Normal Females

Five groups of four birds were taken, 20 females in all. In each group of four, two "red" females were matched with two normal females whose breast feathers had been treated with soap solution (as a control). The four females were placed simultaneously in small aviaries and observed for 4 to 30 days, until a moderately stable hierarchy was formed. Of 20 possible pair contacts between the two "red" and two normal females (4 per experiment in 5 experiments), "red" females were completely dominant in 13 and met only slight opposition in 2. Normal females had more wins in three contacts and approximately equal scores in two. If total winning and losing encounters are compared (Table 3), the supremacy of red-breasted birds

TABLE 3. SUMMARY OF THE TOTAL OF "RED" FEMALE WINS AGAINST NORMAL FEMALES

	Experiment No.					
	1	2	3	4	5	Total
"Red" wins over normal females	75	71	42	92	71	351
Normal wins over "red" females	29	11	25	5	—	70
Percentage of "red" wins of total "red" encounters	72	87	63	95	100	84

TABLE 4.   RESULTS OF FIGHTS BETWEEN "RED" FEMALES AND HAND-REARED FEMALES

	Experiment No.								
	1	2	3	4	5	6	7	8	Total
Number of wins by hand-reared females	5	1	—	—	18	7	—	7	38
Number of wins by "red" females	29	35	28	58	21	34	26	32	263

shows even more emphatically: "red" females won 84 percent of all encounters with normal females. It is evident that female birds with the male red breast have a high chance of success against normal females not previously acquainted with them.

In this experiment the "red" and normal birds were strangers beforehand. We thought perhaps the same thing might also occur with acquainted birds in two flocks of eight female chaffinches that had been together for some time. The two lowest ranking birds were colored red and the effects studied. We found that the chances of success of the subordinate birds increased significantly after coloring and that all rose in their status in the hierarchy. Disguise of a female with the red breast of the male increased her chances of success in aggressive encounters even against previously dominant individuals.

A series of experiments was then designed to ascertain whether or not the response of female chaffinches to the red breast might be the result of conditioning. Eight females reared by hand in the company only of other females were paired with wild-caught "red" females of the same age (Table 4). The "red" females were successful in 87 percent of their encounters with hand-reared females that had had no prior experience with males. In most cases aggressive display was rare since normal females often fled precipitantly on meeting a "red" bird. The success of "red" birds against hand-reared females compared closely with their success against experienced females.

It is possible, however, that for some reason hand-reared birds might be inherently less aggressive than wild birds. In order to evaluate this, another series of experiments was conducted using only hand-reared birds. Of 20 pair contacts, 16 were won by the "red" females with little or no opposition. Of 408 encounters, "red" birds won 78 percent, comparing favorably with the results from the previous series.

In the subsequent year six females were reared in complete isolation from the seventh day after hatching. At an age of about 7 months three pairs were selected at random and one bird in each was colored red. They were placed in new cages and observed during the subsequent month. The overall success of "red" birds was very similar to that in the previous experiments. It is clear that the domination of females by birds with a red breast involves a response that is not dependent on previous experience of being attacked by males.

Females with Green-Dyed Underparts Matched with Normal Females Previously Dominated by "Green" Males

As another approach to the possibility of conditioning, we decided to test whether female chaffinches which had been dominated by green-dyed males would also be dominated by "green" females, either because of learning or because of a fear of strangely colored chaffinches. The results are shown in Table 5. Of the total number of wins, "green" birds had 25 and normal birds 57, suggesting that in fact the abnormal birds may have been at a disadvantage. The attempt to condition females to submit to "green" chaffinches failed.

Matching of "Red" Females with Mixed Groups of Normal Females and Males

Several experiments were run to see whether males would respond differently to "red" and normal females, and to judge whether the presence of males altered the response of normal females to "red" females.

Three groups, each comprising two males, two normal females, and two "red" females were set up and observed in small outdoor aviaries. The same basic pattern occurred in all three experiments, though marred by some irregularity in two of the flocks. In all three, a straight-line hierarchy was formed having the order males > "red" females > normal females. Males dominated both "red" and normal females. "Red" and normal females each received about 50 percent of male attacks. Therefore no effect of the red breast on the frequency of male attacks was revealed. Further, red-breasted females were found to dominate normal females even in the presence of males. The effect was slightly more marked than in the absence of males since "red" females won 93 percent of these encounters as compared with 84 percent in the previous series, where no males were present.

In another attempt to determine whether a red breast could affect the outcome of a female's fights with males, two groups of four females were isolated from males for 4 to 5 weeks and were then colored red and placed in a small aviary with four males. After about 2 weeks, the "red" females were replaced by four normal females, which had also been isolated and thus acted as controls. In one group, "red" females won 57 percent of their fights with males, and the normal females won 8 percent. In the second group "red" females won 32 percent of their encounters with males, and normal control females 7 per-

TABLE 5.   RESULTS OF FIGHTS BETWEEN "GREEN" AND NORMAL FEMALES

	Experiment No.											
	1	2	3	4	5	6	7	8	9	10	11	Total
No. of wins green females		4			7	7					7	25
No. of wins normal females	3		15				7	9	9	10	4	57

cent. As well as encouraging female success against males, the red breast also affected the stability of the hierarchy to such an extent that the relationships between some males and "red" females had not been settled by the end of the observation period.

These results demonstrate that a response can be evoked from males by the red breast under suitable conditions. Presumably, "red" females which have had immediate previous experience of dominant males are unable to exploit the advantage that the red breast gives them over males, because of a learned fear of them. The presence of normal females in the cage may also have an effect by enabling males to perceive that colored females are more similar to normal females than the males.

The Male Red Breast as a Releaser

In the experiments described, female chaffinches disguised with a male red breast won more fights than uncolored birds, against both males and females. There is no reason to think that the process of coloring can have any direct effect on the aggressiveness of a female bird. Since the birds used for coloring were selected at random, we might therefore have expected that an average of half would be dominated by normal females. This was far from the case, for 38 out of 44 dominated their opponents. This success must be a result of the response evoked in other birds by the red breast.

The outcome of every aggressive encounter is a result of mutual interaction, and behavior is strongly affected by that of the opponent. In the present case it appears that the red breast first evokes an avoidance response from other chaffinches; this in turn encourages the red-breasted bird to attack, even though it might otherwise have submitted without fighting at all. With a male opponent, the "red" female's fear usually counters any avoidance response that may be evoked in him by the red breast. Only by taking special steps, which would never occur in the wild state, can male avoidance be experimentally demonstrated.

The red breast of the male chaffinch conforms to the definitions of Konrad Lorenz and Niko Tinbergen of a "social releaser" as "a character peculiar to individuals of a given species and to which responsive releasing mechanisms of conspecific individuals react and thus set in motion definite chains of instinctive actions." It is evidently one of the ways by which chaffinches can distinguish between sexes.

PROXIMITY AS A CAUSE OF AGGRESSION

We had noted that when two nonreproductive chaffinches came into close proximity with each other, either one of them would withdraw or a fight would develop. This occurred during competition for food, water, or favorite perches, and also when birds accidentally alighted close together. We wondered whether it might be possible to measure the distance to which one bird must approach another before fighting occurs.

In these experiments food was used to bring the birds into proximity. Each cage containing eight birds was provided with a pair of food hoppers, at each of which only one bird could feed at a time (Fig. 1). The distance between the adjacent ends of the hoppers could be adjusted, and the effect on the incidence of fighting among feeding birds was noted as the distance was varied. At zero distance it was just physically possible for the birds to feed.

Whenever both hoppers became occupied, an "en-

Figure 1 *Arrangement of food hoppers for testing the distance at which feeding birds would tolerate each other. (Top) One male about to attack another, at 10 cm. (Bottom) Two females threatening each other at zero distance.*

counter" was recorded. Those encounters in which both birds held their positions, with or without aggressive or submissive display, were classed as "nonaggressions"; those in which one bird fled while another remained, as "aggressions." The results were summarized as the ratio of nonaggressions to total encounters, n/t (thus the closer the ratio approaches 1.0, the larger the proportion of nonaggressions); this ratio was plotted against the distance between perches.

Experiments with Normal Birds

In the first series, two flocks of eight females were observed and 669 encounters were recorded for one and 931 for the other. As shown in Figure 2, nonaggressions became less frequent as the distance between hoppers decreased and ceased when the perches were touching. The results may be most clearly reflected by extrapolating on each curve the distance at which nonaggressions make up half of the encounters ($n/t = 0.5$). This "50 percent" distance is about 7 cm in both flocks; that is,

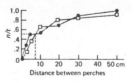

Figure 2 *Graph of the ratio of number of occasions when two birds fed without fighting to total number of encounters (nonaggressions/total = n/t) plotted against distance between the perches, in two flocks of eight females. The 50 percent distance, extrapolated for a value of n/t = 0.5, is 7 cm in each case.*

Peter Marler

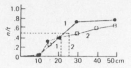

Figure 3 *Results for two flocks of eight males (see Fig. 2). The 50 percent distances are 21 cm and 24 cm.*

when two females alight this distance apart, there is an even chance that one will withdraw.

Knowing that male chaffinches dominate females in the social hierarchy and appear more aggressive, we wished to determine whether this difference was manifest in a difference in sensitivity to proximity. Two flocks of eight males were studied. As shown in Figure 3, non-aggressions became less frequent as hoppers were closer and virtually ceased at 10 cm. The 50 percent distance was 21 cm in flock 1 and 24 cm in flock 2. Thus the distance at which males are as likely as not to fight is about three times greater than the corresponding distance for females. However, the more gradual gradients show that males are more variable in their behavior than females. It should be noted that in neither case does the ratio reach 1.0 even at a distance of 50 cm. This is probably because under these aviary conditions many males move restlessly from hopper to hopper while feeding, however far apart they are; this brings them into proximity with other birds and thus causes more fights than would otherwise occur. Females are less restless and the ratio approaches 1.0 more closely (Fig. 2).

Studies of the dominance hierarchy had revealed that males were more tolerant of females than of other males, and we wished to investigate this difference in relation to proximity. Two mixed flocks were studied, each one consisting of four males and four females. The results were divided into male–female, male–male, and female–female encounters. The 50 percent distance in intrasexual encounters agreed approximately with the findings in the previous experiments: 21 cm and 18 cm for male–male encounters, 10 cm and 12 cm for female–female encounters. In male–female encounters, however, the 50 percent distance was 12 cm for flock 1 and 8 cm for flock 2, or about half the corresponding male–male distance. In fact, a female can approach as close or closer to a male than to another female before fighting or threat develops.

Experiments with Red-Dyed Females

We had shown in earlier experiments that the male red breast is an important stimulus in sex recognition. We now used a similar technique and tested red-dyed females with the distance hoppers. Two flocks of four males were each matched first against four normal females, then against four "red" females, and finally against four more normal females. The same males were used as in the earlier experiment with mixed flocks, where the 50 percent distances for male–male encounters were found to be 21 cm and 18 cm. In the present experiment the 50 percent distance in male–"red" female encounters was 23 cm in flock 1 and 16 cm in flock 2, and in the control experiments with normal females, 12 cm and 6 cm in flock 1, 8 cm and 9 cm in flock 2. Thus while a male would allow a normal female to approach to within about 10 cm of him before attacking, a red-breasted female was attacked at about 20 cm. In fact, a "red" female was treated as though it were a male, if one judges by this criterion.

Since the red breast is also significant to females in

fighting, it is of interest to know at what distance "red" females will tolerate each other. The same experiment showed that "red" females respond to each other as though they were males, with the 50 percent distance estimated at 25 cm for each flock. Experiments designed to study the relationship between "red" females and normal females showed that "red" females responded as both females and males do to normal subordinate females—by allowing them to come very close before attacking them; conversely, normal females approached as close to dominant "red" females as they would to males.

The Effect of Behavior

We may conclude from the experiments with red-dyed females that a chaffinch of either sex will allow subordinates with a red breast to approach up to about 20 cm before it is likely to attack. Those lacking the red breast may approach to within less than half this distance. This leads to the interesting conclusion that, if we use the distance at which others are attacked as a measure of aggressiveness, nonreproductive females are potentially no less aggressive than males. It seems that greater aggressiveness among nonreproductive males may occur primarily because they all carry male colors. Females disguised as males behave similarly. In fact, such females sometimes dominate males in the social hierarchy, which suggests the same conclusion. In both cases, the behavior of disguised females is to be explained not by a direct effect of coloring on their own aggressiveness but by their responsiveness to the changed behavior of others toward them.

It is clear that there is no fixed threshold beyond which another bird is always attacked but rather a gradient, more or less steep in different individuals, along which the *probability* of attack increases. One of the factors making for variability, revealed by qualitative observation, is the effect of the behavior of the opponent. In general, birds showing submissive behavior can come closest, then those showing normal alert behavior, those in escape postures, and, finally, those showing aggressive behavior are attacked farthest away. There is also a suggestion that birds may come closer in the horizontal than in the vertical plane.

"Individual Distance" and Territory

Several investigators have advanced the idea of individual distance as an area around the individual which is kept free of all others, either by attacking intruders or by moving away from them. This clearly corresponds to the phenomenon studied in the chaffinch. Individual distance in the chaffinch was found to be readily measurable, and to vary with the sex, behavior, and individuality of the intruder; in birds in nonreproductive condition it seemed to vary little with changes in internal state.

Many animals defend territories in the breeding season. Then the area defended is usually centered on some object, or collection of objects, often essential for reproduction. It may occupy a fixed position in space, like the breeding territories of most song birds, or it may be centered on something that moves around, such as a host, mate, or young. The obvious difference between territories of this kind and individual distance is that the latter centers on the individual itself rather than on some external object.

Intermediate types of defended area have also been observed. For example, birds on migration may set up small, mobile territories. Other species have a territory centered on the female as a third phase, interposed in time between individual distance and the development of the breeding territory. Male chaffinches in early spring also show an enlarged individual distance that moves with the bird, probably only appearing in the area that will become the territory. This is followed by an intensification of aggressiveness at certain points that subsequently expand into the whole breeding territory.

The most conspicuous characteristic of territorial behavior, the effect of immediate surroundings on aggressiveness, has a parallel in individual distance behavior. An intruder within the individual distance may either be attacked or avoided. One of the factors determining which course will be taken is familiarity with the surroundings, which encourages attack. Thus chaffinches familiar with the cage in winter usually dominate newcomers.

There are good reasons to believe that individual distance is directly related to the defense of other types of area. One might go so far as to suggest that it is the simplest and most primitive form of defended area, perhaps the evolutionary precursor of the other types. It has been observed that the defended area usually provides something essential for life or reproduction, for which the defender is competing with other individuals. Limitation of the supply of some essential commodity will always tend to encourage the aggregation of individuals around the remaining sources, and this in turn will bring individuals into proximity. One can readily imagine the tendency developing, under those circumstances, to defend an area around the individual, as the most economical method of protecting the individual's needs. By placing himself close to the object competed for and attacking all who come too close, the individual increases his chances of retaining it.

This relatively simple mechanism would help in obtaining all the essentials—food, water, mate, nest, or space—and would reduce unnecessary fighting, since the intrusions of individual distance would be rare in conditions of abundance. Once established, it could readily expand or contract in the course of evolution according to the changing requirements of the species. Thus a dispersed food supply would call for a larger individual distance than a concentrated one. Other requirements would be better satisfied by fixing the defended area in space, as, for example, the need for mates to keep in contact for the whole of the breeding season, or to tend eggs or young. Various specialized territories would evolve, culminating in the breeding territory, which is as varied and versatile in function in summer as individual distance is in winter. And since all types result in some degree of uniform dispersal, they may from their very inception have served a multiplicity of functions.

APPETITIVE BEHAVIOR FOR AGGRESSION

Thus far we have dwelt on the external circumstances which trigger aggressive behavior in chaffinches with some brief comments on physiological changes that affect the threshold of responsiveness to those stimuli. Implicit has been the assumption that chaffinches do not seek out opportunities for fighting as such but encounter the stimuli for aggression in the course of other activities. How well founded is this assumption that chaffinches do not normally show spontaneous appetitive behavior for aggression, as they show appetitive behavior for the

consummatory acts of feeding and drinking for example?

In an effort to explore this problem, chaffinches were raised by hand in individual isolation from an early age, and each was then placed with another chaffinch at 100 days of age. After a minute or more, threat behavior appeared, always triggered by proximity occurring during feeding or restless movement, and never preceded by the visible preparations for attack that might identify appetitive behavior for aggression. One could argue however that the inevitable nervousness of birds in these circumstances might have inhibited the tendency to seek out fights.

Study of established flocks revealed that when chaffinches selected one of several food sources to visit, they were more likely than not to choose one already occupied by a subordinate, ousting it from the food before eating. Can this perhaps be construed as seeking for fights? Since the behavior described was most common in individuals occupying intermediate ranks in the dominance hierarchy, an alternative interpretation suggests itself that what we observed was a form of "redirection activity." Once aggressiveness is aroused by fighting it may be turned toward a new opponent. This is especially evident in a peck-right hierarchy, when aggressiveness arising in subordinates only finds expression by the passing on of attack down the hierarchy. Since the despot generally lacks provocation for such redirected aggression, it usually shows no tendency to persecute subordinates.

A false impression that birds are seeking fights is sometimes given by attacks from a distance resulting from considerable individual distance toward some specific individuals. This is most obvious with strangers. Chaffinches, like chickens and many other species, attack strange individuals repeatedly, and at a greater distance than familiar birds. In very close-knit communities of some species this may be the only occasion when individual distance behavior appears. Chaffinches in aggressive posture are also attacked at a greater distance than normal.

Although we are led to the conclusion that association between fighting and other activities of nonreproductive chaffinches is usually only incidental, a more radical association between the two than that mediated by proximity sometimes seems to arise as a result of learning. In experiments involving fighting for food after starvation, one male chaffinch, whose position in the hierarchy was unstable, was strongly persecuted. He developed the habit of making infrequent visits to the food, remaining there longer than normal, and defending himself violently against all who approached. For some weeks after the experiment, as this male became hungry he showed aggressive postures and attacked others at considerable distances. An association seems to have been established between fighting and internal stimuli related to feeding.

CONCLUSIONS ON APPETITIVE BEHAVIOR FOR AGGRESSION

Chaffinches do not show an endogenous tendency to seek fights in the manner in which, for example, they seek for nest material in the breeding season. The first stage of spontaneous appetitive behavior seems to be lacking, in which the animal is "looking for" external stimuli that release further forms of appetitive and consummatory behavior. It is rather by incidental contact with aggressive stimuli that fighting begins. In rare cases, perhaps only in aviary conditions, an association with another activity seems to arise by learning. Apart from this, the seeking out of fights can be shown to be a result of external aggressive stimulation in the near past, the conditions for which are probably also largely confined to captive birds.

As in nonreproductive fighting, field observations do not reveal searching for fighting within the breeding territory that could readily be identified as such, except in parts of it where fighting has occurred in the immediate past such as a contested boundary. Here aggressive postures occur in the absence of a rival, as the bird passes through while appearing to search for intruders. Song in such areas may be accompanied by aggressive postures. When the boundary becomes firmly established, this obvious aggressive behavior ceases.

Consummatory Behavior in Fighting

In studies of fighting it has been assumed that the consummatory act, defined as "an act which constitutes the termination of a given instinctive pattern or sequence," is attacking the opponent. Though quantitative studies have not been made, the evidence suggests that the consummation of fighting in the chaffinch is to be found rather in the elimination of a stimulus situation. Fighting, like fleeing, tends to continue until the stimuli for it are removed, and the movements used are consummatory only in the sense that they help to achieve this removal. The range of movements used is very wide, including for example looking intently at the opponent, flirting of the wings, hopping or flying toward him, swooping or flying noisily at or around him or across his path, displaying with a variety of aggressive postures, making pecking movements, striking with the beak, feet, or wings, as well as calling, singing, or snapping the beak. All may serve to terminate a particular aggressive sequence by causing the opponent to flee.

The cessation of aggressive stimuli is generally achieved by driving the opponent away by movements of this kind, exceptionally by killing him, or alternatively by fleeing from him. The same may be achieved with certain ter-

ritories. A territory centered on the female may be moved by leading her away; but with a fixed territory, fleeing may involve relinquishing part or all of it. Otherwise, fighting generally continues in the presence of the provoking stimuli until the consummatory situation is achieved.

The Function of Fighting

In contrast with other activities, such as sexual, parental, or feeding behavior, spontaneous appetitive behavior for fighting does not normally occur among nonreproductive chaffinches, nor can it be definitely demonstrated in territorial behavior, except in recently contested areas; searching for fights seems to occur as a result of learning, or in an aggressive mood aroused by immediately previous experience. This particular organization of fighting behavior appears to have adaptive value when considered in relation to its function. Fighting is fundamentally a means of competing more efficiently for something in short supply, such as food, mates, nests, or space. Unlike other activities such as eating or copulation, fighting behavior is at best a waste of time and at worst a dangerous and dysgenic activity unless some external commodity is gained by it. An endogenous tendency to seek fights might thus have hazardous consequences. However, certain external stimuli are likely to be present in most or all competitive situations. Thus the danger of wasteful fighting may be avoided by confining it in inexperienced animals to these external stimuli. Searching for fights may develop subsequently by learning, if the animal gains some reward by it.

The distinguished American ethologist Wallace Craig was led by a review of the causes of fighting in animals to a similar conclusion. "Fundamentally among animals fighting is not sought nor valued for its own sake; it is resorted to rather as a means of defending the agent's interest.... Even when an animal does fight he aims not to destroy the enemy but only to get rid of his presence and interference."

SUPPLEMENTARY READINGS

Brown, J. L. The evolution of diversity in avian territorial systems. *Wilson Bull.* 76: 160–169 (1964).

Collias, N. E. Aggressive behavior among vertebrate animals. *Physiol. Zool.* 17: 83–123 (1944).

Craig, W. Why do animals fight? *Intern. J. Ethics* 31: 264–278 (1928).

Crook, J. H. The basis of flock organization in birds. In *Current Problems in Animal Behaviour*, W. H. Thorpe and O. L. Zangwill, eds. Cambridge University Press, New York, 1961, pp. 125–149.

Hinde, R. A. The biological significance of the territories of birds. *Ibis* 98: 340–369 (1956).

Lorenz, K. *On Aggression.* Methuen & Co. Ltd. London, 1966.

Marler, P. Studies of fighting in chaffinches. *Brit. J. Animal Behaviour* 3: 111–117, 3: 137–146, 4: 23–30, 5: 29–37 (1955–1957).

Marler, P., and W. J. Hamilton III. *Mechanisms of Animal Behavior.* John Wiley & Sons, Inc., New York (1966).

Scott, J. P. *Aggression.* University of Chicago Press, Chicago, 1958.

Scott, J. P., and E. Fredericson, The causes of fighting in mice and rats. *Physiol. Zool.* 24: 273–309 (1951).

Seward, J. P. Aggressive behaviour in the rat, I–IV. *J. Comp. Psychol.*, 38: 175–197, 213–224, 225–238; 39: 51–76 (1945–1946).

Part III

COMPARATIVE STUDIES

Editor's Comments
on Papers 12 Through 18

During the past three decades, a substantial number of field and laboratory studies have been concerned with the description of social structures in a variety of vertebrate animals. Many of these studies have focused on the phenomenological aspects of the behavior; a few have attempted to examine the behavior in the broader contexts of ecological considerations. For a recent compilation of this latter approach, the reader is referred to a symposium on the *Ecology and Evolution of Social Organization,* organized and edited by Banks and Willson (1974). A selection of important papers dealing with territoriality is available in Stokes (1974), and a companion volume devoted to dominance hierarchies has appeared (Schein, 1975). Many of the selections in Davis (1974) relate social behavior and ecological factors.

The first paper in this section (Paper 12) was written by the late Bernard Greenberg, a student of Allee at the University of Chicago. It was an early attempt to investigate the complex relationships between two aspects of social organization, social hierarchy and territoriality, in immature green sunfish. The study was experimental and set the pattern for many other studies. An interesting observation emerging from this early work was the social role of "scapegoat" played by omega individuals in these aquarium studies. Recent field work by Myrberg (1972) on the reef fish, *Eupomacentrus partitus*, the bicolor damselfish, amply supports Greenberg's earlier observation.

Paper 13, by Rand, on the adaptive significance of territoriality in iguanid lizards represents a departure from previous essays on this topic. Rand, a leading authority associated with the Smithsonian Tropical Research Institute, formulates a hypothesis and a series of predictions that must follow if the hypothesis cannot be negated. He then fits these predictions in the information bank he has developed through his observations of two species of lizards. The basic assumption of his hypothesis is that "territoriality has evolved through natural selection and is maintained by it." (p. 107). The hypothesis itself relates territorial defense to individual fitness and is now a standard and accepted formulation. Rand's essay is significant in raising the issue, as it relates to lizard sociobiology, and in providing observational evidence in support of his predictions.

Gordon Orians of the University of Washington has been at the forefront of the new wave of students of social structure. He and his students have written extensively in the area of avian social systems with particular emphasis on ecological and evolutionary interpretation. In Paper 14, coauthored with Collier (1963), the question of how variation in social organization between two sympatric species may be related to the energetics of competition as one factor in the evolution of sympatry is elegantly presented. The redwing and the tricolored blackbird have contrasting social systems; the two species overlap in geographic range and differ in certain morphological traits. They also differ in aspects of feeding and parental care behaviors. Paper 14 presents field observations of social interactions between the two species along with a thoughtful interpretation of the evolutionary forces responsible, in part, for the competitive exclusion that has developed between these closely related species.

Richard D. Estes of Harvard University, a leading student of ungulate ethology and ecology published his now classic study (Paper 15) comparing the behavior and ecology of two species of gazelles in 1967. In addition to documenting major differences and similarities in a variety of social behaviors between Grant's and Thomson's gazelles, Estes articulated such distinctions as feeding preferences and water requirements into an evaluation of major distinctions in social structure.

More recently, Estes (1974) has completed a general evaluation of the social organization of the African Bovidae in terms of ecological and evolutionary factors. This work, in conjunction with an ecologically sophisticated examination of the African antelope by Jarman (1974), represents some of the most penetrating analyses of social organization in a major taxon of mammals.

Studies in rodent social behavior and organization are legion. One of the more comprehensive investigations is that of John Eisenberg of the U.S. National Zoo, Smithsonian Institution. In the work presented here (Paper 16), Eisenberg utilizes the comparative methodology fostered by the early conceptual proposals of Lorenz (1957) and Tinbergen (1951). The rodent family, Heteromyidae, is the object of a behavioral, ecological, and evolutionary analysis. The full gamut of social, maintenance, locomotory, and comfort movements and group formation behaviors is subjected to a comparative analysis. The selective factors operating on the evolution of the patterns is articulated in those instances in which the data are available. The article is a useful source of seminal ideas regarding the effects of ecological factors on social behaviors, and stands as a model that might well be followed by current phylogenetically oriented studies. Because of space limitations, the 25 tables forming an appendix to the paper have not been included in this reprinted version.

Alverdes (a German social biologist of the early twentieth century), Deegener, Espinas, and Wheeler were the intellectual leaders of their time in the field of animal sociology. Alverdes was firm in his belief that the fundamental criterion of social life was the possession by species members of a social instinct. Alverdes writes (1927) "In short, no social instinct, no society." This statement was written in the context of the problem early students of general sociology had in discriminating between associations and societies. The former were believed to consist of accidental congregations produced by external factors alone, e.g., insects collected around a light source. For Alverdes, social instincts were organized around sexual relationships. Thus, in the article reproduced here on mammalian packs and herds (Paper 17), he systematically defines and provides examples of mateship systems. Alverdes also distinguished between closed and open groups, a concept later sharpened by Eisenberg (1966). The topics of hierarchy formation and role differentiation, e.g., guards and leaders, lend quite a modern tone to Alverdes' writing.

The final paper of this section deals with the social structure of non-human primates. Thomas Struhsaker of the New York Zoological

Society and the Rockefeller University is a leading field ethologist. He has studied a variety of mammalian taxa and in Paper 18, on the African Cercopethicine monkeys, he concentrates on the relation between environment and social organization of these rain-forest primates. His studies are characterized by careful regard for clear definitions of terms and rigorous acquisition of measurements. In addition to providing a wealth of observational data on the social behavior and structure of the cercopithecines, Struhsaker questions some of the conventional wisdom that has emerged from other primate studies. Past generalizations regarding group size and territoriality, habitat preference and group structure, have been based primarily on studies of taxa inhabiting the savannah areas of Africa. His studies of rain-forest species lead to the suggestion that when the evolution of grouping tendencies and social structures are under consideration, the phylogenetic relationships of the groups being compared should not be overlooked. In their attempt to establish broadly conceived classification schemes relating social organization to ecological factors, some have, according to Struhsaker, neglected to weigh in the balance the impact of the phylogenetic heritage of the species to which these classifications have been applied. The paper is of significance because of the extraordinary mix of empirical findings and theoretical interpretation that Struhsaker provides. Other recent essays on the relationships between primate social structures and ecological factors include those of Crook and Gartlan (1966); Crook (1970); and Eisenberg, Muckenhirn, and Rudran (1972).

REFERENCES

Alverdes, F. 1927. Social Life in the Animal World. Kegan Paul. London. 216 pp.

Banks, E. M., and Willson, M. F., eds. 1974. Symposium on the Ecology and Evolution of Social Organization. *Amer. Zool.* 14(1): 7–264.

Crook, J. H. 1970. The socio-ecology of primates. In *Social Behaviour in Birds and Mammals,* ed. by Crook, J. H. Academic Press. London. Pp. 103–168.

Crook, J. H., and Gartlan, J. S. 1966. Evolution of primate societies. *Nature* London **210**: 1200–1203.

Davis, D. E. 1974. Behavior as an Ecological Factor. Dowden, Hutchinson & Ross. Stroudsburg, Pa. 390 pp.

Eisenberg, J. F. 1966. The social organization of mammals. *Handbuch der Zoologie* 8(39).

Eisenberg, J. F., N. A. Muckenhirn, and R. Rudran. 1972. The relation between ecology and social structure in primates. *Science* **176**: 863–874.

Estes, Richard D. 1974. Social Organization in the African Bovidae. Paper No. 8 (166–205). In *The Behaviour of Ungulates and Its Relation to Management,* Vol. 1. Eds. V. Geiost and F. Walther. IUCNNR, Morges, Switzerland.

Jarman, P. J. 1974. The social organization of antelope in relation to their ecology. *Behaviour* **48**: 215–267.

Lorenz, K. Z. 1957. Methoden der Verhaltensforschung. *Handbuch der Zoologie* **8** (pt. 10, sec. 2): 1–22.

Myrberg, A. A. 1972. Social dominance and territoriality in the bicolor damselfish, *Eupomacentrus partitus. Behaviour* **41**: 207–231.

Schein, M. W. 1975. *Social Hierarchy and Dominance.* Dowden, Hutchinson & Ross, Stroudsburg, Pa. 401 pp.

Stokes, A. W. 1974. *Territory.* Dowden, Hutchinson & Ross. Stroudsburg, Pa. 398 pp.

Tinbergen, N. 1951. The study of instinct. Oxford Press. London. 228 pp.

12

Reprinted from *Physiol. Zool.*, 20(3):267–282, 285–299 (1947)

SOME RELATIONS BETWEEN TERRITORY, SOCIAL HIERARCHY, AND LEADERSHIP IN THE GREEN SUNFISH (LEPOMIS CYANELLUS)[1]

BERNARD GREENBERG[2]

Whitman Laboratory, University of Chicago

THERE is a basic and widespread tendency for living things to aggregate, more often than not with mutual benefit (Allee, 1931). Integration of vertebrate aggregates is influenced by three major principles of behavior—territoriality, hierarchy, and leadership. Many species appear to organize in accordance with but one of these principles; nevertheless, in most instances traces of the operation of all three may eventually be recognized.

The modern concept of territory was first presented by Altum (1868) and Howard (1907–14). Although many observations have been gathered since Howard's (1920) stimulating book, few detailed studies of territorial behavior are yet available. Noble (1939a) gives a simple, inclusive definition of territory as "any defended area." Nice (1941) reviews the literature on birds and classifies a number of types of territory with regard to mating, nesting, feeding of young, roosting, etc. Territorial behavior seems to be most varied in birds but is also prevalent in the bony fishes, reptiles, and mammals.

Social hierarchies in domestic fowl were described by Schjelderup-Ebbe (1922, 1935). Small flocks of hens maintain a linear system of dominance, which has come to be called the "peck-right."

The dominant bird has the right to peck all the others without, in turn, being pecked; the next in rank can attack all except the first; and so on down the line. Masure and Allee (1934a, b) confirmed the existence of this social order in chickens but found that certain other species of birds have a more flexible type of relation, which they named "peck-dominance." The dominant individual delivers more blows than it receives, but the subordinate often retaliates, and the outcome of any single contact is not strictly predictable. Masure and Allee observed that certain members of a group are more aggressive in one place than in another, and Allee (1942) suggests that this territoriality may be an underlying factor in the peck-dominance type of society.

Still other kinds of relations between territory and hierarchy have been proposed by Diebschlag (1941) for pigeons and by Greenberg and Noble (1939, 1942, 1944) for the American chameleon, *Anolis*. Diebschlag discovered that pigeons defend both residence points of limited area and "influence spheres," which may extend to an entire laboratory cage. Individuals were observed to engage in peck-dominance exchanges with others that held residence areas within the formers' spheres of influence. Greenberg and Noble recorded similar situations in *Anolis* lizard groups; however, the relations between the territory-holding males were of the peck-right type.

Leadership-followership phenomena

---

[1] This study originated from a suggestion by Dr. W. C. Allee, to whom the author is deeply indebted for friendly guidance. Dr. C. M. Breder, Karl P. Schmidt, and Clifford H. Pope read the manuscript and contributed much helpful criticism.

[2] Present address: Department of Biology, Roosevelt College, Chicago, Illinois.

219

are well known in birds and mammals, but few thorough studies have yet been made. It is difficult to define leadership so as to exclude social facilitation, which Allee (1947) considers to be "any increment in frequency, intensity or complexity of behavior of one individual resulting from the presence of another." A common example of this effect is the consumption of more food by animals in groups than when isolated. Leadership seems to be a special case of facilitation, in which one animal sets the pace of group activity or initiates changes in it of various kinds.

' An urgent problem in this area of sociology is the accumulation of data concerning the social behavior of lower vertebrates. On the well-grounded assumption that mammalian and avian behavior patterns stem from those of reptilian ancestors and that homologies can be drawn with the inheritable behavior of amphibians and fishes, it would seem to be of fundamental importance to trace the social instincts by comparisons between the vertebrate classes. It is difficult and sometimes impossible to distinguish homologous from analogous behavior, but valid generalizations of broad significance can often be discerned (Allee, 1938, 1945; Noble, 1938, 1939a, b; Collias, 1944).

Early in the present study it was discovered that small groups of immature sunfish (*Lepomis cyanellus* Raf.) establish hierarchies but may also become organized territorially. A simple kind of leadership or facilitation of behavior was noted. Therefore, it has been possible to undertake an experimental analysis of the relations between these types of social organization.

THE ANIMALS AND THEIR TREATMENT

The experiments reported here were designed to investigate (1) the types of organization established by small groups of immature sunfish, (2) some of the factors that determine position in a hierarchy or other organization, and (3) the effects of certain formalized complications of the physical environment upon organization. In addition, control observations were made of the adult fighting and courtship patterns.

All the sunfish used in this study were collected from a small permanent pond in the Tinley Park Forest Preserve in Illinois. A wide range of sizes and probably of ages was found during the summer of 1944, and in the fall a stock was maintained in large aquaria. These fishes were used throughout the following year until early summer, when immature fishes again became abundant. Collections were made also in July and August, 1946.

Hubbs and Cooper (1935) state that green sunfish ordinarily do not reach sexual maturity until they are about 3 inches in length; they found very few of the second-summer fish to be mature. For my experiments, individuals were selected that ranged from about 0.75 to 1.5 inches in standard length. It is almost certain that none of these was more than 1 year old. Sex could not be distinguished externally but was determined later at autopsy. Individuals could be identified through the circumstance that they had encysted parasites (trematodes) imbedded in skin and muscle and, in fact, all through the body; the parasites formed distinctive patterns of black-pigmented dots. The behavior of these fishes differed in no discernible way from that of noninfested sunfish collected from Hyde Lake, Illinois. Groups were selected in which the members could be identified readily and did not vary greatly in size.

Some of the experimental aquaria were situated in the Whitman Laboratory greenhouse, but the greater part of the study was performed in a large laboratory room, where the aquaria were kept under fluorescent lighting. Screens were arranged so as to minimize outside disturbance. The water used was Lake Michigan tap water, either heated or run through a charcoal filter to remove the toxic chlorine; it was changed in the aquaria every 4–7 days. The fishes were fed daily with enchytraeid worms, tubifex worms, or a cooked mixture of corn and soy meal, supplemented with liver or egg yolk. They grew best on live food but also did well on the substitutes.

In the experiments involving manipulation of environmental relations, single or multiple galvanized-wire partitions were set into the aquaria, each with a 2-inch square opening at the bottom. Details of procedure in the various experiments can most conveniently be given with the results.

BEHAVIOR PATTERNS
AGGRESSIVE BEHAVIOR

Sunfish that have settled their dominance relations do not ordinarily fight in the sense of showing the full fighting pattern. Instead, the dominant individuals drive their subordinates, that is, they make quick dashes at them, and the latter swim away, usually without being nipped. Occasionally, two may threaten by spreading their gill-covers and moving head-on toward each other; the threatened individual turns broadside and, with body held rigid, vibrates in a weaving fashion. An encounter seldom goes any further, unless the two are strangers.

The full fighting pattern of the immature sunfish cannot be distinguished in any particular from that of the adults. In both there is the same remarkable set of special pigmentation. The bony opercular flap has a large black spot, mar-gined with salmon-red on the flexible portion. When fighting, the fish spreads its gill-covers, and this gesture carries the red-margined patch of black, which roughly resembles an eye, out to a position where it appears to the observer that the head has suddenly become very much enlarged. Another concentration of black pigment is present on the flexible portion of the dorsal fin and less conspicuously also on the anal fin. These spots darken and lighten with the apparent excitation of the fish. An actively fighting individual almost always has these spots intensified, but, if it loses the fight, the spots disappear. An interesting series of changes in eye color occurs during fighting. The iris is normally lightly pigmented, with a variable amount of red around the pupil. During an encounter, the red becomes more prominent, and a dark, inverted triangle appears above the pupil. If the fish loses, black pigment spreads to the rest of the iris and the red is obscured. A "black eye" is here an unfailing sign of defeat.

During the breeding season, male green sunfish develop fringes of white on the dorsal, caudal, anal, and pelvic fins. The latter have thickened anterior rays, although this does not seem to be connected with mating behavior. All the various tints of the male are strengthened at this time. The female, on the other hand, becomes less conspicuous. Females, and also subordinate individuals of either sex, exhibit a dorsoventral striping or banding of light and dark. There is some overlap in regard to the sex-specificity of these color patterns. Immature females often show the white margining of the fins but to a lesser extent than the males, and subordinate males are somewhat banded.

The initial move in the fighting pattern consists of a head-on advance toward the opponent, with gill-covers erected; the latter responds with a seem-ingly defensive measure, namely, a weav-ing motion that appears to help in either avoiding or mitigating the force of a blow. The fishes take turns in advancing and weaving, but soon begin to nip furiously at head or body. Often they will catch onto each other's jaws and hold until one breaks away. When the fight is lengthy, the movements tend to become formalized, and as the oppo-nents tire, they may move side by side, making weaving motions, or alternate in threatening, weaving, and nipping.

Females show the same fighting pat-tern as the males except when ready to spawn, at which time they are very sub-ordinate.

REPRODUCTIVE BEHAVIOR

During the winter and spring (1944–45) a number of sunfish grew to mature

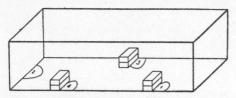

Fig. 1.—Breeding aquarium. Bricks were piled up to provide nest sites; the shorter sides of the aquarium are opaque.

size in the laboratory, and in May an at-tempt was made to induce them to breed. Two large aquaria ($8 \times 2 \times 1.3$ feet) were arranged for this purpose, as shown in Figure 1. The bottom was covered with about 1 inch of coarse sand, and bricks were placed at intervals, as indicated, to provide the kind of shielding barrier used by some species of sunfish (Breder, 1936).

In the first trial, two large males and what appeared to be a female were placed in one of the aquaria. After several days, a male dug an oval nest at position 1 and, sometime later, the other male nested at position 2, at the opposite end of the aquarium. On May 23, Male 1 and the female spawned at position 1, after

which the female paid no further atten-tion to the nest, while the male continued to guard the eggs. On June 2 the female was seen on the second nest, responding to courtship but not laying any eggs. On June 6, tiny fishes were detected; the adults were immediately removed in order to rear the young. Meanwhile, the same males were placed in the second aquarium, together with two other equal-ly large males. To make conditions more nearly like the pond environment, 7 smaller fishes were added that were about half the size of the males. The first two males that had established nests at 1 and 2 in the first aquarium repeated their performance in the closely similar habitat in corresponding places. In ad-dition, the two new large males built nests at positions 3 and 4 (Fig. 1). The group of smaller fishes did not nest but engaged in nipping and chasing among themselves; all were strictly subordinate to the large males.

Nest 3 was the least secure, and the male on it was subject to drives by Male 1 and Male 4. The latter especially exer-cised dominance over Male 3 in "nip-right" fashion so that for more than a month a kind of "territorial" hierarchy (Greenberg and Noble, 1944) could be detected, with Male 3 able to defend his nest only against the smaller fishes.

There was no sizable female available, and after a time it was observed that the borders of nest 1 were being progressively enlarged, almost to the extent of de-struction of the nest. Finally, on July 9, a small female was observed to spawn with Male 2; she was about half the size of the male, but the behavior of both was the same as in the mating seen earlier, and eggs were laid. Nesting behavior on the part of the males continued until the middle of August, when they began to move about more freely, with less atten-tion to the nests.

The pattern of courtship and mating of *L. cyanellus* has not hitherto been recorded in the literature; it was very similar in all observed instances. While the male attempts to drive the female toward the nest and will make repeated excursions in her direction to do so, it is the female that determines when spawning will occur. The driving male behaves in the same manner as toward a male rival, but with much less intensity. The female finally circles the nest and comes to rest with head toward the angle made by the bricks and the glass. Here the male

the corner. At intervals, he nips her and she leaves the nest, only to return after a short time. When the female is out of the nest, the male enters the corner head-first, and makes vibratory fanning movements with his tail.

During the various experiments, the immature fishes never showed the complete courtship pattern. In a single instance, observed from July 26 to August 12, 1945, a small male behaved in a peculiar fashion on a territory, and some phases of sexual behavior were detected such as circling a particular spot, driving

TABLE 1

SUMMARY OF DATA CONCERNING 44 GROUPS OF 4 SUNFISH EACH, KEPT IN SMALL, BARE AQUARIA (10×8×8.5 INCHES) CONTAINING 4 LITERS OF WATER

	Expt. I	Expt. II	Expt. III	Expt. IV	Expt. V	Totals
Number of aquaria.........	8	8	8	8	12	44
Duration in calendar days..	19	19	25	22	22	....
Number of observation days	15	17	17	15	14	....
Length of observation period (in minutes)........	10	10	15	15	15	....
Total recorded drives..	4,258	3,733	7,707	5,592	10,244	31,534
Av. drives per group/hour.	240*	164.6	226.6	186.4	244	214.6

* Group 5 with four territories was omitted because too few drives were recorded (62).

moves to a position directly over her. As he does so, the pair surge together into the corner, the male pressing on the back of the female; she turns ventral side up, rising upward; as her belly approaches that of the male, eggs are seen to be emitted from the genital opening. The pair move back and forth in the corner; at each movement the female presses her belly upward, close to the male's cloacal opening, and small numbers of eggs are expressed, to fall on the sides of the nest, on the plants bordering it, and in the nest itself. Sperm could not be seen, but fertilization undoubtedly occurred, since the eggs later hatched. The pair may stop and circle the nest, with the male at the outside, herding the female toward

others toward it, and moving above another fish. This was an exceptional observation.

TYPES OF SOCIAL ORGANIZATION

The first series of experiments involved forty-four groups of four sunfish each, in small, bare aquaria (10 × 8 × 8.5 inches) containing 4 liters of water. These groups were observed for 19–25 days, between October 2, 1944, and January 14, 1946 (Table 1).

In the majority of the groups (27) the fishes developed a dominance order or hierarchy. A single individual became dominant and was able to move unhindered over the whole aquarium and freely drive all the others, whereas a second fish could drive the remaining two, and a third drove only the fourth or "omega" fish. Such relations were often established by the very first day of observation and persisted, except for temporary reversals, during most of the observation period.

In the remaining seventeen aquaria, two or more individuals defended territories. In four cases, two fishes held territory, and the others were not only subordinate to them but were also organized in a hierarchical relationship. In twelve cases, three of the four held territories, while the fourth was subordinate to all. In the one unusual case, all four mem-

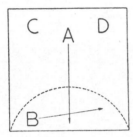

FIG. 2.—"Partial" territory. *A* is dominant in the whole aquarium, while *B* defends a small area against all but *A*.

bers of a group succeeded in holding territories; the lowest-ranking fish was the last to establish itself. Further details concerning these territories will be given below.

It was impracticable to distinguish cases of a single fish holding territory from those in which a despot ruled the entire aquarium. With a floor space of about half a square foot, the dominant fish frequently darted across the entire area to attack a subordinate. Territoriality was clear enough, however, when two fishes threatened each other, neither dominant, and subdivided the available space, although with a somewhat indefinite boundary. Perhaps all hierarchies in small laboratory aquaria should be considered as single territories

held by the despot, but the concept of territory as "any defended area" implies boundaries, defended against trespass. In the case of a single territory, therefore, it would be necessary to show that the boundaries lie within the aquarium walls, i.e., that the dominant fish voluntarily restricted its defensive reactions to only part of the total area. Presumably, in nature, boundaries are set by interaction between neighboring territory-holders, and available space is thus parceled out. Experimental conditions would seem to rule out the category of "single territory" from consideration in this series of experiments.

The twenty-seven hierarchies were complicated by "incipient" or "partial" territoriality. This was not difficult to recognize in seventeen instances; the fish ranking next to the despot was observed to take up residence in a corner or side of the aquarium and defended this area against trespass by all but the despot. The first-ranking fish could enter and drive this territory-holder, but the latter would often resist and would resume its defense when the despot left the immediate area. In the remaining ten cases this behavior was less evident and was not recorded as such.

The concept of "partial" territory differs somewhat from Diebschlag's (1941) terminology of "resting place" and "influence sphere." Figure 2 may serve to illustrate the kind of relationship observed in green sunfish. Here *A* is the despot, but the second-ranking fish, *B*, defends the area indicated by a dotted line whenever *A* does not interfere. Individuals *C* and *D* do not defend any specific place but rest more often in the indicated corners than elsewhere, and they can be said to have resting places. In this case there are no limits to the "influence sphere" or territory of *A*; the artificiality of aquarium conditions eliminates territory-holding neighbors and boundaries which would be delimited by mutual adjustment. Hence A does not hold the same kind of territory that exists when two or more fishes subdivide the aquarium space.

The partial territory of B may represent a step toward later equality with A, or it can result from a change in the status of B, which might previously have held full territory. In either circumstance, partial territory may be considered as intermediate between hierarchy and territory.

Even under our laboratory conditions, with oversimplification of habitat and limitations of space, territoriality is an outstanding phenomenon in the social life of immature sunfish. The territorial nesting behavior of adult male sunfish during the breeding season was known to naturalists long before its description by Reighard (1902), but there is no mention in the literature of territorial behavior in the immature fish of any species. The present experiment demonstrates that first-year green sunfish show aggressive behavior and defend territories, although not yet hormonally stimulated to exhibit reproductive behavior.

DESCRIPTION OF HIERARCHIES

Behavior on first encounter was variable, and usually the fishes spent some time becoming accustomed to the aquarium before reacting to one another. Soon one fish threatened another, eliciting either avoidance or counterthreat. Dominance was often decided by the early submission of a nonaggressive or smaller individual. If the fishes were fairly well matched, vigorous fights occurred at any time within the first hour and sometimes continued intermittently throughout the day. The winner showed the right to "drive" its subordinate, and the latter retreated. In most instances the sub-

ordinates became well trained to retreat and did so at the first move in their direction. As a result, the dominant fish seldom spread its gillcovers or showed other signs of the fighting pattern and rarely came close enough to nip. Such a "drive-order" contrasts with the "nip-order" described by Braddock (1945) for *Platypoecilus maculatus*.

Within a few days after the establishment of the hierarchy, its members settled down to incessant aggressive activity. In twenty-six groups,[3] the most dominant member drove its three subordinates at the average rate of 44.3 attacks per hour, the next-ranking individual drove its subordinates an average of 29.2 attacks per hour, the third-ranking fish drove its subordinate at a rate of 14.7 attacks per hour, and the lowest-ranking fish resisted its superiors at the average rate of 1.7 drives per hour (Table 3). In spite of this high frequency of aggressive interaction, the hierarchies were fairly stable. In seven of the above twenty-six groups, the rank order that pertained on the first day of observation remained unchanged, except for one or two temporary reversals. In another case the two upper ranks reversed after a few days and thereafter relations remained constant. A permanent upset of the two intermediate positions occurred in nine groups, whereas in four others the lowest fishes either exchanged places or never completely settled the dominance order. In four groups a single fish died and was replaced, with consequent changes in the hierarchy. In the remaining case, early instability of the three lower ranks was followed after 4 days by settling of relations.

The total drives made and received by fishes occupying each of the four ranks

[3] One group (No. 6, Expt. V), has been omitted from the total figures because the rank order was reversed several times (Table 7).

are compared in Table 2. The data for each hierarchy represent the longest stable interval during the observation period, and the nature of any major changes that took place is indicated. Table 3 presents the averages for each rank in terms of drives per hour of observation; the fishes occupying the alpha, or highest, positions drove their subordinates quite equally, on the average, regardless of rank. The beta, or second-ranking, fishes showed a tendency to drive their lower-ranking subordinates more often, although this was not statistically significant. The number of drives given are directly correlated with rank, whereas the drives received are inversely correlated.

A total of 582 reverse drives (subordinate driving superior) is recorded in Table 2. The fishes in each rank category were resisted in direct proportion to the rank of the subordinate. The orderly progression of numbers shown in Tables 2 and 3 suggests that the members of a hierarchy occupy steplike levels in dominance. The most dominant fish not only drives all the others but drives more frequently; resistance to the alpha fish is also weakest, as shown by the low frequency of return drives that it receives. The second-ranking fish drives its subordinates more actively than does the third-ranking fish and is resisted more actively by the latter than by omegas or lowest-ranking fish. The third-ranking individuals, while decisively dominating the omegas, also receive a large number of drives in return.

The majority of the hierarchy relations were of the stable "peck-right" type. Table 4 presents the frequency distribution of reverse drives in each of the six contact-pair relations. In sixty-four instances, the subordinate did not return the drives of its superior, while in thirty-five other cases it did so only once or

twice. Relations between the despot fish and its three subordinates were more decisive than among the others; in the lower ranks, reciprocal driving was seen much more frequently (Table 2). In such cases one can consider the relation to be similar to the flexible "peck-dominance" of pigeons (Masure and Allee, 1934a). At any given time, either the dominant or the subordinate member of the contact-pair might be driving, with a better chance that it will be the dominant.

The stability of the relationships can also be shown by analyzing their duration in terms of percentage of total observation time. Table 5 gives the longest duration in calendar days of each of the six contact-pairs, arranging the members of the group by their rank assigned in Table 2; next to the duration is the percentage of the observation period that it represents. It may be seen that dominance was quite stable; the alpha rank was most consistently maintained, remaining unchanged for an average of 97 per cent of the observation period. In preparing this table a change was considered to have occurred whenever a subordinate drove its superior two or more times in excess of the number of times it was driven during the observation interval (10–15 minutes). Such changes were seldom permanent except during the early stages of hierarchy formation and for a few days afterward. The changes usually again were reversed in favor of the consistently dominant fish.

Table 6 presents an analysis of the duration of the entire hierarchy and of various subrelationships within it. The single pair-contacts lasted, on the average, 88 per cent of the observation period. Alpha dominated its subordinates very consistently (97 per cent of the time, on the average), whereas beta and gamma were not quite so effective (70 and 74 per cent, respectively). The hierarchy itself lasted

TABLE 2

TOTAL DRIVES GIVEN AND RECEIVED IN EACH OF THE 6 PAIR-CONTACTS COMPRISING A HIERARCHY OF 4 MEMBERS

Group Number	Pair-Relations A—B	A—C	A—D	B—C	B—D	C—D	Days Duration	Type of Change*
Experiment I								
1	79— 1	69— 0	49—0	35— 1	17— 2	6— 2	9	c
3	46— 1	128— 0	233—0	61— 2	150— 4	60— 1	19	a
4	111— 1	96— 0	97—0	46— 1	92— 6	12— 3	15	c
6	87— 2	185— 0	130—0	41— 5	87— 1	50— 10	19	a
7	82— 0	187— 0	163—0	49— 11	88— 9	34— 12	12	c
8	36— 3	79— 0	75—0	139— 0	191— 0	9— 0	19	a
Experiment II								
17	41— 2	32— 0	30—0	59— 37	52— 12	30— 4	10	c
20	71— 5	81— 1	99—0	53— 11	42— 16	21— 14	15	c
21	64— 6	44— 3	40—4	52— 2	47— 2	7— 1	9	b
22	45—10	26— 0	62—1	38— 11	37— 4	11— 2	10	c
23	71— 0	111— 0	59—0	10— 7	23— 3	25— 8	14	e
24	32— 5	18— 1	30—0	43— 1	111— 1	6— 4	19	d
Experiment III								
6	187— 0	96— 0	11—0	43— 10	23— 6	56— 9	16	c
12	364— 2	231— 0	199—0	137— 18	223— 1	169— 11	25	a
15	121— 6	206— 3	226—0	77— 4	83— 8	43— 15	25	a
18	237— 6	99— 0	198—3	105— 6	109— 34	33— 16	25	a
Experiment IV								
3	172— 0	141— 0	96—0	27— 5	29— 7	23— 22	19	d
15	63— 0	54— 1	62—0	34— 3	43— 0	80— 0	8	c
Experiment V								
5	139—10	93— 1	64—0	129— 7	117— 0	26— 0	16	f
7	173— 0	167— 0	106—0	19— 11	41— 1	33— 18	17	d, f
8	178— 0	157— 0	118—0	54— 9	35— 2	9— 2	22	a
18	97— 0	89— 0	59—0	37— 0	14— 9	35— 3	21	d
19	90— 1	145— 0	135—1	41— 12	31— 6	37— 15	15	f
20	154— 0	89— 0	176—0	23— 5	37— 4	14— 8	15	c
25	117— 0	150— 1	135—0	118— 2	122— 0	104— 2	15	d
26	126— 0	90— 0	87—0	91— 0	120— 0	17— 0	11	f
Totals	2,983—61	2,863—11	2,739—9	1,561—181	1,964—138	950—182		

* a, No permanent changes in hierarchy; b, A—B pair-contact reversed permanently; c, B—C pair-contact reversed permanently; d, C—D pair-contact reversed or unsettled; e, early instability of lower three ranks, stable after 4 days; f, one f was replaced, leading to changes in the hierarchy.

unchanged for an average of 54 per cent of the observation period.

In general, the hierarchies appeared to become more stable with time. There were some exceptions, in which permanent reversal of ranks occurred well along in the observations. In one case (Group 6, Expt. V), the fish ranking third at the start worked its way up to first and, in the process, brought about a series of changes in the hierarchy, as shown in Table 7. As L rose in rank, M dropped to omega. There was a parallel rise in rank of N to second place, and, as a result, O dropped to a fairly consistent third. All these fishes were females, and they were very closely matched in size at the start. At the end of the experiment, however, it was found that L and N had grown considerably more than M and O, so that a decided size discrepancy resulted.

This introduces the whole matter of position in the social hierarchy and growth, a subject that is reserved for a joint communication later with others

TABLE 3

AVERAGE RATES OF DRIVING PER HOUR OF OBSERVATION
IN 26 HIERARCHIES OF FOUR MEMBERS

RANK OF DRIVING FISH	RANK OF FISH DRIVEN			
	A	B	C	D
A..........	....	44.6±2.4	44.1±1.4	44.1±3.9
B..........	0.97±0.23	....	26.8±2.9	31.5±3.7
C..........	0.19±0.08	3.3±1.0	....	14.7±2.1
D..........	0.15±0.09	2.2±0.5	2.7±1.1	....

TABLE 4

FREQUENCY DISTRIBUTION OF NUMBER OF REVERSE DRIVES IN EACH
OF THE 6 CONTACT-PAIRS OF 26 HIERARCHIES
(Pair-Relations Are in Order of Rank)

NUMBER OF BACK-DRIVES	FREQUENCY OF OCCURRENCE						TOTAL
	A—B	A—C	A—D	B—C	B—D	C—D	
0...............	11	19	22	3	5	4	64
1...............	4	5	2	3	4	2	20
2...............	3	0	1	3	4	4	15
3...............	1	2	1	1	1	2	8
4...............	0	0	0	1	3	2	6
5...............	2	0	0	3	0	0	5
6...............	3	0	0	1	3	0	7
7...............	0	0	0	2	1	0	3
8...............	0	0	0	0	1	2	3
9...............	0	0	0	1	2	1	4
10...............	2	0	0	1	0	1	4
11...............	0	0	0	4	0	1	5
12...............	0	0	0	1	0	1	2
Over 12.........	0	0	0	2	2	6	10
Total pair-contacts.......	26	26	26	26	26	26	156

TABLE 5

(Expressed in Calendar Days and Percentage of Total Observation Period)

Group No.	A—B Days	A—B Per Cent	A—C Days	A—C Per Cent	A—D Days	A—D Per Cent	B—C Days	B—C Per Cent	B—D Days	B—D Per Cent	C—D Days	C—D Per Cent
Experiment I												
1	19	100	19	100	19	100	7	37	19	100	19*	100
3	19	100	19	100	19	100	19	100	19	100	19	100
4	19	100	19	100	19	100	15	79	12	63	9	47
6	19	100	19	100	19	100	19	100	19	100	10	53
7	19	100	19	100	19	100	9	47	11	58	11	58
8	13	68	19	100	19	100	19	100	19	100	19	100
Experiment II												
17	19	100	17	89	19	100	9	47	10	53	19	100
20	19	100	19	100	19	100	15	79	9	47	9	47
21	17	89	19	100	19	100	18	95	15	79	11	58
22	9	47	19	100	19	100	9	47	10	53	9	47
23	19	100	19	100	19	100	15	79	12	63	15	79
24	9	47	19	100	19	100	19	100	19	100	14	74
Experiment III												
6	25	100	25	100	25	100	9	36	15	60	25	100
12	25	100	25	100	25	100	11	44	25	100	25	100
15	25	100	25	100	25	100	25	100	25	100	18	72
18	21	84	25	100	25	100	25	100	10	40	9	36
Experiment IV												
3	22	100	22	100	22	100	22	100	22	100	10	45
15	22	100	22	100	22	100	8	36	22	100	22	100
Experiment V												
5	14*	100	14*	100	14*	100	14	64	22	100	22	100
7	22	100	17*	100	22	100	11*	65	22	100	5*	30
8	22	100	22	100	22	100	22	100	22	100	22	100
18	22	100	22	100	22	100	22	100	16	73	19	87
19	22	100	22	100	16*	100	22	100	13*	59	13*	59
20	22	100	16	73	22	100	22	100	22	100	15	68
25	22	100	22	100	22	100	22	100	22	100	16	73
26	22	100	22	100	13*	100	22	100	13*	100	13*	100
Average duration in percentage	94		98.5		100		79		83		74	

* One member of group died.

who have been and are associated in this particular phase of the inquiry. Although the evidence is by no means sufficiently collected to allow a definitive statement, it is known that size is an important factor in determining social dominance in the green sunfish.

In connection with joint experiments on growth, a test was made to secure further evidence on the stability of these fish hierarchies. Eight groups were maintained without observation of their behavior for 33 days (March 23–May 4, 1945). After this period, the fishes in each aquarium were watched for nine daily observation periods, each lasting 10 minutes. Tables 8 and 9 give analyses of the drives recorded and the duration of the contact-pair relationships. Relatively few reverse drives were seen and most of the contact-pairs remained unchanged throughout the 10-day period.

The general picture of activity within the sunfish hierarchy is fairly clear. After a time, relations tend to become formalized. The alpha fish, if its subordinates are all unaggressive, will drive at random any fish it comes across, and the omega member of the group simply avoids the approach of any of the others.

TABLE 6

DURATION OF THE 26 HIERARCHIES AND OF VARIOUS
INTERNAL SUBRELATIONSHIPS

Duration of Observation Period (Per Cent)	One Fish Dominates One Other	Alpha Dominates Three Others	Beta Dominates the Two Others	Gamma Dominates Delta	Four Fishes Maintain Same Hierarchy
1– 10	0	0	0	0	0
11– 20	0	0	0	0	0
21– 30	1	0	1	1	3
31– 40	4	0	3	1	4
41– 50	11	2	4	3	5
51– 60	10	0	5	5	6
61– 70	5	1	2	1	3
71– 80	10	1	1	4	2
81– 90	5	2	0	1	0
91–100	110	20	10	10	3
Total cases	156	26	26	26	26
Average duration in percentage	88.1	97	70	74	54

TABLE 7

DAILY RANK ORDER IN A HIERARCHY WHICH SHOWED SEVERAL REVERSALS

Rank	1/14	1/16	1/19	1/20	1/21	1/24	1/25	1/27	1/28	1/29	1/31	2/2	2/3	2/4
1	M	M	M	M	L	L	L	L	L	L	L	L	L	L
2	O	O	O	L	O	N	N	*	†	N	N	N	N	O
3	L—N	L	L	O	N	O	O	*	†	O	M	O	O	N
4	L—N	N	N	N	M	M	M	*	†	M	O	M	M	M

* Triangle situation: M
　　　　　↗ ↘
　　　N ← O

† Only one contact seen (N drove O once).

If the second-ranking fish happens to be active and aggressive, it may be as much or more of a tyrant than its superior. It then begins to approach equality with the latter, which may drive it infrequently.

Conditioning up and down the scale has been observed within the hierarchy. Frequently, a second- or third-ranking fish, after repeatedly driving a subordinate, may then challenge a superior, usually with disastrous results. Conversely, repeated attacks by alpha may weaken the resistance of beta or gamma to drives by the others, leading to temporary changes in position.

An interesting example of conditioning is the phenomenon here called the "substitution" effect. In several cases the lowest-ranking fish was observed to take

TABLE 8

TOTAL DRIVES RECORDED IN EACH OF THE PAIR-RELATIONS OF 8 HIERARCHIES
MAINTAINED FOR 43 CALENDAR DAYS AND OBSERVED ONLY
DURING LAST 10 CALENDAR DAYS (EXPT. VI)

Group No.	PAIR-RELATIONS IN ORDER OF RANK					
	A—B	A—C	A—D	B—C	B—D	C—D
1	78—1	49— 0	36—0	29—13	28—2	62— 4
2	56—1	63— 0	74—0	64— 3	108—0	31— 0
3	44—0	83— 0	62—0	10—14	18—0	91— 1
4	38—1	46— 0	37—0	21— 6	70—0	33— 0
5	113—3	72— 0	47—0	34— 0	24—0	2— 1
6*	14—0	89— 0	68—0	54— 0	45—0	13— 4
7	75—0	70— 0	77—0	49— 0	46—3	9— 4
8	10—2	45—10	32—0	4— 3	6—0	8— 0
Totals	428—8	517—10	433—0	265—39	345—5	249—14

* Two territories, held by A and B.

TABLE 9

LONGEST DURATION OF PAIR-RELATIONS IN HIERARCHIES OF TABLE 8

Group No.	A—B		A—C		A—D		B—C		B—D		C—D	
	Days	Per Cent	Days	Per Cent	Days	Per Cent	Days	Per Cent	Days	Per Cent	Days	Per Cent
1	10	100	10	100	10	100	3	30	10	100	10	100
2	10	100	10	100	10	100	10	100	10	100	10	100
3	10	100	10	100	10	100	4	40	10	100	10	100
4	10	100	10	100	10	100	9	90	10	100	10	100
5	10	100	10	100	10	100	10	100	10	100	10	100
6*	10	100	10	100	10	100	10	100	10	100	5	50
7	10	100	10	100	10	100	10	100	10	100	9	90
8	10	100	10	100	10	100	10	100	10	100	10	100
Average duration in percentage	100		100		100		82.5		100		92.5	

* Two territories, held by A and B.

advantage of the tendency of beta and gamma to flee the approach of the despot; it would dash at them, and they would run, unless they turned around to fight back, and then they would drive omega away with ease. Some of the reverse drives recorded in Table 2 undoubtedly can be traced to this kind of substitution, which also was performed by the others.

FACTORS IN DOMINANCE

Among the factors that can be postulated as affecting the outcome of a fight or the place of an individual in a rank order, sex, size, and previous experience or conditioning lend themselves most readily to experimentation. Although the sex of the immature fish could not be told externally, it was possible to distinguish the sex composition of the groups at autopsy by macroscopic inspection of the gonads (method checked by cross-section of gonads). In Experiments II–V, distribution of the sexes was determined and found to be 78 males to 66 females. Additional data on the effect of sex on hierarchy position were also secured from Experiment VI and from two other sets of ten groups each, observed under the same conditions as the others, but for a shorter time—August 6–August 10 and August 29–September 6. These two sets were controls for an experiment on the effect of screen partitions on the form of social organization (to be reported below). From these eight sets of experiments, totaling seventy-two groups, sixteen can be selected that fulfil the following conditions: first, they were hierarchies, not complicated by established territory; and, second, they consisted of two males and two females. In these sixteen critical cases, the alpha rank was held by fourteen males and two females, the beta rank by eleven males and five

females, the gamma rank was held by four males and twelve females, and the omega or lowest rank by three males and thirteen females. This indicates a correlation between maleness and high social rank.

During the early stages of formation of a hierarchy, the relative size of the fish is an important factor. Large sunfish dominate much smaller individuals without resistance by the latter; even within the minor range of size differences that were present, the larger fishes regularly occupied the higher ranks. Since both sex and size are factors in dominance, the question arises as to how they interact. To what extent does larger size overcome the handicap of femaleness? Some indication can be secured by re-examining the sixteen cases already discussed as critical for the sex factor. There are ninety-six pair-relations represented, of which two-thirds are heterosexual and a third unisexual. In sixteen male-male contact-pairs, the dominant individual was the larger in eleven cases; the dominant female was the larger in nine of the sixteen female-female pair-contacts. In sixty-four heterosexual contacts, the dominant male was larger in forty-four of fifty-three cases, while the dominant female was larger in seven of eleven cases. The fact that so many of the males that dominated females were also larger somewhat confuses the picture. It is possible, although not probable, that this may be a systematic error arising from the composition of the field collections. Hubbs and Cooper (1935) found that male sunfish in nature grow faster than females but are not significantly larger until their third summer. Our laboratory experiments made in another connection demonstrate a significantly faster rate of growth of males over females in the first and second years.

Hubbs and Cooper suggest that "the

increased growth of the males has been of selectional significance, enabling them to ward off enemies from the nests which they guard so pugnaciously." This hypothesis may be restated by defining enemies to include intra-specific male rivals; and the selection would then be in part also intra-specific sexual selection of the Darwinian type (Huxley, 1938).

The effect of conditioning on the members of a hierarchy has already been mentioned. The members of a group develop stereotyped reactions to one another and to the aquarium; place conditioning is especially prominent and may lead to territory; but, even when no definite territories exist, each fish tends to stay more often in a particular place than in others. The subordinates have certain positions in which they are safer from attack than elsewhere and also develop movement habits that take them out of the path of their superiors in rank.

When a new fish is introduced to replace a member of the group, it is quickly relegated to the bottom rank, unless it is much larger. The newcomer transgresses upon all the relations that have developed and at first is driven by each resident fish. After a while, however, the newcomer begins to assert itself and may advance in rank. Thus, in Group 5, Experiment V, a fish introduced as replacement for the omega rose to the alpha position in a few days.

From May 9 to May 11, 1945, the dominant individuals in the hierarchy groups of Experiment VI were interchanged to see what rank a long-established alpha fish would take in a strange hierarchy. The alphas that were exchanged differed in size by little more than 1 mm. in three instances but were somewhat unequal in the fourth case. Since these individuals were long accustomed to be dominant, it was of interest to see whether they could substitute

mechanically for each other or whether they would meet with resistance.

After the exchange of the dominant fish in Groups 1 and 2, both were observed together for $1\frac{1}{2}$ hours. For the first 15 minutes, the second-ranking fish in Group 2 took over, and the newcomer went to the bottom of the hierarchy; soon, however, it began to return challenges and within another 20 minutes it had succeeded in defeating all the resident fish and became the despot. In the other aquarium the newly introduced fish at first made persistent efforts to get through the glass of the aquarium. During this time, it was repeatedly driven by the second-ranking fish and occasionally by the others. After a while it began to resist attack, first defeating the second-ranking fish and then dominating the group.

In the second exchange, between Groups 3 and 4, events proceeded in a much similar fashion. The alphas, when placed in strange aquaria, at first did not resist attack and were driven by the residents. Within 45 minutes, however, each had won to the top.

The alpha fishes of Groups 5 and 6 were unequal in size, measuring 47.3 and 39.7 mm., respectively. When they were interchanged, the smaller went to the bottom of the new hierarchy and stayed there for 2 hours and 15 minutes of observation. In the other group the large newcomer was not active, and, although he was driven a few times and later drove once in return, no conclusive results were secured.

In the fourth exchange the two alphas were of about the same size. After 11 minutes one of them was dominant in its new group; by 31 minutes, the other had asserted itself and had become dominant. Neither encountered much opposition.

Thus in three cases the dominant fishes of two groups were exchanged and replaced each other, rising to top rank in 35, 51, 45, 45, 11, and 31 minutes. In the remaining case one of the alphas was too small for its new group, becoming omega, and the other was unaggressive in the new situation. All these fishes, when returned to their old places, almost immediately resumed their original positions. During their absence fights occurred among the subordinates, but there were no reversals in rank.

This apparent ability of a long-dominant fish to retain its dominance even in a strange group probably resulted as much from the conditioning of the subordinates as from its own previous experience. These tests are too few to be taken as approaching conclusive results, but they do indicate the importance of the factor of experience or conditioning.

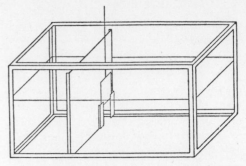

Fig. 3.—Maze used in tests of relations between leadership and hierarchy. Adapted from Welty (1934). Aquarium measures 16×10×10 inches.

[*Editor's Note:* Material has been omitted at this point.]

DESCRIPTION OF TERRITORIES

Seventeen of the forty-four groups comprising Experiments I–V (Table 1) became organized along territorial lines. In nine of these seventeen groups, some or all of the areas were claimed before the start of observation. A detailed record of the manner of organization of territories was secured in two cases (Groups 27 and 28, Experiment V), and a less complete account can be given of six others. The history of certain of these groups is presented in some detail because it provides an insight into the nature of territoriality.

The four members of Group 28 engaged in round-robin fights, B defeating D, and C dominating A. Soon B emerged as the despot by winning an encounter with C, the latter retaining the ability to drive A. Next, C was observed in a corner of the aquarium, which it defended against all except the despot, B. This behavior has already been defined as partial territory (p. 272). A half-hour passed, with no discernible change, and then B and D again fought; neither could dominate, but D tended to withdraw to one corner, while B remained at the opposite side (Fig. 5). Then there occurred a change in dominance between A and C; the former defeated C and took its place in the corner. After a while B challenged A, and they fought vigorously; but neither seemed to give way, and A retained its corner position. A and D challenged each other repeatedly and several times exchanged nips, neither achieving mastery. Soon it became clear that three territories existed in the aquarium and that C was omega or subordinate to all.

An interesting change was seen the

following day. The three territories still remained, but D had become the most aggressive, with the largest territory at one side of the aquarium (Fig. 6), while A and B had shifted so that they occu-

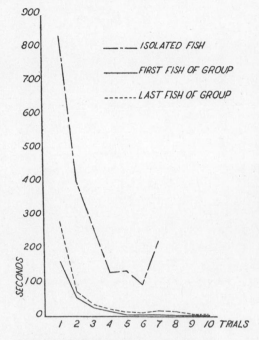

FIG. 4.—Average learning curves of isolated and grouped fishes in maze shown in Fig. 3.

pied adjacent corners opposite D. These positions were kept for the rest of the 22 days of observation. The later history of this group is discussed in part in another section (p. 288). Fish A was the least aggressive of the territory-holders, and, as a result, C was repeatedly driven into A's territory and tended to find an insecure resting place there. Thus A drove the omega actually more often than did the other fishes; nevertheless, by January 24, C was resting mainly in A's territory when exhausted, and D often invaded that area to attack omega. A few days later, it was seen that D's habit of invading had worked to the detriment of A, since D now also drove A; A and B were still in territorial relation to each other at the time. On February 2, A was also driven by B whenever forced into B's area. The omega died on February 4 and thereafter A was, in effect, the lowest-ranking member of the group.

At first, Group 27 followed a similar course. Individuals C and D fought very actively soon after they were placed to-

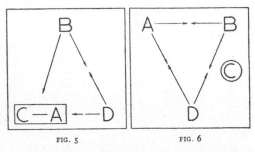

FIG. 5   FIG. 6

FIGS. 5 and 6.—Changes in territories of Group 28 (Expt. V). B and D were first to establish a mutual, territorial relation; then A established itself in the lower left-hand corner (Fig. 5). Next D became more aggressive than the others and caused a shift of the territories, as shown in Fig. 6. Apposed arrows show territory; unapposed arrows indicate dominance. C became omega (encircled).

gether, and C was the winner. In turn, C defeated A and B and became the despot, in control of a large part of the aquarium. Next, B withdrew to a corner and defended it against the others. This corner was directly opposite the usual residence of C, and it appeared that the location of this territory was determined by the prior position of the despot. The following day, D had displaced B and thereafter held the corner territory.

Subsequently, the two territory-holders became very aggressive and kept A and B in one corner; they remained there until January 26, when an abrupt change took place: A drove B out of the corner, chasing it persistently (forty-nine times in

15 minutes); A was relatively unmolested (5 drives). Thereafter, A gradually won a small territory in the corner; on the other hand, B was driven relentlessly, and on February 6 it was dead.

The sequence of events is similar in these two cases. First, vigorous fights resulted in an unstable dominance order. The resting place occupied by the most dominant individual seemed to determine an opposite pole, at which several fish alternated, and ultimately one established territory there. When the third territory was seen, it was at an intermediate position.

Seven other groups were observed initially when they still were hierarchies; in five of these, territory was eventually taken up strictly in accordance with rank in the hierarchy. The beta individual became equal in rank with the alpha, and the two restricted their movements to separate areas. When a third territory also was held (in five of the total of eight groups), the gamma fish occupied it. In one exception the two highest members of the hierarchy first held territories; 9 days later, there occurred a reversal in rank of the two subordinates, and very soon thereafter the winner also took up a territory. Another exception arose when the lowest-ranking member of a hierarchy died and the new fish that replaced it succeeded in maintaining an area against the residents. The original alpha and the newcomer both drove the other two fishes, but not each other, and stayed in separate parts of the aquarium.

This behavior of a replacing fish was seen in two other instances. In Group 9 (Experiment III), B and D were first observed to hold territories, and C was omega. The next day, C was dead and was replaced with a fish that was 4 mm. larger than A and $\frac{1}{2}$ mm. larger than B and D. That same day the newcomer was driven by both territory-holders but not by A; the following day, the new-comer had become equal in rank with D, and 3 days later it held a territory along-side B and D. On the days immediately before the new fish acquired its territory, it drove omega twenty-eight and twenty-three times in two 15 minute observation intervals, chasing it into the paths of B and D, which, in turn, drove it repeated-ly. It was my impression that in this manner the nonresident fish was building up an immunity from attack.

A comparable set of observations was made with Group 3 (Experiment III). At the very first observation, three terri-tories were seen, with D as omega. Two weeks later, B lost its position and at the same time became subordinate to D. On that day, B was driven a total of one hundred and three times in 15 minutes of observation; it was dead the next day. The fish replacing B was the same size as the residents. It was attacked persistent-ly by all, and evidently could not find a resting place in the aquarium. It was not reactive to the social relations that al-ready existed, for example, the invisible boundaries and neutral areas, whereas D, also subject to attack, could retire to a certain corner and be safe for a while. The new fish was driven almost con-stantly by A and C, while D remained unmolested and even took part. On the following day, however, the newcomer had ousted D from its resting place in the former territory of B and was in process of taking over this area. Again, it was seen that in the short observation periods on consecutive days the newcomer drove omega twenty-six, twenty-nine, and sev-enteen times, seeming to force D out in front of it and thereby gaining a measure of immunity from attack. It held this territory until the end of observation, 5 days later.

Group 5 (Experiment I) was found to be organized into three territories at the start of observation; A, B, and D each stayed in a corner of the aquarium and sallied out to drive C or to challenge each other. Fish C found a resting place only in the remaining unoccupied corner. Three days later, C had succeeded in re-pulsing drives by first A and then B, on each side of it. From then on, the four fish were all equal. They challenged but seldom drove each other, except when one trespassed on another's territory. Each stayed close to its corner and ac-tivity tended to be at a minimum, com-pared with other groups.

By way of summary, the following three generalizations can be made:

1. Territories may be established at any time during the course of a hierarchy and are usually taken up in order of rank. When low-ranking individuals establish territories, they first defeat their immedi-ate subordinates.

2. Territories are, in large measure, de-termined by the shape of the aquarium, i.e., the corners of a square aquarium are much used as resting places. A second territory is set up at the opposite pole from an already existing one, the third is equidistant from the two others, and the rare fourth territory is located in a neu-tral area with relation to the three others.

3. Territories are stable but may un-dergo changes such as elimination of one, transfer of ownership, shifts in position and boundaries, and variations in rela-tive aggressiveness of the possessors.

In the foregoing accounts there is a suggestion of an important social func-tion of the omega fish. Especially in the case of territories, the omega seems to be a sort of buffer, releasing tension between the high-ranking individuals. When a territory was about to be taken, attacks upon the omega may have been an ef-fective part of the readjustment of the group.

This suggested role of the omega was

investigated in the following series. of tests. At the end of Experiment IV (December 25), Groups 18 and 21 were retained because there were three firmly held territories in each. On January 1 the omegas were removed permanently, to see whether this would affect the territories; the groups were observed for 15 minutes immediately afterward. The territories were not disrupted, but there was a rise in the frequency of challenges. On the last day of previous observation (December 25) sixteen challenges were recorded in Group 18; after the removal of omega, there were twenty-seven. In Group 21 there were three challenges on December 25 and twenty-six after the removal of omega. The omegas were kept out of the aquaria overnight, and on the following day the two groups were again observed for 15 minutes each. In Group 18 there were twenty-five challenges, four full fights, and two drives, and in Group 21 there were thirty-eight challenges and five drives.

A new fish was introduced into each group, and it was observed for 15 minutes, after a 2-minute interval for adjustment to disturbance. There were eight challenges and four drives among the three resident members of Group 18, whereas they combined to drive the nonresident fish one hundred and fifty-seven times. In Group 21 the newcomer was driven two hundred and twenty-five times in 15 minutes, keeping the residents so busy that they challenged each other only six times and drove each other seven times, occasionally perhaps by error.

These tests show that either the omega or the nonresident fish draws upon itself the attacks of the territory-holding animals and thereby reduces tension between them. In the absence of other objects of attack, fishes that have territories close together, as in our aquaria, challenge each other, and tension may increase.

A further test was made with Groups 27 and 28 of Experiment VI. On February 4, the last day of that experiment, the omega of Group 28 died, and A, which earlier had held a territory but was now subordinate to D, came in for more than its usual share of attacks. Two days later, Group 28 was observed for 15 minutes, and during this time A was driven thirty-six times by B and eighty-two times by D. There were also eleven challenges between B and D. A new fish that had been isolated was added. The newcomer was in the same size range as the other three fishes. The group was observed for 30 minutes; during the first 15 minutes, D drove A only nine times, and B drove A just once, while B and D challenged each other six times. This was a considerable advance for A and a reduction in reactions between B and D. The latter drove the newcomer thirty-nine times, and B drove it thirteen times. The nonresident fish was forced to stay in the same corner with A, where it was driven least by B and D. In the next 15 minutes, A drove the newcomer twenty-one times and also vigorously resisted D, challenging it six times and once engaging in an active, though losing, encounter with both D and B. D drove A thirteen times, and B drove it once. There was no tension in the form of challenging between B and D. The new omega was driven ninety-six times by D and twenty-three times by B.

The omega was left in the aquarium for 24 hours, and the group was observed again for 15 minutes. The situation had improved even more for A, and it had won back its territory. It was driven only once by D and twice by B. The omega was driven thirty-three times by A, nineteen times by B, and fifty-eight times by

D. It appeared to the observer that A was driving the newcomer out in front and that it became a buffer between A and the aggressive D. Often, when D drove omega near to A, the latter, in turn, drove it out toward D's territory, and D's attention was taken up with omega, leaving A unmolested. The results, as described, indicate an important social function of the omega fish, in territorial organizations at any rate.

On February 6 the situation in Group 27 was similar to that described for Group 28; the omega had just died, and when the group was observed for 15 minutes, A was found to be the subordinate. Another fish (B') of about the same size was added; it had been kept isolated (in an adjacent aquarium). The newcomer was immediately attacked by D, but it resisted vigorously; the two were equally matched, and after 8 minutes they separated, both exhausted. The new fish drifted near A, and the latter challenged and attacked it. B' fought back and won, driving A out of its resting spot. Then B' attacked C, and the resulting fight was indecisive. A short while later the newcomer attacked D, and another vicious fight ensued, with neither able to overcome the other. After 19 minutes, the newcomer had taken over A's territory and was on an equal footing with C and D. In the next 25 minutes, B' alternately fought with C and then with D, always holding its own, and between fights it drove A as often as the others did. After a total of 35 minutes of observation, B' was firmly established on A's territory, and because of B''s belligerence D was displaced from its resting area, moving over to the adjacent corner. This situation still existed 24 hours later.

In this case the newcomer was very aggressive from the start and succeeded in withstanding the attacks of the two territory-holders. By ousting the weakened A, it gained a territory and was even able to extend it at the expense of D. Here the introduction of a new fish resulted in a marked deterioration of the position of A, but the same event in Group 28 brought about an improvement for a fish in a similar situation. The critical difference was the relative aggressiveness of the two newcomers, in proportion to that of the residents.

FACTORS INFLUENCING THE ESTABLISHMENT OF TERRITORY

When the space for territories is limited, adjacent individuals may divide this space unequally but still remain in balance at the boundaries. Aggressiveness diminishes with distance from the territorial center, and this phenomenon may explain, in part, the paradox of social equals in possession of disparate areas.

Territorial balance may be achieved at the first fighting encounter or shortly afterward. If established later, within a hierarchy, the territories are taken in order of rank and result from a rise in status of the highest-ranking subordinate to equality with the despot. In such cases, prior conditioning to an area may be involved in rise in dominance level. Another distinct possibility is that the tyrant may withdraw from a portion of its range, perhaps becoming in time less aggressive, and this may enable the second-ranking individual to establish a territory.

The factors which determine whether two animals will enter upon a territorial relation are those that decide their relative aggressiveness. The larger will usually dominate; this has been repeatedly observed. Sex may be critical, since males dominate females, but not necessarily decisive. All other matters, such as health, strength, or previous experience —in fact, all the factors involved in dominance (Allee, 1945)—are also important in the establishment of territory. The aggressiveness of each animal is the result of the complex interactions of these factors, and, since territory is a balance, variations on each side are significant for the territorial relationship.

In fifteen of the seventeen cases of territory observed in Experiments I–V, sex was determined at autopsy; the ratio was thirty-three males to twenty-seven females. Twenty-one of the males and fifteen of the females held territory, and this was roughly in the same proportion as the relative frequency of the sexes in the total population. Seven of the fifteen females that held territory were the largest in their groups, whereas only two of the twenty-one males were the largest. This suggests that size was a more important factor for females than for males. That females of heterosexual groups held territory at all is evidence that sex alone is not decisive in this kind of dominance.

In five groups there was, at first, a hierarchy with a female as alpha. In each case the next-ranking fish was a smaller male that subsequently rose to equality with the female, and the two established territories. The rise in rank of the beta fish may have been correlated with maleness; but the female alpha fish, by virtue of its experience as alpha, may have retained sufficient advantage to prevent its complete overthrow. Conversely, in three cases, second-ranking females that were the largest in their groups rose to equality with the smaller alpha males but did not overthrow them and, instead, established territories. These cases point to the importance of previous experience in relation to territory.

There were two cases in which a smaller female initially held territory alongside two males but was ousted from her position and became omega. But in the majority of cases, territories held by males or females were stable.

Further evidence is derived from the twenty groups observed between August 5 and September 6, 1946, in connection with an experiment to be described below. Analysis of the data from these groups, which had a sex ratio of exactly 50:50, supports the points already made. Twenty-four males and eighteen females held territory; ten of the eighteen females and only seven of the twenty-four males were the largest in their groups. The relations between territoriality and sex or size held as a rule; the exceptions were due to undetermined factors.

Certain external influences modify territorialism; some of these are the configuration, subdivison, and amount of available space and the numbers of fishes present. A corollary is the factor of size; an area may be crowded for two large fishes but adequate for four much smaller ones. Preliminary experiments have been carried out with special attention to the factor of spatial subdivision.

With size of aquarium and number and size of fishes held constant, a wire-screen partition (with a 2-inch square gate at one corner) was introduced into the aquaria to divide them into adjoining compartments; in a series of twenty-seven experiments, the partition was placed about one-third of the way to one side. Twenty of the twenty-seven groups developed territories, and nineteen of them established two territories, in conformity with the arrangement of the aquarium. This is a considerably higher ratio than prevailed in Experiments I–V, with undivided aquaria. Since the groups in partitioned aquaria were maintained under somewhat different conditions of light and temperature (in a greenhouse, under natural illumination), another set of experiments was performed, this time under the same conditions as in Experiments I–V. Two series were arranged, each consisting of ten aquaria with a partition set exactly in the middle, and alternating with these were ten control aquaria without the partitions. The first series was observed from August 5 to August 10, until stable relations had apparently been established; the second series was observed from August 29 to September 6. Table 12 presents the data with regard to types of social organization established. Fifteen of twenty groups in partitioned aquaria developed territories while only five of twenty control groups established territories.

The principal effect of the partitions was in relation to defense of boundaries. The gate became the site of interaction between the most aggressive of the fishes, alpha and beta occupying separate compartments, and the subdivision into territories appeared to be facilitated.

When the partition was set one-third of the way across the aquarium, the aggressiveness of the two territory-holders may have been correlated with amount of space. Frequently, the fish dominant in the large section kept all the others from entering it. In these cases the smaller compartment contained a typical hierarchy. Whenever the territory-holding fish in the smaller area drove a subordinate through the gate, it was promptly chased back again. After a time, movements tended to become stereotyped; the subordinate was driven with regularity and usually retired to a neutral position; but when it was evicted through the gate, the other territory-holder would be ready and waiting to drive it back. Each fish of the group played its role smoothly, and there seemed to be little friction; after a time the two territorial individuals seldom challenged each other and even dom challenged each other and even

came to rest close together on opposite sides of the gate.

Not all territorial relations remained so stable; occasionally the more aggressive individual became bolder and trespassed upon the territory of the other, at first entering a short distance to drive a subordinate; the two dominant animals threatened each other, and the intruder then retreated. In several cases the more aggressive fish also began to drive its neighbor and by stages reduced the

TABLE 12

EFFECT OF SUBDIVIDING THE AQUARIA WITH PARTITION AND GATE UPON TYPE OF SOCIAL ORGANIZATION IN GROUPS OF 4 SUNFISH

SERIES	HIER-ARCHIES	TERRITORIES	
		Two	Three
1 { Experimental......	3	6	1
{ Control..........	7	2	1
2 { Experimental......	2	8	0
{ Control..........	8	2	0

latter's territory to a partial one. The despot kept one compartment clear for itself, but also entered the other compartment and attacked all the others. The subordinates could seldom be driven from the smaller space, because there they were relatively safer than in the larger space.

In the majority of cases the groups were first organized as hierarchies and then changed to territories after the second-ranking fish had become equal in status and dominant in one of the compartments. The second in rank resisted attack more and more vigorously, especially at the gate, upon entrance of the dominant fish, and soon the latter challenged only at the boundary. Usually the individual resident in the smaller space thus gained a territory, but occasionally it was the other way around. That is, in some cases a fish kept the smaller space as its residence area but invaded the larger space to drive all the others. Then the second-ranking fish would progressively limit these invasions by challenges and fighting, until there existed two territories.

Essentially the same relationships prevailed in aquaria divided into two equal compartments. Here, too, it was evident that the more aggressive fish stayed alone in its compartment but often made

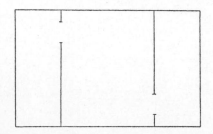

Fig. 7.—Wire-screen partitions divide aquarium into three compartments. Aquarium measures 16× 10×10 inches.

minor forays into the adjacent area. There was an interesting difference in orientation of the territories; the control groups generally held territories dividing a front from a rear area, but the groups in experimental aquaria followed the left-right division provided by the partition.

At the end of this series of experiments each group was observed for 15 minutes, the partition was gently raised, and observations were taken for another 15 minutes. In a typical case, one of the two territory-holders (C) had kept its side clear of subordinates and, in consequence, had very little to do to maintain its area. It drove only four times (during the first observation period), as contrasted with the activity of its neighbor (B), which drove the two lower-ranking fishes forty-six times. After removal of

the partition, C became very active and was able to drive the subordinates, since it now could reach them without trespassing at the door. C drove seventy-six times, and B drove sixty-six times; altogether, the total number of drives within the group rose from sixty-six to one hundred and fifty-seven in equal 15-minute periods of observation. At the same time, tension between B and C increased, as shown by eight challenges and eight drives between the two, as against none before. The territories were not abolished, but C proved itself to be more aggressive, continuing to keep its area free from subordinates, yet repeatedly invading B's area to drive them. The group returned to normal within a few minutes after the partition had been replaced.

Removal of the partitions in no case disrupted the territories, but there generally was an increase in challenging and driving between the territorial fishes. In four of the six groups there also followed an increased driving of subordinates by the two territory-holders. Other minor changes in status were observed; on replacing the partitions, the relations within the groups returned to normal.

The success of this method of influencing territoriality with partitions led to some preliminary trials with more complicated systems. In a set of larger aquaria (16 × 10 × 10 inches) two wire screens were introduced to provide three areas, the middle one being the largest (Fig. 7). Sixteen groups of four were observed in such aquaria; fourteen of them developed territories, an equal number being two- and three-territory situations. Hence it appears that provison of three compartments and more space resulted in establishment of a larger number of territories.

When there are three territories, the omega fish is at a decided disadvantage and may not survive, being shunted from place to place and kept continually on the move. In one typical instance, it was seen to rest only when it retreated to a corner of the middle compartment, at the rear, and buried its head between the screen and the glass, facing away from its tormentor. Even there, it was sought out after a while and driven by all three territory-holders. The death or removal of the omega did not disrupt the territories, but subtle changes indicating an increase in tension could be noted. The fish threatened each other more often, and invasions increased in frequency. As in previous observations, the omega appeared to act as a buffer or releaser of tension between the territory-holding fishes.

There were interesting variations in the pattern of territorial division of space. Sometimes an aggressive individual took for itself two adjacent compartments and kept the others in the third. Here both a territory and hierarchy of three existed, as described for the two-compartment aquaria. In other, instances the territorial fish remained in the side sections and the subordinates stayed in the middle one. The territory-holders would take turns entering the middle space to drive or would enter simultaneously while ignoring each other; in such cases individual recognition was obvious, for they often came to rest side by side, only to dart at another fish.

Another type of arrangement of partitions is shown in Figure 8. Only four groups were observed in these aquaria. In two cases, one fish held adjacent compartments, and two other fishes each held one area; the remaining fish was thus an omega. In another case one fish was so aggressive that it excluded all the others from three compartments, while a second fish maintained a territory in the fourth compartment. The fourth group was a hierarchy, with two fishes

holding partial territories, subject to invasion by the dominant individual.

As shown in Figure 9, partitions may also be arranged so that four triangular niches open into a central square. Two groups of four fishes each, when placed in such an environment, established three and four territories. The fishes stayed in the outer niches, entering the central space either to threaten a neighbor or to drive the omega, in the case of the triple-territory situation.

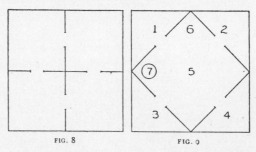

FIG. 8.

FIG. 9.

FIG. 8.—Arrangement of wire-screen partitions to provide four communicating chambers. Square aquarium measures 10×10×12 inches.

FIG. 9.—Arrangement of wire-screen partitions with gates to provide multiple niches as a stimulus to territory. Square aquarium measures 10×10×12 inches.

Three trials were made with groups of seven fishes in such aquaria. The first trial yielded the most interesting example of territoriality so far recorded. Each corner was occupied by one individual, while a fifth was able to maintain itself in the middle section and resist the threats of the other four. A sixth fish held a partial territory within the angle made by two partitions (Fig. 9) but was subordinate to No. 5, as well as to the four that held corner territories.

The seventh fish (encircled) was omega and moved around the aquarium, driven from all directions. At the end of 3 days, the position of No. 5 had become untenable, and only the four residents in the side compartments still held territory. The three fishes in the center were driven by these but still maintained a hierarchy. The group was almost constantly in motion; the territory-holders threatened each other, but it was obvious that they were inviolate in their own corners. First one and then another was the most aggressive in the central chamber and entered it farthest. Each fish appeared to be able to distinguish between a territorial and a nonterritorial fish, as well as perhaps individuals themselves, and drove the subordinates unerringly. Here, too, the impression was gained that the presence of subordinates

TABLE 13

EFFECT OF SPACE ON TYPE OF SOCIAL ORGANIZATION IN GROUPS OF 4 SUNFISH

Size of Aquaria (In Inches)	Territories	Hierarchies	Total Drives
4.5 × 4.5 × 4.5	2	8	2,864
16 × 10 × 10	6	4	3,560
24 × 24 × 16	3	5	657

lessened tension and diverted the attention of the territory-holders from each other. The group was under daily observation from August 27 through September 3, 1944, and less constantly through the rest of September; no essential change was observed.

Two other groups of seven under the same conditions established only three territories per group. In both cases an aggressive individual held two adjacent corners, while two other fishes each held one corner; none of the remaining fishes held territory.

Although these experiments with more complex arrangements of niches are preliminary, they indicate that the condition of the physical environment with regard to subdivision into niches helps determine the location and extent of territories.

The amount of space available to a group is probably a factor in determining whether territories will be established, as shown by Shoemaker's (1939) study with canaries. A preliminary test of this factor in the sunfish was carried out from September 9 through September 29, 1945. From a stock that had been kept in the greenhouse for several weeks in extralarge aquaria (8 × 2 × 1.4 feet), twenty-eight groups (of four each) were selected, composed of fishes that were all about the same size. Ten groups were placed in 1-liter jars, ten others in medium-sized aquaria, and eight groups in much larger aquaria. Six rounds of fifteen minutes each were made between September 12 and September 29. Table 13 gives the total number of territories, hierarchies, and drives recorded under each of the three space conditions. In the matter of drives, there is a suggestion of an optimum space for the greatest number of reactions. Similarly, with regard to territory, the jars were decidedly too small, and the intermediate-sized space seemingly was most effective. These results suggest that for a given size of fish, there may be an optimum aquarium space for territoriality and/or frequency of aggressive interactions.

The effect of numbers present per unit of space has been investigated by Hixson (1946). Miss Hixson found an increase of frequency of territories in correlation with size of population, in the same volume of water. In eight groups of two fishes each, no territories were seen; in each of five groups of three, a single territory could be detected; in each of four groups of four, two territories were recorded; in three groups of six fish, two became organized as four territories, the other two as three territories; one of two groups of eight fish developed six territories, while the other developed five territories. This surprising correlation needs further study.

DISCUSSION

The literature on sunfish courtship and nesting behavior has been summarized in excellent fashion by Breder (1936). He finds that the species that have been studied are confined to a similar type of habitat and display essentially the same reproductive methods. *Lepomis cyanellus* has not been studied in the field, but in our laboratory its mating behavior followed closely the family pattern. An oval nest was constructed by the males and guarded both before and after eggs were laid, while the females participated only in egg-laying. As shown in Figure 1, all the nests were behind some opaque barrier. This closely resembles the nesting of *L. auritus* in a flowing stream, as reported by Breder (1936). Since no current existed in the aquaria, one must ascribe the position of the green sunfish nests to some other factor, possibly orientation to direct window lighting. The males entered the nests so that their heads pointed toward the angle made by the aquarium glass and the barrier, and they made sweeping movements with their tails to clear out an oval nest.

Reighard's (1902) brief discussion of the sunfish fighting pattern mentions the spreading of the gill-covers and the role of color display as an aid in driving away intruders, as well as perhaps in attracting the female. Sex differences in pigmentation and their significance in courtship were analyzed by Noble (1934) with *L. gibbosus* and by Breder (1936), with *L. auritus*. Although agreeing on the details of behavior, the authors differed with regard to the functions of male display in Darwinian sexual selection. Noble maintained that a "true" sexual selection exists in sunfish, based on the conspicuous colors of the males, which attract the attention of the females. Breder and Coates (1935) object to this hypothesis on the grounds that the sex

ratio in sunfish is nearly equal and that hence such selection would be of no practical effect. Laboratory studies by Noble and Curtis (1939) with jewel fish seem to confirm the existence of a selection mechanism in which the color display of the males attracts the females. Greenberg and Noble (1944), experimenting with the lizard *Anolis*, also support the Darwinian hypothesis in modified form. A. E. Emerson suggests that natural selection acts not upon the individual male but upon the mating pair as a unit; the brilliant male display functions to integrate the pair more effectively. However, the Darwinian concept is concerned rather with which one of several competing males enters into the mating relation.

Adult male centrarchids have long been known to hold territories during the breeding season. Abbott (1894) observed that sunfish nests are often in a row or cluster: "The occupants of the several nests do not molest each other, and never intrude beyond the limits of their own 'homes.' " Lydell (1902) recorded the fact that male black bass, preparing to nest, fight actively; in a clear artificial pond an aggressive male will prevent any other bass from building a nest within 25–30 feet. However, nests of the smallmouth bass in the "natural water" were frequently built against a stone or log so as to be shielded on one side and no farther apart than 4 feet. Lydell constructed rectangular nest frames, placed them in spaced rows, and thereby increased the number of effective breeders. This method has become standard practice in bass culture (Langlois, 1936); it utilizes one factor in determining size of territory—the physical complexity or subdivison of an area.

Much of the literature on territorial behavior of fishes has been reviewed by Noble (1938). Defense of nesting site is common among oviparous fishes. That territorial organization may not necessarily be associated with nesting is indicated by the observations of Meyer-Holzapfel (1941). She studied eleven South American Characinidae (*Hemigrammus caudovittatus*) of unknown sex in a large aquarium (9 × 24.4 × 10 inches). The specimens averaged 7 cm. in length, presumably being adults. All except four established tri-dimensional territories; the four subordinates were relegated to "neutral areas," not defended by the other fishes. Contrary to Wunder (1930) and Breder (1936), who have found territoriality only during the mating season and held only by males, Meyer-Holzapfel suggests that there might have been a female among her territorial fishes and that they do not build nests. She concludes that territory is much more fundamental than hitherto thought.

It may be objected that, inasmuch as there were some fishes that did not have a territory, these may have been females. However, the present set of experiments with immature green sunfish of known sex supports her conclusions. Territoriality in the sunfish is independent of overt sex behavior or the genetic sex of immature individuals, although sex is a factor in deciding dominance or territorial relations in a contact-pair.

The factors influencing establishment of territory have already been discussed in part (p. 289). Underlying the behavior of adjacent territory-holders is a psychological balance at boundary lines, which manifests itself, in the sunfish, in the form of challenges or even nips but without any change in status. Although the fish must be equal at the boundaries, their territories may be decidedly unequal in extent, and this can be used as a measure of relative aggressiveness. Meyer-Holzapfel (1941) found that size of individuals was not important (within

limits of $\frac{1}{2}$–1 cm.), in the observed social organization of *Hemigrammus*. The sunfish seem to be more sensitive to size differences, but a minor disparity in size can be overruled by other elements of a complex combination of factors. These factors are the same whether the relation is territorial or hierarchical, except for additional influences, such as the effect of residence in an area, which may reinforce the dominance drive of one or both individuals.

The hierarchy had not been investigated in any centrarchid; in fact, for teleosts generally, published accounts of hierarchies are limited to partial summarizing statements by Noble (1939a, b), two abstracts by Noble and Borne (1938, 1940), and the investigation by Braddock (1945) of a viviparous poecilid. There is substantial disagreement between Noble and Braddock, although the species involved, *Xiphophorus helleri* and *Platypoecilus maculatus*, may be interbred in the laboratory. The most important discrepancy concerns the stability of hierarchies; Noble stated that rank orders could remain unchanged for several weeks, whereas Braddock's published results show the average duration of a hierarchy of four members to be little more than a day. Although both authors claimed that the organization of their species was of the same nature as the "peck-right" society of chickens, Noble's data have not been published in full and an examination of Braddock's (1945) tables leaves the decided impression that "peck-dominance" would better describe the shifting hierarchies of *Platypoecilus*.

The social hierarchy of the green sunfish is not so stable as that of chickens. Nevertheless, my findings agree with those of Noble and Borne; some hierarchies lasted without change for more than 3 weeks, and the majority of contact pair relations were stable (Table 5). The frequency of "reverse drives," i.e., the subordinate, in turn, driving its superior, was generally low; in ninety-nine of one hundred and fifty-six contact-pairs, representing twenty-six full hierarchies, not more than two return drives were recorded during the entire period of observation (Table 4). Such one-sided dominance can be called the relatively absolute, or "peck-right," type, but there were frequently instances among the lower ranks in which subordinates returned a large number of drives. This condition of instability may arise from partial or incipient territory, the "confusion" effect (p. 279), unsettled dominance early in group formation, interference by the despot in the settling of dominance between the subordinates, and other irregularities. "Peck-dominance" in fishes may straighten out in time to "peck-right," as seen in Experiment VI; but, even after a long period, pair-contacts may still be indecisive.

In all but two of the one hundred and seventy-two groups that I observed, there was at least one subordinate individual; thus, where territory was held, hierarchal relations also existed, and the fishes defending areas drove the subordinates but not one another. Even in the groups organized solely as hierarchies, the drive to hold territory could be detected in the form of the partial territories held by one or more of the subordinates. Such semi-territories were seen in seventeen of the twenty-seven hierarchy groups of Experiments I–V.

These facts indicate that the principles of hierarchy and territory are not sharply separate but interplay in a variety of ways to shape the form of sunfish organization. It is highly probable that hierarchal relations of some sort exist in every instance of territory, but there is reason to believe that certain species that

exhibit hierarchy are not at all territorial (Braddock, 1945). Whether territoriality is entirely absent even in these cases, or perhaps can be detected in modified form, needs to be further investigated.

A simple form of leadership can be discerned in groups of sunfish. It expresses itself in facilitation of various forms of behavior, especially feeding and the driving of subordinates. Leadership implies initiative and control of the pace of an activity, usually by one individual or alternately by several. A test for its presence in the sunfish was made by putting six groups through a learning situation to determine whether the most dominant individual in a group is also the leader. It was impracticable to secure the learning time for each individual; the six dominant fishes "led" in 47 per cent of the total trials (60), whereas the chance expectation is approximately one-half of this. Also a comparison of the average group performance with the learning shown by isolated sunfish (Fig. 3) indicates that there was a marked facilitation of learning in the groups.

Welty (1934) demonstrated that grouped goldfish are markedly superior to isolated ones in similar learning tests. It is possible to ascribe part of the group advantage to leadership, but it was observed also that isolated sunfish are more easily startled and become conditioned against elements of the test situation. Thus several isolated fishes gave erratic learning performances, while others did almost as well as the groups. The elimination of such variations may be part of the group benefit.

The tendency of the dominant fish to be the leader has far-reaching implications for further study. Settled hierarchies in chickens have greater survival value than groups kept constantly in reorganization (Guhl and Allee, 1944). If the dominant individual were also the leader, it could contribute to group survival; this would balance some of the destructive individual advantages of aggressive drive.

The outstanding group value of aggressiveness in the sunfish probably centers about the importance of territory-holding in egg-laying, and the defense of eggs and young by the males. Fishes strung out along the shore on individual nests constitute questionable groups, but from an ecological point of view they are part of an integrated breeding aggregation. Territoriality spaces the nests and may make for more effective breeding.

The role of hierarchy in breeding survival is less obvious. A clue is afforded by the observations centering about the omega individual in a territorial situation. When three members of a group defend areas, the fourth almost always is severely punished, being driven at an excessive rate and often killed. Removal or death of the omega was followed by a rise in tension and increased frequency of aggressive contact among the territory-holders. Introduction of a newcomer centered attention upon it and reduced other aspects of intra-group friction. Thus it appears that these subordinates have a definite function in integrating the group.

SUMMARY

Groups of four immature sunfish (*L. cyanellus*), observed for periods of 19–25 days in small bare aquaria, maintain hierarchies or territories or both. Hierarchies are fairly stable and become progressively more so with time. Males usually dominate females, and the larger individuals tend to dominate smaller ones.

Among these fishes, hierarchy represents distinct levels of aggressiveness, whereas territoriality arises from a balance between almost equally aggressive individuals. This balance may be struck at first encounter; but, when it occurs later within a hierarchy, territories are first established by the two highest-ranking individuals, then by the third, and, rarely, by the lowest. Seventeen of forty-four groups kept in bare aquaria developed territoriality. In twelve cases, three of the four fishes held territories, while the fourth was subordinate to all. In one group each member held territory; here few aggressive contacts were observed.

Subordinates frequently defended areas against all but the most dominant fish. This "partial" territory seems intermediate between hierarchy and territory.

Groups tested in a simple maze-learning situation were superior to isolated controls. Leadership through the maze shifted at almost every trial, but the dominant individual tended also to be the "leader."

Aquaria divided into compartments by means of wire-screen partitions with square gates favored formation of territories, in direct proportion to the complexity of the subdivision.

Subordinates appear to lessen tension among territory-holding fishes; removal of the omega from aquaria with three territories markedly increased aggressive activity among those remaining, whereas the introduction of a new fish led to their attacking it instead of one another.

LITERATURE CITED

ABBOTT, CHARLES C. 1894. A naturalist's rambles about home. New York: D. Appleton & Co.

ALLEE, W. C. 1931. Animal aggregations: a study in general sociology. Chicago: University of Chicago Press.

———. 1938. The social life of animals. New York: W.W. Norton Co.

———. 1942. Social dominance and subordination among vertebrates. Biol. Symp. 8:139-62.

———. 1945. Human conflict and co-operation: the biological background. In: Approaches to national unity: fifth symposium of the Conference on Science, Philosophy and Religion, ed. L. Bryson, L. Finkelstein, and R. M. MacIver, chap. xx, pp. 321-67. New York: Harper & Bros.

———. 1947. General sociology. Article in Encyclopaedia Britannica. (In press.)

ALTUM, B. 1868. Der Vogel und sein Leben. Munster: Niemann.

BRADDOCK, J. C. 1945. Some aspects of dominance-subordination in the fish *Platypoecilus maculatus*. Physiol. Zoöl., 18:176-95.

BREDER, C. M., JR. 1936. The reproductive habits of the North American sunfishes (family Centrarchidae). Zoölogica, 21:1-48.

BREDER, C. M., JR., and COATES, C. W. 1935. Sex recognition in the guppy, *Lebistes reticulatus* Peters. Zoölogica, 19:187-207.

COLLIAS, N. E. 1944. Aggressive behavior among vertebrate animals. Physiol. Zoöl., 17:83-123.

DIEBSCHLAG, E. 1941. Psychologische Beobachtungen über die Randordung bei der Haustaube. Zeitschr. f. Tierpsychol., 4:173-88.

FISCHEL, W. 1927. Beiträge zur Soziologie des Haushuhns. Biol. Zentralbl., 47:678-95.

GREENBERG, B., and NOBLE, G. K. 1942. Dominance, social order and territory in the lizard. Anat. Rec., 84:58 (abstr.).

———. 1944. Social behavior of the American chameleon (*Anolis carolinensis* Voigt). Physiol. Zoöl., 17:392-439.

GUHL, A. M., and ALLEE, W. C. 1944. Some measurable effects of social organization in flocks of hens. Physiol. Zoöl., 17:320-47.

HIXSON, G. A. 1946. The effect of numbers on the establishment of hierarchies and territoriality in the green sunfish, *Lepomis cyanellus*. Master's thesis in library of University of Chicago.

HOWARD, H. E. 1907-14. The British warblers, a history, with problems of their lives. 6 vols. London: R. H. Porter.

———. 1920. Territory in bird life. London: Murray.

HUBBS, C. L., and COOPER, G. P. 1935. Age and growth of the long-eared and the green sunfishes in Michigan. Papers, Michigan Acad. Sci. Arts and Letters, 20:669-96.

HUXLEY, J. S. 1938. The present standing of the theory of sexual selection. In: Evolution: essays on aspects of evolutionary biology, edited by G. R. De Beer. Oxford: Clarendon Press.

LANGLOIS, T. H. 1936. A study of the small-mouth bass, *Micropterus dolomieu* (Lacepede) in rearing ponds in Ohio. Ohio Biol. Surv., Bull. 33, 6: 191-225.

LYDELL, D. 1902. The habits and culture of the black bass. Trans. Amer. Fish. Soc., 31:45-73.

MASURE, R., and ALLEE, W. C. 1934a. The social order in flocks of the common chicken and the pigeon. Auk, 51:306-27.

———. 1934b. Flock organization of the shell parakeet, *Melopsittacus undulatus* Shaw. Ecology, 15:388-98.

MEYER-HOLZAPFEL, M. 1941. Das Territorium als Grundlage der sozialen Organization bei einer Gruppe von Schwanzbandsamlern (*Hemigrammus caudovittatus* E. Ahl). Rev. suisse zool., 48:531-36.

NICE, M. M. 1941. The role of territory in bird life. Amer. Mid. Nat., 26:441-87.

NOBLE, G. K. 1934. Sex recognition in the sunfish, *Eupomotis gibbosus* (Linne). Copeia, 151-55.

———. 1938. Sexual selection among fishes. Biol. Rev., 13:133-58.

———. 1939a. The role of dominance in the life of birds. Auk, 56:263-73.

———. 1939b. The experimental animal from the naturalist's point of view. Amer. Nat., 73: 113-26.

NOBLE, G. K., and BORNE, R. 1938. The social hierarchy in *Xiphophorus* and other fishes. Bull. Ecol. Soc. Amer., 19:14 (abstr.).

———. 1940. The effect of sex hormones on the social hierarchy of *Xiphophorus helleri*. Anat. Rec., Suppl., 78:147 (abstr.).

NOBLE, G. K., and CURTIS, B. 1939. The social behavior of the jewel fish, *Hemichromis bimaculatus*. Bull. Amer. Mus. Nat. Hist., 76:1-76.

NOBLE, G. K., and GREENBERG, B. 1939. Social behavior and sexual selection of the Florida chameleon. Bull. Ecol. Soc. Amer., 20:28 (abstr.).

REIGHARD, J. 1902. The breeding habits of certain fishes. Science, N.S., 15:380-81.

SCHJELDERUP-EBBE, T. 1922. Beiträge zur Sozialpsychologie des Haushuhns. Zeitschr. f. Psychol., 88:225-52.

———. 1935. Social behavior of birds. In: Murchison's Handbook of social psychology, chap. xx, pp. 947-73. Worcester, Mass.: Clark University Press.

SHOEMAKER, H. H. 1939. Effect of testosterone propionate on behavior of the female canary. Proc. Soc. Exper. Biol. and Med., 41:299-302.

WELTY, J. C. 1934. Experimental explorations into group behavior of fishes: a study of the influences of the group on individual behavior. Physiol. Zoöl., 7:85-128.

WUNDER, W. 1930. Experimentelle Untersuchungen am dreistachligen Stichling (*Gasterosteus aculeatus* L.) während der Laichzeit. Zeitschr. f. Morph. u. ökol. d. Tiere, 16:453-98.

13

Reprinted by permission from *Lizard Ecology: A Symposium*, W. W. Milstead, ed., University of Missouri Press, Columbia, Mo., 1967, pp. 106–115

The Adaptive Significance of Territoriality in Iguanid Lizards

A. STANLEY RAND

INTRODUCTION

TERRITORIAL behavior is one of those untidy subjects that lie on the boundary between two fields of study. What one says about territoriality depends largely on whether it is approached from an ethological or an ecological point of view. Dr. Carpenter has presented a paper that emphasizes the behavioral aspects of territoriality. I will try to balance this with a consideration of its ecological importance.

Territorial behavior has been described in a number of iguanid lizards (among others: Blair, 1960; Carpenter, 1962; Evans, 1951; Fitch, 1956; Greenberg and Noble, 1944; Greenberg, 1945; Milstead, 1961; Noble, 1934; Oliver, 1948; Schmidt, 1935; Stebbins, 1944; and Tinkle et al., 1962. Evans, 1961, provided an extensive bibliography). There has been little speculation about its adaptive significance in lizards compared with the amount that has been published on this topic in other groups of animals (C. R. Carpenter, 1958; Hinde, 1956; Nice, 1941; Noble, 1939; Tinbergen, 1939, 1957).

The usual approach to this sort of problem is the inductive one of collecting all the possible cases, grouping them into a pattern, and then drawing such generalizations as are appropriate. A major difficulty with this approach is that what is usually collected is not factual material but conclusions of various investigators. In this paper I shall try a deductive approach. I shall begin with an hypothesis and a list of predictions that must follow if it is true, and then compare these with my field observations on *Anolis lineatopus* and *Iguana iguana*. In framing the following operational definition I have used suggestions of Emlen (1957) and Willis (personal communication): a territory is an area or space within which a particular individual dominates certain categories of intruders who dominate it elsewhere. An individual dominates an intruder if it drives it away, excludes it, or supplants it at will. This constitutes successful defense of the territory. The behavior patterns by which a territorial holder dominates intruders I will call territorial defense. In discussing the function of territory it is important to exclude those factors, such as familiarity with an area, associated with home range or site attachment (Tinbergen, 1957). These are undoubtedly important, but would occur even if the area was not a territory.

BASIC ASSUMPTION

One basic assumption underlies the hypothesis I want to discuss: territoriality has evolved through natural selection and is maintained by it. The amount of energy spent in territorial behavior, the increased conspicuousness to predators of displaying and fighting lizards, and the possibility of injuries in fighting, all would provide strong selection against territorial behavior. Territorial behavior can persist only if there are still stronger selective pressures for it.

Selection for territorial behavior need not operate at all times and in all places. The following are among the cases where a selective advantage might be absent:

1. In small populations in unusual habitats or at the periphery of the range, where gene flow prevents precise adaptation to local conditions.

2. In recently occupied areas and in areas where conditions have changed so recently that counterselection has not had time to have an effect.

3. Selection need not operate continuously and it may be seasonal or even operate only in exceptional years.

These exceptions to the assumption that territorial behavior in lizards is of selective advantage probably are rare and for this paper can be ignored. I want now to present an hypothesis on what the possible selective advantages may be.

HYPOTHESIS

The hypothesis is: Successful defense of a territory gives a selective advantage to the individual by increasing the chances of: (1) its securing a necessary share of environmental resources, and/or (2) mating, and/or (3) survival of its offspring.

Few would argue that this is never so, and most of the suggested functions of territoriality in birds fall into these categories, but several authors have said that it need not always be true. Especially those who argue for the importance of territorial behavior in control of population density, such as Wynne-Edwards, would disagree. I will return to the relation of territories and population densities later.

QUALIFICATIONS

Before further discussion, the hypothesis must be qualified. Behavior, like morphology, is the result of a balance between selective forces operating in different directions and these set limits beyond which territorial behavior cannot evolve:

1. No individual can defend an area in such a way that mating is prevented (for example, a male cannot exclude all females from his vicinity during the mating season).

2. No defense can be elaborated to the point where survival of the offspring is endangered.

3. Defense can be carried only to the point where the advantages of winning, the probability of winning, and the cost of winning considered together outweigh the disadvantages of losing and the probability of losing. If there is a great deal to be gained by winning, an animal may attempt to defend its area even if the chances of winning are small and those of being injured great. Alternatively, it may not attempt to defend if the prize is small even though the chances of winning are good and the danger minimal.

PREDICTIONS

In order to test the hypothesis we must derive from it the most specific predictions possible and compare these with the observed data. These predictions can be arranged as the answers to five questions: Who defends? What is defended? Against whom? When? With what result? These questions must be answered independently for each of the three selective advantages postulated: share of environmental resources, mating, and survival of offspring.

1. Environmental resources

Only critical resources, those in absolute short supply (in the sense of Milstead, 1961), will be involved. If more than enough of any resource is available for all of the population, then there is no advantage in protecting a share of it. If the use of the resource by one individual does not interfere with its use by others, there will be no defense. For a lizard, the most important resource that fulfills these conditions is food. Others that may sometimes be important are certain kinds of hiding places, basking sites, etc.

In order to illustrate the predictions, I have compared them with my observations on the population of *Anolis lineatopus* that I watched for ten months near Kingston, Jamaica. The females and juveniles defend their territories as do the adult males.

Who defends?

Those individuals for whom the resource is potentially in short supply. Though I know of no case in lizards, it is possible, where there are closely knit social groups, for the defense to be done by certain members. It would be possible, if the food habits change radically, for one age group to take a superabundant food and not defend, while another depends on a less abundant food and defends.

Example: In *A. lineatopus,* all need food, and all defend.

What is defended?

An area containing the critical resource. It need not contain all the individual uses, nor need the individual have exclusive use of the resource within his area, but it must contain at least some important fraction, and the defending individual must have some priority on its use. It should be stressed here that the critical resource itself may not be defended. Consider what would happen if a hungry lizard tried to defend a grasshopper.

The larger a territory, the more difficult it is to defend, so the size of the defended area should be such that an increase in it would increase the cost in effort necessary to defend beyond the value of the additional resource gained. Thus, size of the defended area should vary with the amount and distribution of the critical resource, relative to the requirements of the individual. Since the amount of defense necessary presumably also increases when the number of lizards to be defended against is larger, one would expect a reciprocal relationship between territory size and population density. The upper limit of this relationship is an area containing all of the critical resource an individual could ever possibly use and the lower limit an area containing the absolute minimum.

Example: An *A. lineatopus* gets most of its food in the defended area and it does not defend areas where it does not feed. The relationships between population density and territory size are unknown, but I have the impression that territories are smallest where population densities are greatest. Generally, larger *lineatopus,* with presumably greater food requirements, have larger territories.

Against whom?

Competitors for the critical resource. The more similar the requirements of the individuals involved, the more vigorous the defense. Since members of the same species are usually the most serious competitors, one would expect them to be the objects of most defense, but when other species are competitors interspecific defense could occur. Those members of the same species that do not compete would not defend against one another.

Example: A. *lineatopus* females and juveniles defend against other anoles of about their own size, regardless of species (adult males are larger than any conspecific females). Within *lineatopus* there is a correlation between size of food and size of lizard so that individuals of the same size are in more direct competition for food than are those of different sizes.

When?

During the period when demands on the critical resource approach the supply; possibly also in the period immediately preceding this.

Example: A. *lineatopus* defend during the day when feeding and looking for food but not in the morning and evening when on the way to and from sleeping sites.

With what result?

The successfully defending individual should be able to supplant its competitors with respect to the critical resource itself or with respect to some part of the environment necessary to secure the resource. Where there is variation in the quality of the resource, the defending individual should be able to secure more than its share of the most desirable manifestations of the resource.

Example: Within their territory A. *lineatopus* supplant lizards of their own size on the perches from which they watch for food. I found that successfully defending individuals were not starving but, also, that they were always hungry enough to eat food I offered them. I would suppose that those individuals who had not established territories to be less well fed than those who had, and even when they were getting enough to eat, that they would likely be taking a larger proportion of less desirable prey (more distasteful perhaps) than the individuals with established territories. I have no evidence on either of these two points.

It is striking that in two well-studied herbivorous species, *Ctenosaura pectinata* (Evans, 1951) and *Amblyrhynchus cristatus* (Schmidt, 1935), the females rarely fight. The same seems to be true of the largely herbivorous *Dipsosaurus dorsalis* (Norris, 1953). At the other extreme, in one population of *Sceloporus merriami* that Milstead (1961) concluded to be limited by competition for food, the females were almost as pugnacious as the males.

2. Mating

The predictions discussed below as following from the second adaptive advantage are those applicable to the iguanid mating system, where polygamy and promiscuity are either usual or occasional.

Who defends?

Adult males. A male can increase the number of young it will father by copulating with several females, but a female will lay no more eggs if she copulates with more than one male. Thus only the adult males gain a mating advantage by defending unless the sex ratio is so unbalanced that there are more females to be fecundated than the available males can service.

Example: Adult *A. lineatopus* males defend, using more elaborate displays and morphological structures than either the females or the subadult males.

What is defended?

The area where pair formation occurs. If the pair bond is weak and short-lived, that is, exists only just before and during copulation, then the place where copulation occurs will be defended. If the pair bond is strong and exists for a longer period prior to copulation, then the area where the pair bond is formed is more important than where copulation itself occurs. Where females are sedentary where is an advantage to a male in holding as large a territory as possible and in having as many females living in it as possible.

Example: In *A. lineatopus* male territories are much larger than the territories of the females. The territory of a male includes the territories of one or more females, and both pair-bonding and copulation occur within it.

Against whom?

Competitors for mates: adult males of the same species. No interspecific defense would be expected. Even males of the same species might be ignored if they did not attempt to court or copulate. Those that did attempt to participate would be the most vigorously driven out or supplanted.

Example: *A. lineatopus* adult males are much more tolerant of adult males of other species than the juveniles and females of *lineatopus* are of juveniles and females of other species, although the *lineatopus* adult males are less tolerant intraspecifically than are the juveniles and females. Displaying males elicit more attacks than those that do not display.

When?

During the breeding season. One might expect that territorial defense would be associated with levels of reproductive hormones.

Example: A. *lineatopus* breed year-round and also defend year-round.

With what result?

A male will attempt to defend only when doing so increases his long-range chances of contributing to the gene pool. This may explain the natural formation of male hierarchies in certain lizards (Evans, 1951; Fitch, 1940). If the chances of defeating a larger male are small, while the chances of living longer than he and inheriting his area and harem are fair and are improved by not challenging him and losing, one would expect the smaller male not to attempt to drive out the larger but rather to avoid him and to concentrate on driving away other small males that would be competitors for the inheritance. A similar explanation has been suggested for why young males in certain birds seldom breed (Selander, 1965).

The hypothesis predicts that the males most successful at holding territories, particularly the more dominant ones, form pair bonds and copulate with more females and perhaps larger females.

Example: A. *lineatopus* males who do have females living within their territories are usually larger than those males who do not. These larger males are also the most successful at obtaining and defending territories. Of all the copulations seen, only one involved a male who was not the largest and most dominant in the area.

The published accounts of male iguanid antagonistic behavior accord quite well with the predictions. Of particular interest is the report of Evans (1951) on a colony of *Ctenosaura pectinata* in which the only courtship activities involved the dominant male, and the report of Blair (1960) in which it was concluded that yearling males of *Sceloporus olivaceous* mate much less frequently than larger males, presumably because of the antagonistic behavior of the latter.

3. Survival of offspring

The third adaptive advantage mentioned in the hypothesis is associated with parental care. Iguanids, like most lizards, have evolved little of the parental care so highly developed in birds, mammals, and some fishes, but choosing a nest site, constructing a nest, laying the eggs, and covering them is parental care of a sort. I might omit this category from discussion except that I have recently observed a case of defense of nest site in a group of colonial nesting *Iguana iguana* on Barro Colorado Island, Canal Zone.

Who defends?

Relatives; the closer the relationship, the more likely they are to defend. In lizards, only mothers defend, but in other animals, fathers and even siblings may defend.

What is defended?

An area around the offspring themselves (eggs or young), or around some environmental resource in short supply that is necessary to the offspring.

Against whom?

Competitors for the critical resource, or predators on the offspring.

When?

During the breeding season.

With what result?

The chances of survival of the offspring are improved.

Example: In the colony of nesting *Iguana,* females defended nest burrows while they were digging and filling them, driving off other females. A female that dug into the nest of another while making her own nest, kicked out the earlier eggs, which were then devoured by waiting black vultures. Inasmuch as the *Iguanas* showed a preference for digging in a hole that had already been started, a female able to keep others away until she had her nest filled and the ground leveled decreased the chances that her eggs would be dug up. The only other example that I know concerning iguanid females fighting over nesting sites is that just described by Dr. Carpenter in *Amblyrhynchus cristatus:* the females compete for nesting space on the sandy beaches.

Conclusions and Implications

To the question of why iguanid lizards defend territories, I have suggested that it is because of the selective advantage gained by a successfully defending individual, and I have discussed the hypothesis that successful defense of a territory gives a selective advantage to the individual by increasing the chances of (1) securing a necessary share of environmental resources, and/or (2) mating, and/or (3) survival of its offspring. The data I have examined on iguanid behavior are consistent with this hypothesis. Until more data are presented, it should not be rejected.

There is one final point, mentioned in the introduction but not yet discussed: regulation of density is frequently cited as *a* function, sometimes *the* function, of territoriality. It is sometimes implied that the behavior has evolved because of this function, that is, as the result of selection at the population level.

There is little question that territoriality usually limits the number of individuals that can live in the same area and therefore limits population density. This control may benefit the population as a whole. Certainly the trait can not persist if it is a severe handicap to the population. An effective means of population control may sometimes give a species an important edge in interspecific competition or enable it to survive times of environmental stress when it might otherwise become extinct.

In view of the large number of species that have become extinct through geologic time and in view of the evidence for interspecific competition among species living today, I will not deny the possibility of selection at the species level. In fact, I believe it to explain why territoriality is so widespread among animals. The hypothesis I have presented demands that territoriality can evolve and be maintained only when it is of selective advantage to the individual.

A species does not evolve territoriality because of the benefit it could receive from the control of its population density, but the species in which territoriality has evolved may be able to survive because of it.

In summary, territoriality is evolved and maintained in any one species because of the advantage gained by the successfully defending individuals but the reason it occurs in so many species is that it provides an effective mechanism of population control.

LITERATURE CITED

Blair, W. F. 1960. The rusty lizard. A population study. Univ. Texas Press, Austin. 185 p.

Carpenter, C. C. 1962. Patterns of behavior in two Oklahoma lizards. Am. Midland Naturalist 67: 132-151.

Carpenter, C. R. 1958. Territoriality: A review of concepts and problems, p. 224-250. *In* A. Roe and G. G. Simpson, [ed.], Behavior and evolution. Yale University Press, New Haven.

Emlen, J. T., Jr. 1957. Defended area?—A critique of the territory concept and of conventional thinking. Ibis 99: 352.

Evans, L. T. 1951. Field study of the social behavior of the black lizard, Ctenosaura pectinata. Am. Mus. Novitates 1493: 1-26.

Evans, L. T. 1961. Structure as related to behavior in the organization of populations in reptiles, p. 148-178. In W. F. Blair, [ed.], Vertebrate speciation. Univ. Texas Press, Austin.

Fitch, H. S. 1940. A field study of the growth and behavior of the fence lizard. Univ. California Publ. Zool. 44: 151-172.

Fitch, H. S. 1956. An ecological study of the collared lizard (Crotaphytus collaris). Univ. Kansas Mus. Nat. Hist. Publ. 8: 213-274.

Greenberg, B. 1945. Notes on the social behavior of the collared lizard. Copeia 1945: 225-230.

Greenberg, B., and G. K. Noble. 1944. Social behavior of the American chameleon (Anolis carolinensis Voigt). Physiol. Zool. 17(4): 392-439.

Hinde, R. A. 1956. The biological significance of the territories of birds. Ibis 98: 340-369.

Milstead, W. W. 1961. Competitive relations in lizard populations, p. 460-489. In W. F. Blair, [ed.], Vertebrate speciation. Univ. Texas Press, Austin.

Nice, M. M. 1941. The role of territory in bird life. Am. Midland Naturalist 26: 441-487.

Noble, G. K. 1934. Experimenting with the courtship of lizards. Nat. Hist. 34: 3-15.

Noble, G. K. 1939. The role of dominance in the social life of birds. Auk 56: 263-271.

Norris, K. S. 1953. The ecology of the desert iguana Dipsosauris dorsalis. Ecology 34: 265-287.

Oliver, J. A. 1948. The anoline lizards of Bimini, Bahamas. Am. Mus. Novitates 1383: 1-36.

Schmidt, K. P. 1935. Notes on the breeding behavior of lizards. Zool. Ser. Field Mus. Nat. Hist. 20: 71-76.

Selander, R. K. 1965. On mating systems and sexual selection. Am. Naturalist 99(906): 129-141.

Stebbins, R. C. 1944. Field notes on a lizard, the mountain swift, with special reference to territorial behavior. Ecology 25: 233-245.

Tinbergen, N. 1939. The behavior of the Snow Bunting in spring. Trans. Linn. Soc. New York 5: 1-94.

Tinbergen, N. 1957. The function of territory. Bird Study 4: 14-27.

Tinkle, D. W., D. McGregor, and S. Dana. 1962. Home range ecology of Uta stansburiana stejnegeri. Ecology 43: 223-229.

14

Reprinted from *Evolution*, 17:449–459 (Dec. 1963)

COMPETITION AND BLACKBIRD SOCIAL SYSTEMS

GORDON H. ORIANS[1] AND GERALD COLLIER[2]

Received November 1, 1962

The principle of competitive exclusion, variously known as Gause's Law, the Volterra–Gause Principle, or Grinnell's Axiom, has emerged as an important ecological generalization notwithstanding the reservations of some ecologists (Cole, 1960) and the active antagonism of others (Andrewartha and Birch, 1954). The evidence in support of competitive exclusion, derived from six main sources, strongly suggests that interspecific competition has had an important influence on the evolution of contemporary community structure despite the apparent relative abundance of resources with respect to the sizes of the populations utilizing them (Hutchinson, 1957). Unfortunately, most of the supporting evidence is indirect.

Closely related sympatric species have attracted the most attention, and whenever they have been carefully investigated, important ecological differences have been discovered (Lack, 1944, 1945, 1946; Carpenter, 1952; Diver, 1940; MacArthur, 1958; Gibb, 1954). However, since identical species are theoretically unlikely, differences are to be expected whether competition is manifest or not. This evidence would, therefore, be less convincing if it were not for evidence of the second type, namely, that closely related species may be more different morphologically and ecologically in areas of sympatry than in areas of allopatry (Lack, 1947, 1949; Vaurie, 1951; Brown and Wilson, 1956).

A third source of evidence is the structure of the incomplete communities of isolated mountains and remote islands, where those few species present nearly always occur in a wider range of habitats than in areas with a full faunal complement (Moreau, 1948; Lack and Southern, 1949; Crowell, 1962). Fourth, though the problems of defining the boundaries of arbitrary communities are formidable, Elton (1946) and Moreau (1948) found fewer closely related species occurring together than would be expected if species were randomly distributed. Fifth, studies of breeding bird censuses (MacArthur, 1957, 1960) have revealed a rather consistent logarithmic relationship between the relative abundances of species. MacArthur (1960) has shown that this distribution is to be expected if niches are contiguous and non-overlapping. No distributions are currently known that fit assumptions of either completely separated or completely overlapping niches. This implies that, for birds at least, food determines the abundance of all species since it is the only unsharable, completely utilizable resource (Hairston, 1959).

The sixth and last source of evidence is direct observation of interactions between species under natural conditions. Unfortunately, direct evidence of this sort is more difficult to obtain than indirect evidence, but in many bird species interactions can be observed readily, provided sufficient time is spent (e.g., Pitelka, 1951). Moreover, evidence from behavioral interactions is potentially of great theoretical interest since the utilization of some form of social organization to aid environmental exploitation and the development of complex behavior patterns to mediate population interactions are fundamental features of vertebrate evolution. The simple interaction of a population with its food supply as found in *Daphnia*, for example (Slobodkin, 1954), does not exist among birds and mammals. Moreover, it

[1] Museum of Vertebrate Zoology, Berkeley, California. Present address: Department of Zoology, University of Washington, Seattle.

[2] Department of Zoology, University of California at Los Angeles. Present address: Department of Zoology, San Diego State College.

is characteristic of many manifestations of social behavior, such as territoriality, that they usually result in each species depressing its own population more than the populations of other species, a vital condition of coexistence (MacArthur, 1958). In addition, the operation of a social system is consuming of both time and energy (Orians, 1961), since a territorial bird requires more food but has less time in which to obtain it than if it were not engaged in aggressive behavior.

The purpose of this paper is twofold. First, to present data on interactions between the Red-winged Blackbird (*Agelaius phoeniceus*) and the Tricolored Blackbird (*A. tricolor*) in California, and second, to develop from them a theoretical framework for thinking about the energetics of competition as an evolutionary factor in the establishment of sympatry. The data for this paper were gathered during intensive studies of social organization among the two blackbirds from 1958 to 1961. Orians worked in north-central and Collier in southern California, each in a wide variety of habitats, so that the observations reported here are probably representative of interactions between the species wherever they occur together. We shall refer to all aggressive encounters between the species as interactions, reserving until later a consideration of their competitive significance.

COMPARATIVE SOCIAL ORGANIZATION

The widespread Redwing is sympatric with the Tricolor throughout the narrow range of the latter in the lowlands of California and adjacent Oregon and Baja California. Morphologically they are very similar, but the Tricolor has narrower and more pointed wings and a more slender bill, and the males have prominent white borders to the middle wing coverts. Behaviorally, however, the species are strikingly different (Orians, 1961; Orians and Christman, MS; Collier, MS). In California male Redwings defend moderately large territories of 500–30,000 square feet, usually in marshes, within which several females establish subterritories and build nests. The males take no part in the construction of the nest or incubation and normally do not feed the nestlings until after they have fledged, if at all. The females gather food partly on the territory and partly on the adjacent dry land, the pattern of exploitation varying with the vegetation types involved.

In contrast, the nomadic Tricolor is the most highly colonial of North American passerines. Colonies of less than 50 nests are rare, and in favorable habitat, such as the Californian rice fields, they may be as large as 200,000 nests (Neff, 1937). Territories average about 35 square feet in area but are difficult to measure accurately in congested areas. One to three females construct nests within these small territories. The males do not assist in nest building or incubation and may leave the marsh while the females are on the eggs. However, once the eggs hatch, the males take an active role in the feeding of the young. No food is gathered on the small territories and the radius of exploitation around a large colony may extend as far as four miles and include more than 30 square miles of utilized land.

STUDY AREAS

Observations were concentrated at several localities in north-central and southern California (fig. 1). Extensive field work was carried out at Lake Sherwood, an artificial body of water (formed in 1905) in the Santa Monica Mountains located 25 miles east of Ventura, Ventura County, at an elevation of 955 feet. At its western end is a marsh about five acres in extent, divided into north–south halves by an east–west land bank. The northern half is occupied by bulrushes (*Scirpus*), the southern half by cattails (*Typha*). The quarter-acre Hidden Valley Marsh, 200 yards distant, is separated from the northwest corner of Lake Sherwood by a low ridge. Blackbirds nest in both areas. Another area of concentrated field work was the East Park Reservoir, located in the North Coast Ranges of Colusa County

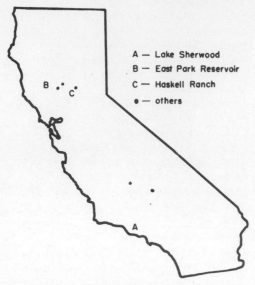

A — Lake Sherwood
B — East Park Reservoir
C — Haskell Ranch
● — others

FIG. 1.   Blackbird study areas.

at an elevation of 1,200 feet. It is also an artificial body of water whose level fluctuates markedly during the year, as the water is used for irrigation in the Sacramento Valley. It is surrounded by heavily grazed blue oak parkland–chaparral, and, at the south end, by wheat and alfalfa fields. The Haskell Ranch marsh, eight miles southeast of Marysville, Yuba County, is surrounded chiefly by irrigated pastures and owes its existence to runoff water from these pastures. Observations were also made in others areas in the

Sacramento and San Joaquin valleys and in southern California.

INFLUENCE OF ENVIRONMENTAL CONDITIONS ON BREEDING SUCCESS

If the effects of interactions between the species are to be understood, they must be separated from the influence of environmental fluctuations upon reproductive success. It has been shown (Orians, 1961) that the colonial system of the Tricolored Blackbird is more demanding of energy than the territorial system of the Redwing, primarily because of the enormous amount of energy expended in flying back and forth from the distant feeding areas when gathering food for the young. From this it follows that the Tricolor requires rich food supplies which can be obtained rapidly once the feeding ground is reached. Therefore, this species has an unpredictable breeding distribution and has poorer reproductive success than the Redwing in unfavorable years.

Of the years of this study, 1958 was very wet, with heavy rains continuing into April in all areas of the state. In contrast, 1959, 1960, and 1961 were drought years (table 1). Adequate spring rains did not occur on any of the study areas in 1959. No rain fell at Lake Sherwood between February 21 and April 25. In 1960 spring rains were well distributed in northern California, but the southern part of the

TABLE 1.   *Monthly precipitation and seasonal totals for study areas in north-central and southern California*

| | Precipitation in inches | | | | | | |
| | Lake Sherwood | | | | Sacramento | | |
Month	1957–1958	1958–1959	1959–1960	1960–1961	1957–1958	1958–1959	1959–1960
September	none	0.16	0.10	none	1.35	0.12	1.54
October	1.33	0.02	0.02	none	1.35	0.42	T
November	0.40	0.44	none	4.15	0.33	0.16	0.01
December	6.30	0.13	1.96	0.16	3.07	0.72	1.28
January	3.13	3.22	3.92	1.54	5.38	4.62	3.25
February	8.64	4.65	2.06	none	9.13	3.64	2.91
March	7.39	none	0.38	0.54	5.93	0.46	1.62
April	6.89	0.52	2.26	0.34	4.41	0.30	1.26
May	0.05	none	none	0.02	0.72	T	0.41
Seasonal total	34.13	9.14	10.70	6.75	32.37	10.44	12.28
Long-term average		19.04 inches				16.32 inches	

state remained abnormally dry. Nearly all of the spring rain at Lake Sherwood fell during a heavy storm April 27.

Response of Tricolors to drought is of three main types: failure to breed, desertion, and poor nesting success. At Lake Sherwood in 1960, the second year of the drought, about 400 Tricolors established territories on March 27 and engaged in territorial and reproductive behavior for two days. However, by March 30, all signs of breeding behavior had ceased, and two weeks later the marsh was completely deserted. Redwings, however, bred successfully in the marsh that year. In 1959, about 150 Tricolors moved into the Hidden Valley Marsh April 2, and 65 nests were started. However, it was evident by April 10 that the number of birds was sharply dropping and only 20 nests were completed. Eggs were laid in only 7, and young hatched in only 3 nests. A similar situation was observed at the East Park Reservoir in 1959. Several thousand birds were present when the colony started April 20, and complete clutches were laid in most nests, but in early May desertions began. On May 7, there were actively incubated eggs in only one-fourth of the nests, and by May 16, only 15 females were feeding young. Redwing breeding was successful that year, but fewer young were fledged per nest than in the wetter spring of 1960 (Orians, 1961). The season of 1960–1961 was the driest on record for Lake Sherwood. Grain crops withered before the plants exceeded a height of ten inches and Tricolored Blackbirds made no attempt to breed either at Lake Sherwood or other breeding areas in the Los Angeles region. Massive desertion of full clutches and poor survival of fledglings (also correlated with poor environmental conditions) characterized the two fall colonies in the Sacramento Valley in 1959 (Orians, 1960).

INTERACTIONS

Prenesting situation.—Since the Tricolored Blackbird is a nomadic species, the situation prior to the start of nesting varies considerably. In some localities both species roost in the nesting marsh months before breeding begins, while in other localities the Tricolors may appear suddenly in marshes from which they have been absent for months. Neither species consistently breeds earlier than the other, but since male Redwings establish and maintain territories up to several months prior to the commencement of nest building, Redwing territories are always occupied at the time the Tricolors start breeding.

Certain sites regularly used by colonies of Tricolors are avoided by Redwings during the period of territory establishment, but it is difficult to be certain that those areas are in fact suitable for Redwings. It has become apparent to us that the Redwing is primarily a species of the edges of marshes and that large expanses of unbroken cattails in California are normally occupied only on the periphery. Thus, the failure of Redwings to occupy the site of the Tricolor colony at the Capitol Outing Club in Colusa County (Orians, 1961), and the large bulrush section of the main marsh at Lake Sherwood, may have nothing to do with the presence of Tricolors.

However, there remains one case which is difficult to dismiss on these grounds. The Haskell Ranch, Yuba County, has been the site of a large Tricolor colony for at least eight years and probably much longer. The colony has always been located in the southwest half of the marsh though its extent varies considerably (figs. 2 and 3). Redwings do not establish territories in that portion of the marsh even prior to the arrival of the Tricolors, despite the fact that the marsh is nowhere broad in extent and is surrounded by similar irrigated pastures its entire length. It is difficult to know how an avoidance of Tricolor areas by Redwings prior to the arrival of the Tricolors could be selected for and maintained in the Redwing population. Even if adult Redwings with previous experience avoided these areas, birds from other areas and individuals breeding for the first time would presumably settle

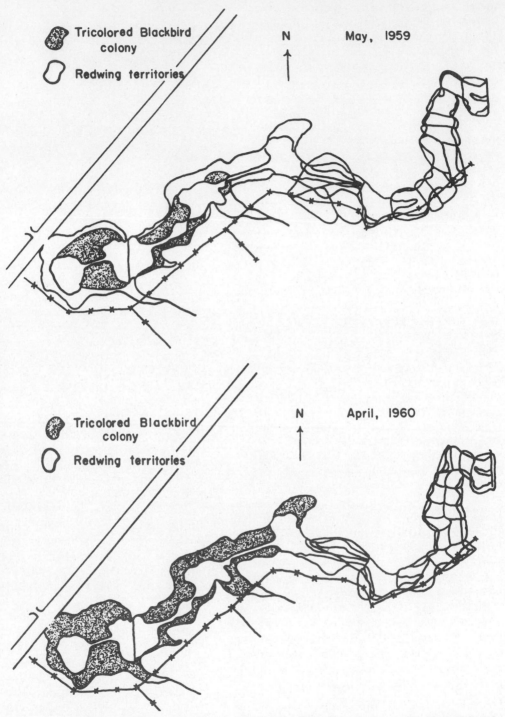

FIG. 2. Location of Redwing territories and Tricolored Blackbird colony at the Haskell Ranch, Yuba Co., California in 1959.

FIG. 3. Location of Redwing territories and Tricolored Blackbird colony at the Haskell Ranch, Yuba Co., California in 1960.

there, particularly since territories are known to be in short supply (Orians, 1961). Perhaps these areas are unsuitable for Redwings in some undetected manner.

Colony establishment period.—Interactions between Redwings and Tricolors are most common at the time colonies are formed. At the East Park Reservoir, in 1959, Redwing territories, established during February and March, occupied the entire marsh. A large flock of several thousand Tricolors roosted there nightly but they were completely absent during the day until the first week of April when they began to shift from site to site, as if prospecting for the best location for the colony, before leaving in the morning. These shifting flocks were continually attacked by the resident male Redwings, but without success, even though aggression was actually never countered or resisted.

Suddenly, on April 21, the Tricolors remained all day, the colony site was established, and by afternoon nest building was under way. Aggression by male Redwings whose territories were invaded was intense but ineffectual and gradually subsided. During the period of territory occupation by the Tricolors these male Redwings were still present on their territories, but they stopped singing and were no longer aggressive. When mass desertion of Tricolors took place during the first week of May, the male Redwings again resumed singing and became aggressive towards those few female Tricolors still incubating. However, the female Redwings had deserted these territories, leaving unfinished nests behind them, and no new nestings were attempted.

At Hidden Valley Marsh in 1959, there were 16 male Redwings defending territories. However, the intermittent presence of Tricolors disrupted the system and by the time the Tricolors began to breed, nine of the Redwings had deserted their territories leaving seven holding territories on the periphery of the marsh while the Tricolors occupied the center (fig. 4). The remaining peripheral male Redwings har-

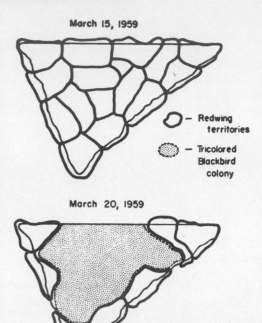

FIG. 4. Interactions between Redwing and Tricolored blackbirds at the Hidden Valley Marsh, Ventura Co., California in 1959.

assed incubating female Tricolors and birds of both sexes returning to feed young. Even though the Tricolors were largely unsuccessful, no additional Redwing territories were established later in the season.

When the Tricolor colony at Lake Sherwood was established in 1958, Redwings already had young in some of their nests. Dead and mutilated nestlings were found in several nests soon after the Tricolors took over the area. They may have been pecked by the Tricolors, but there is no direct evidence for this.

We have additional evidence from colonies less well studied that invasion of a marsh by Tricolors may result in nest desertion by Redwings. In early May, 1959, thousands of Tricolors began nesting in a marsh five miles southeast of Marysville, Yuba County. We had not visited the area prior to the establishment of the colony, but deserted Redwing nests with eggs were found in several parts of the colony. Territorial male Redwings remained only on the fringes of the colony and only a few broods of young were successfully reared, whereas nesting success in adjacent marshes, with no Tricolors, was normal. In 1960, the entire marsh was occupied by Redwings, and there was no large Tricolor colony. At Lake Sherwood in 1957, incubating female Tricolors were harassed by immature male Redwings when the latter came to roost in the evenings. Many fights were observed, but no cases of actual nest destruction were discovered.

Nesting period.—Normally, by the time the nestling Tricolors hatch, the period of interaction is over unless Tricolor colonies have become very small due to desertions and/or failures. By the time the Tricolor eggs hatched at the East Park Reservoir in 1959, only five females and no males remained in that part of the colony which had been most intensively studied. Females returning to the nests with food were vigorously attacked by the male Redwings and, while this did not completely prevent their feeding the young, the delay was at times considerable. Furthermore, whenever one of the Tricolors attempted to gather food at the edge of the marsh, she was chased by the male Redwing in whose territory she foraged. This completely prevented the Tricolors from foraging adjacent to the marsh in the areas in which female Redwings gathered food. The probable result was a reduction in the rate of food delivery to the nestlings.

During the afternoon of May 16, one of the Tricolor nests was destroyed. A male Redwing was observed behaving strangely near the nest site, disappearing into the cattails near it while uttering the guttural notes normally given when attacking other birds. When visited a few minutes later, the nest was empty and the nestlings could not be found in the dense vegetation. Whether the Redwing actually tossed them out or whether he had been attracted by some predator, such as a snake, could not be determined. At Lake Sherwood in 1959,

when the Tricolor colony was reduced to a few active nests, the females also had difficulty in returning to incubate and feed the young because of Redwing aggression.

The almost constant aggressive interactions between Redwings and Tricolors in the field wherever colonies of the latter are bounded by territories of the former thus have conspicuous adverse effects upon the reproductive success of one or the other of the species. Which species suffers depends upon the particular situation, but the behavior of the two species is remarkably and consistently different. The Redwing exhibits strong territorial aggression towards Tricolors, but field observations clearly indicate that mistaken identification is not involved. These attacks are unsuccessful not because of their low intensity but because of the numbers of Tricolors involved. Individual Redwings are always dominant to individual Tricolors, but attacks by single Redwings on flocks of Tricolors are ineffectual. When repeated attacks on the Tricolors fail, the male Redwings respond either by continuing to occupy the territory but ceasing to display there, or by completely deserting the area. Females also desert and no successful Redwing nest has been found by us in a Tricolor colony except at its very edge. On the other hand, it is possible that Redwings may actually destroy Tricolor nests if the number of Tricolors is greatly reduced. In this situation attacks upon those few remaining Tricolors are successful.

In contrast, the Tricolors normally exhibit no signs of interspecific aggression; the possible case of nest destruction at Lake Sherwood in 1958 is perhaps an exception. Rather, the birds merely fly a few feet and land again, only to be chased again by the male Redwing. But while some birds are being evicted, others move in to take their place so that despite lack of overt resistance by individuals, continued occupation of the desired site by the population is maintained if there are enough birds.

DISCUSSION

Since the interactions between the two species of blackbirds have such striking effects upon their reproductive success, it is to be expected that the existence of such interactions through time has had an influence upon the evolution of their social behavior. The Tricolored Blackbird probably evolved in California at a time when the pancontinental Redwing population was split into at least two groups, perhaps by Pleistocene glaciation, but possibly earlier. There has been a long period when both species have coexisted in California, during which time the Redwing has developed several well-marked races, including those in which there has been character displacement in the male plumage from that of the Tricolor. On the other hand, there is little to suggest that the major features of social organization of either species have been substantially altered by interspecific contacts. The colonial system of the Tricolored Blackbird appears to be well adapted to the alternating flood and drought conditions prevailing in the lowlands of California and probably was established in approximately its present form when the Redwing reinvaded the area (Orians, 1961). It is possible, however, in view of the success of Redwing aggression when the number of Tricolors is small, that continued interspecific contacts have served to reinforce the advantages inherent in close synchrony in Tricolor colonies. In general, however, data support the conclusion that the different social systems have evolved in response to ecological conditions independently of interspecific contacts (Orians, 1961).

Since aggressive behavior must entail disadvantages as well as advantages, the level of aggression which evolves must be a compromise between their relative importance. The expenditure of excessive amounts of time and energy would result in increased mortality and neglect of offspring, which would be uncompensated if all necessary environmental requisites could be provided by a lesser time and energy expenditure. On the other hand, an insufficient expenditure of time and energy could seriously impair the reproductive output if, as a result, some requisites were in short supply. The particular spacing system which evolves will therefore depend upon many variables, among them the ecology of the species, its previous history, and the nature of the environment which it exploits.

Whereas territorial aggression is usually restricted to conspecific individuals, species which are ecologically very different may defend mutually exclusive territories when they come into contact. However, natural selection should quickly act to eliminate interspecific aggression *no matter what the degree of divergence in species-specific characters*, because a mutation causing its possessor to ignore individuals of the other species would immediately save a great deal of time and energy without concurrent disadvantages. Conversely, if ecological divergence has not been sufficient for the species to be truly compatible, selection should favor the continuance of interspecific aggression, again no matter what the degree of differences in species-specific characters. Interspecific aggression is by no means restricted to those species which are morphologically similar, nor are similar species necessarily interspecifically territorial. A review of interspecific territoriality in birds as related to the age of the species groups, structural diversity of vegetation, and methods of exploitation will be given elsewhere (Orians and Willson, in prep.).

On this view, certain problems currently considered in the literature on speciation and isolating mechanisms can be given at least partial resolution. For example, ornithologists have questioned whether species coexist because of plumage differences evolved after contact or whether these plumage differences, evolved during their allopatric evolutionary histories, permit the sympatry (Hamilton, 1962). We suggest that the problem of sympatry is basically independent of the degree of divergence in species-specific plumage patterns although the possibility of further divergence in such plumage patterns following the establishment of sympatry is not thereby discounted. In fact, the character displacement in male Redwing plumage is attributable to this cause. Ecological compatibility, the prime requirement for sympatry, is relatively independent of species-specific plumage patterns, but rather is strongly influenced by such behavioral attributes as feeding behavior which vary remarkably in morphologically similar species (Hartley, 1953; MacArthur, 1958).

Similarly we suggest that the strong segregation of congeneric species into different habitats where they are sympatric is primarily related to the inability of the species to achieve ecological sympatry rather than to the problems of hybridization and ethological isolation. To deny access to suitable habitats for one's self and offspring to achieve ethological isolation is highly disadvantageous and should at best be only temporary. Failure to achieve sympatry may result from swamping of peripheral adaptations by gene flow from the center of range of the species (Hamilton, 1962; Mayr, 1954; Snow, 1954), or because the environment is structurally too simple to afford opportunities for alternate patterns of exploitation.

However, divergence of ecological as well as species-specific plumage characters is to be expected following the establishment of sympatry provided peripheral adaptations are not completely swamped. In the zone of sympatry there will be strong selective advantage to any individuals which are ecologically more divergent from their congener because of their lesser competitive impact. By this means divergence in excess of that required to make interspecific aggression energetically disadvantageous is possible and likely.

Behavioral patterns are important in species interactions because they may decrease the incidence of misidentification and may result in different foraging techniques and hence important ecological differences. In the case of the Redwing and Tricolored blackbirds the impact of behavior is not primarily upon foraging technique, which appears to be identical in the two species, but upon the places in which the individuals are foraging. Whereas Redwings are primarily restricted to feeding areas within 300 yards of their nests, Tricolors regularly forage as far as four miles from the nesting colony.

Evolution of interspecific aggression in blackbirds.—If we postulate that the essential differences between Redwings and Tricolors were in existence at the time contact between the species was established, it follows that territorial aggression was initially low in Tricolors. In view of the impact of interspecific aggression it might be argued that selection should have favored the evolution of increased aggression towards Redwings. However, it has been noted that aggression by Redwings is successful only when Tricolor colonies have been seriously reduced for other reasons, at which time reproductive success of the remaining individuals may be low in any case. Moreover, under normal circumstances non-aggressive resistance by Tricolors is completely effective because of the overwhelming influence of numbers. A similar relationship between Noddy and Sooty terns on the Seychelles Islands has been reported by Vesey-Fitzgerald (1941), but his observations were fragmentary. A mathematical model for such relationships has been developed (Hutchinson, 1947) but good substantiating field evidence has not previously been available.

The Redwing, on the other hand, because of its strongly territorial system, would initially have been aggressive to Tricolors as well as other members of its own species. Because of the devastating impact of a large Tricolor colony upon the reproductive activities of the Redwings whose territories they invade, selection should favor the continuance of this interspecific aggression. If the size and location of Tricolor colonies were more predictable from year to year, it would be expected that avoidance on the part of Redwings of traditional sites of Tricolor colonies would evolve. Since this is apparently not possible, Redwings continue to occupy all suitable areas and remain aggressive to Tricolors attempting to take over their territories, despite the low degree of success of such aggression.

Since most objections to competitive exclusion as a factor of importance in nature are based upon the indirect nature of most of the supporting evidence, it is vital that direct evidence bearing upon the problem be gathered whenever possible. If by careful observation competitive interactions cannot be demonstrated, theories relegating the importance of competition to the past could rightly be viewed with greater skepticism. But we suggest that such evidence will be easily obtained in many bird groups. Moreover, if competition has been important in determining the structure of natural communities, it should be demonstrable today. In all sexually reproducing species enormous amounts of new variability are produced each generation, the extremes being continually selected against. Moreover, in nature many bird species enter the breeding season with larger populations than the usual habitats can support. Given these facts, it should be impossible to completely avoid competition by specializations to habitat, and continued interactions are to be expected among closely related species wherever they come into contact.

SUMMARY

Evidence for competitive exclusion, largely indirect, is supplemented by observations of interactions between the closely related Redwing and Tricolored blackbirds which differ strikingly in their social organization and, where they are sympatric in California, interact strongly. Male Redwings establish territories early in the winter in California so that marshes are usually fully occupied at the time the nomadic Tricolors start to establish their breeding colonies. Nesting success in both species is influenced by amount and distribution of rainfall during the winter and spring, but the Tricolor, because of the greater demands of its system, suffers more acutely in drought years.

When large numbers of Tricolors move into a marsh inhabited by Redwings, there is strong aggression on the part of the male Redwings, but through superior numbers, Tricolors are successful without offering any counteraggression. Redwings either desert their territories, or if they remain on them, cease defending them. No successful Redwing nests have been found within large Tricolor colonies. Redwing aggression is successful against Tricolors only when the numbers of the latter are small, thus providing evidence in support of Hutchinson's model.

The evolutionary consequences of aggressive interactions are considered in the light of the probable action of natural selection upon time and energy expenditure. Levels of aggression are assumed to evolve as a compromise between the pattern of spacing which provides maximum reproductive success through the securing of the nest sites, mates, and food supplies, and the increasingly large expenditure of time and energy needed to maintain the system. Interspecific territoriality is to be expected when two species come into contact, but if they are ecologically compatible, selection should eliminate interspecific aggression. Conversely, in the absence of sufficient ecological divergence, continuance of interspecific aggression is likely. The Redwing and Tricolored blackbird have diverged primarily through their social systems which permit different spatial utilization of environmental resources in heterogeneous environments. Observations of direct interactions are of particular importance since most of the evidence for competitive exclusion is indirect.

ACKNOWLEDGMENTS

The work of Orians in north-central California was supported by a National

Science Foundation Graduate Fellowship for the academic years 1958–1960. Travel expenses were in part defrayed by a grant from The Museum of Vertebrate Zoology. Collier was supported by a Frank M. Chapman Memorial Grant from the American Museum of Natural History during 1957 and by a National Institute of Health Predoctoral Fellowship (E-9555) for the academic years 1959–1961. Costs of travel were borne by the Department of Zoology, University of California, Los Angeles. We wish to thank Frank A. Pitelka and Thomas R. Howell under whose guidance our work was carried out and Nicholas E. Collias and Alan J. Kohn who offered valuable suggestions during the preparation of the manuscript.

Literature Cited

Andrewartha, H. G., and L. C. Birch. 1954. The distribution and abundance of animals. Univ. Chicago Press, Chicago.

Brown, W. J., Jr., and E. O. Wilson. 1956. Character displacement. Systematic Zool., 5: 49–64.

Carpenter, C. C. 1952. Comparative ecology of the common garter snake (*Thamnophis s. sirtalis*), the ribbon snake (*Thamnophis s. sauritus*), and Butler's garter snake (*Thamnophis butleri*) in mixed populations. Ecol. Monog., 22: 235–258.

Cole, L. C. 1960. Competitive exclusion. Science, 132: 348–349.

Crowell, K. L. 1962. Reduced interspecific competition among birds of Bermuda. Ecology, 43: 75–88.

Diver, C. 1940. The problem of closely related species living in the same area. *In* Huxley, J. (ed.), The new systematics. The Clarendon Press, Oxford.

Elton, C. 1946. Competition and the structure of ecological communities. J. Anim. Ecol., 15: 54–68.

Hamilton, T. H. 1962. Species relationships and adaptations for sympatry in the avian genus Vireo. Condor, 64: 40–68.

Hartley, P. H. T. 1953. An ecological study of the feeding habits of the English titmice. J. Anim. Ecol., 22: 261–288.

Hutchinson, G. E. 1947. A note on the theory of competition between two social species. Ecology, 28: 319–321.

——. 1957. Concluding remarks. Cold Spring Harbor Symposia on Quantitative Biology, 22: 415–427.

—— and R. H. MacArthur. 1959. On the theoretical significance of aggressive neglect in interspecific competition. Amer. Nat., 93: 133–134.

Lack, D. 1944. Ecological aspects of species formation in passerine birds. Ibis, 86: 260–286.

——. 1945. The ecology of closely related species with special reference to Cormorant (*Phalacrocorax carbo*) and Shag. (*P. aristotelis*). J. Anim. Ecol., 14: 12–16.

——. 1946. Competition for food by birds of prey. J. Anim. Ecol., 15: 123–139.

——. 1947. Darwin's Finches. Cambridge Univ. Press, Cambridge.

——. 1949. The significance of ecological isolation. *In* Genetics, paleontology and evolution: 299–308. Princeton Univ. Press.

—— and H. N. Southern. 1949. Birds on Tenerife. Ibis, 91: 607–626.

MacArthur, R. H. 1957. On the relative abundance of bird species. Proc. Nat. Acad. Sci., 43: 293–295.

——. 1958. Population ecology of some warblers of northeastern coniferous forests. Ecology, 39: 599–619.

——. 1960. On the relative abundance of species. Amer. Nat., 94: 26–35.

Mayr, E. 1954. Change of genetic environment and evolution. *In* Huxley, J. (ed.), Evolution as a process.

Moreau, R. E. 1948. Ecological isolation in a rich tropical avifauna. J. Anim. Ecol., 17: 113–126.

Neff, J. 1937. Nesting distribution of the Tricolored Redwing. Condor, 39: 61–81.

Orians, G. H. 1960. Autumnal breeding in the Tricolored Blackbird. Auk, 77: 379–398.

——. 1961. The ecology of blackbird (*Agelaius*) social systems. Ecol. Monog., 31: 285–312.

Pitelka, F. A. 1951. Ecologic overlap and interspecific strife in breeding populations of Anna and Allen hummingbirds. Ecology, 32: 641–661.

Slobodkin, L. B. 1954. Population dynamics in *Daphnia obtusa* Kurz. Ecol. Monog., 24: 69–88.

Snow, D. W. 1954. The habitats of Eurasian tits (*Parus* spp.). Ibis, 96: 565–585.

Vaurie, C. 1951. Adaptive differences between two sympatric species of nuthatches (*Sitta*). Proc. X Int. Orn. Congr. 1950: 163–166.

Vesey-Fitzgerald, D. F. 1941. Further contributions to the ornithology of the Seychelles Islands. Ibis, 5 (14): 518–531.

15

Reprinted from *J. Mamm.*, 48(2):189–190, 192, 194, 196–209 (1967)

THE COMPARATIVE BEHAVIOR OF GRANT'S AND THOMSON'S GAZELLES

RICHARD D. ESTES

[*Editor's Note:* Figures have been omitted because of space limitations.]

ABSTRACT.—The comparative behavior of two associated gazelles, *Gazella granti* and *G. thomsonii* was observed in Ngorongoro Crater, Tanzania during a 2½-year antelope behavior study. Though superficially alike, the two species actually differ so much in details that morphological characters alone are probably sufficient to prevent interspecific confusion. Ecologically, the gazelles have quite different requirements: *thomsonii*, a grazer, prefers short-grass steppe and requires water; *granti* is mostly a browser and water independent. They share the same six predators in Ngorongoro, of which two jackals are considered the most important, purely on the basis of fawn predation. The same elaborate system for concealing newborn fawns is found in both gazelles and mothers sometimes cooperate in pairs to drive away pair-hunting jackals. Adult *thomsonii* are preferred prey of East African wild dogs, and territorial male gazelles may be most vulnerable to their hunting methods. Behavioral interaction between associated ungulates is discussed and found to extend no further than mutual response to warning signals, probably even in the gazelles. The gazelle warning signals are practically identical, featuring snorting, stamping, twitching of the side stripe, and most important, stiff-legged bounding (stotting). Gazelles and many other gregarious, territorial antelopes have the same basic social system. Differences between the gazelles may be largely based on their habitat preferences; for example, *granti* males defend much larger territories than do *thomsonii* males. Grant's nursery herds are correspondingly widely spaced and also smaller, on the average. The two seem to differ most in territorial behavior, including fighting styles. Male *thomsonii* are more vigorously territorial, and besides scent-marking by urination-defecation like *granti*, mark extensively by means of a secretion from preorbital glands, which are less developed in *granti*. Agonistic encounters between *granti* males are usually settled by means of a neck intimidation display, whereas conflicts between *thomsonii* males routinely end in fighting. Neck development is very important in the *granti* fighting style, where males lock horns and push, attempting to twist one another out of position—this is the basis for the intimidation display. In *thomsonii* natural selection has operated on horn configuration and fighting style to produce a relatively safe type of parry-thrust combat, thus obviating the need for a display substitute. Compared to territorial behavior, epigamic behavior in the gazelles appears remarkably similar. However, subtle-looking differences in emphasis, sequences, and combinations of the basic displays turn out to be possibly as divergent as the epigamic displays that serve to sexually isolate some other genera. But in the absence of any demonstrated tendency for these gazelles to interbreed, morphological differences alone may be considered adequate species-specific isolating mechanisms. In this respect, olfactory differences may well be more important than visual characters: *thomsonii* have and *granti* lack strong-smelling inguinal glands.

As congeneric, sympatric species frequently found associated in East African steppe-savanna, *Gazella granti* Brooke and *Gazella thomsonii* Günther are two of the most interesting of the numerous plains antelopes for comparative study. This article is intended to point out some of the striking similarities and equally striking differences in their morphology, ecology, and behavior.

DESCRIPTION OF STUDY AREA AND METHODS

The study was made in Ngorongoro Crater, Tanzania, a caldera enclosing an area of approximately 120 sq miles which is predominantly open grassland lying at an elevation of 5800–7000 ft. With comparatively abundant rainfall and fertile volcanic soil, the Crater carries a resident population of about 25,000 herbivores, including 15 ungulate species. The dominant five and their approximate numbers are: wildebeest (*Connochaetes taurinus*), 14,000; Burchell's zebra (*Equus burchelli*), 5000; Thomson's gazelle, 3500; Grant's gazelle, 1500; and eland (*Taurotragus oryx*), 350.

From January 1963 to July 1965 I lived in the Crater and studied the wildlife. The behavior and life history of the wildebeest was the principal subject of the study (Estes, 1966); the gazelles were studied for comparison with wildebeest and one another, with emphasis on behavior. Most observations were made with binoculars from a vehicle, by means of which all species except eland could be watched undisturbed from within 100 yards. Seven adult male and three adult female *granti* were marked by a former Game Warden, D. Orr, the Ngorongoro Game Biologist, J. Goddard, and myself. They were immobilized with succinylcholine chloride delivered in a syringe projectile by a CO_2 powered Cap-Chur gun, branded, and ear-tagged. Five yielded useful information on territory and home range, respectively. Marking attempts were discontinued when it was found that a 2cc syringe hit with sufficient force to break the legs even of adult males.

RESULTS

Recognition Characters

Grant's and Thomson's gazelles are superficially so alike that newcomers almost invariably confuse them, although when they are seen side by side some differences are obvious. First of all, *granti* is a much bigger animal, adult males weighing 150–180 pounds compared to 45–60 pounds for *thomsonii* males. Overall coloration too is usually dissimilar, the fawn-colored *granti* looking faded beside *thomsonii* with its rich cinnamon coat, and the markings may be seen to differ in detail. Even postures and movements are clearly unlike: *granti* carries its head more erect both when walking and running and appears generally more stately; movements of *thomsonii* look abrupt and nervous by comparison. But when they are seen apart such differences are hard to judge. Coloration and markings, moreover, are subject to considerable geographic and individual variation (Brooks, 1961: 94). Yet certain differences in their markings are diagnostic and, once mastered, assure positive identification under all conditions.

The most reliable and easily seen difference for human observers is in the rump patch. While both species have white underparts and rump, the disc in *granti* is larger and more conspicuous (Fig. 1, A and B). Set off by a dark pygal stripe on either leg (often faint or lacking in Ngorongoro *thomsonii*), it continues over the tail rostrum, from either side of which two white arrows project forward onto the hips; these give the disc a pronounced T-shape when viewed from behind. This difference alone is sufficient to distinguish the two species, including females and young, where the chances of confusion are greatest.

Other subtler differences in posterior markings may be equally important to the animals. In *thomsonii* the part of the tail overhanging the rump patch is solidly brown or black and comparatively full (Fig. 3, A). In *granti* the tail is thin, the dorsal surface white with a black tip, the ventral surface black and hairless (Fig. 1, A). Thomson's gazelles are famous for continually wagging their tails; the function has not been determined, but may be a social facilitation signal comparable to tail-flipping in some gregarious passerines (Andrew, 1956). Tail-wagging in *granti* is less pronounced, if not less constant, due to the minimum contrast between tail and rump disc. But when the tail is held out, it becomes conspicuous. Furthermore, the raising of the tail reveals the contrasting perineum, black and hairless, of the adult female *granti* (Fig. 2, G); in fawns of both species and usually also in adult female *thomsonii*, the comparatively long white hair of the disc conceals the perineum (Fig. 1, D). Thus when urinating (Fig. 2, G), and at the stage of courtship when the females raise their tails, two strikingly different designs are presented—a black line over a white disc in *thomsonii*, a white line over a white disc bisected by the black perineum in *granti* (the difference in pygal stripes may or may not hold for other sympatric populations). These markings could well have specific releasing value for epigamic behavior.

The black side stripe so characteristic of gazelles is normally very conspicuous in all adult *thomsonii* but is generally lacking in adult male *granti*, and may be fully, partly, or not at all developed in female *granti*. The last is commonly the case in Ngorongoro. The idea that character displacement may be operative in sympatric populations is supported by the fact that both sexes of *Gazella granti notata* Thomas, an allopatric population found in western mid-Kenya, possess the side stripe (Haltenorth, 1963). On the other hand, female *granti* may generally retain the mark in other sympatric populations, for instance on the Athi Plains near Nairobi, Kenya. While the actual causation awaits investigation, it would not be surprising if character displacement caused the side stripe to disappear at an accelerated pace in Ngorongoro, where the two species are permanently associated at relatively high densities and are perhaps largely isolated from other populations.

When horn development is compared, it is female *thomsonii* which seem to be in the process of losing this character. The horns are thin, no more than 4 to 5 inches long, commonly malformed, and not infrequently absent altogether (Fig. 3, F and H). Of 89 heads collected by Brooks (1961), only 24% had evenly paired, normal horns. The horns of *granti* females are usually symmetrical, if thin, and up to 1 ft long (Fig. 2, G). The males of both species have well-developed horns, black and heavily ringed almost to the tips, but very different in shape (Fig. 1, A and B). The horns of *granti* are proportionally very large and lyre-shaped, though subject to great variation. Horns of *thomsonii* are straighter, closer set, and less variable. The significance of these differences is dicussed in the section on fighting.

Facial markings in the two species, even more superficially alike than body markings, are again seen to differ when closely examined. In *thomsonii* a bold dark cheek stripe tapers from the eye to the mouth (Fig. 3, E); this stripe is thin and often incomplete in *granti* (Fig. 1, A). Conversely, the latter has a black patch over each eye which is absent in *thomsonii*. In both species, the forehead and snout are usually darker than the general body color; but the forehead may be quite grizzled in *thomsonii* (Fig. 1, B), and the bordering white lines tend to be less distinct. Finally, a dark nose chevron distinguishes even newborn *granti* fawns (Fig. 1, C); this mark is either lacking or amorphous in *thomsonii*.

When the two species are compared in detail, considerable morphological differences are evident. Indeed, it would be hard to find two *Gazella* that are more unalike. Obscure as the differences may appear to the casual observer, the different markings, posture, and locomotion alone may well provide sufficient isolating mechanisms insofar as the animals are concerned.

Ecological Separation

Compared with most common African ungulates, *thomsonii* is highly restricted both in range and in habitat preferences, occurring only in the steppe-savanna from central Kenya to northern Tanzania (but the dubiously distinct *G. albonotata* and *G. pelzelnii* occur in southern Sudan and Somalia, respectively—Brooks, 1961: 112). As Brooks noted, the preferred habitat provides short, preferably green grass and dry ground in the vicinity of water. When on green pasture, however, it can go without drinking—I have never seen one drink in Ngorongoro. Both Brooks (1961) and Talbot (1962) observed that *thomsonii* is predominantly a grazer, with legumes, herbs, and shrubs composing not more than 10–20% of the diet. Grant's gazelles, on the other hand, are mixed feeders (40% grass, 60% browse, according to an analysis of 10 stomachs by Talbot, 1962), are probably entirely independent of surface water, and are adapted to habitats ranging from tall grassland and light bush to the desert regions of southern Sudan.

Considering how restricted is *thomsonii* in range and habitat, it is noteworthy that it is often the most numerous herbivore within its centers of distribution. In an aerial count of western Masailand, for instance, Stewart and Talbot (1964) estimated the gazelle population at 480,000–800,000, compared to 239,500 wildebeest, the next most abundant species. Grant's gazelles were not counted separately, but reportedly made up a very small percentage of this total, probably less than 60,000.

The status of Thomson's gazelle in Ngorongoro Crater, where it ranks third behind wildebeest and zebra, is thus atypical. The estimate of 1500 *granti* and 3500 *thomsonii* is the average of two gazelle ground counts carried out in October 1964 and May 1965. While *thomsonii* outnumbers *granti* by more than two to one, in fact there is apparently a greater biomass of *granti* in the Crater.

Ngorongoro is seemingly marginal habitat for *thomsonii*, at least during the rains. From November to May, the main population was concentrated in an area of only 16–25 sq miles, on alkaline soils bordering the Soda Lake and the western sector of the Crater, where the grasses are naturally short. None was known to have migrated from the Crater. The rest of the Crater floor, which has a total area of approximately 104 sq miles, is dominated by Bermuda grass (*Cynodon dactylon*), Rhodes grass (*Chloris gayana*), and bluestem grass (*Andropogon greenwayi*) that reach heights of 1 to 2 ft during the rains. Only a few isolated *thomsonii* males were to be found there at this time. During the dry season, when the grass was reduced to a height of less than 1 ft, the species extended its range to an area of 55 sq miles or more.

Grant's gazelles, by contrast, range an area of about 112 sq miles throughout the year, which includes more or less the entire area that is utilized by plains game. They are particularly abundant on the hillsides, where red-oat grass (*Themeda triandra*) predominates and is never short enough to suit *thomsonii* except when burnt. In the Ngorongoro Crater *granti* appears to browse much of the time, selecting legumes and low, thorny shrubs such as *Indigofera* spp., and it generally favors areas where such browse is most abundant. As Talbot pointed out (1962:136): "In spite of the availability of other food species, the principal species of grass and other foods chosen by the Grant's gazelle are those which invade or become dominant in overgrazed, abused grasslands. . . ." In short, what appears to be poor pasture for cattle, for wildebeest or other pure grazers, may be perfect for Grant's gazelle and such browsers as the black rhinoceros (*Diceros bicornis*), and thus an essential part of an environment that supports a broad spectrum of herbivores.

Interspecific Associations

Mixed gazelle herds are commonplace. For that matter, nothing is more characteristic of African wildlife than interspecific herbivore associations. While ecological interactions between members of such associations have been demonstrated (Vesey-FitzGerald, 1960), some observers have inferred a degree of behavioral interaction that is unjustified on the basis of the known facts. Before examining the relations between the gazelles, it is therefore desirable to discuss the causation of interspecific associations and the kinds of behavioral interactions that may occur between the members.

It must be noted, first of all, that no obligatory associations between ungulates are known, and that any existing association is usually purely temporary. Two types may be defined: "neutral" or chance associations, and "positive" associations. When two or more species are brought together in time and space by environmental factors (Fig. 2, G), as when game concentrates on a green pasture in the dry season, it is a neutral association. Likewise, a Grant's and a Thomson's gazelle may feed side by side on the Crater floor, one grazing, the other browsing on inconspicuous shrubs, purely due to chance.

A positive association is demonstrated when isolated members of one species join members of another species. Zebra, wildebeest, topi (*Damaliscus lunatus*), hartebeest (*Alcephalus buselaphus*), waterbuck (*Kobus defassa*), the gazelles, and numerous other ungulates are all known to associate with one another on this basis. The simplest and most probable explanation is that every gregarious species is led by its need for social contact to seek the companionship of other herbivores when isolated from conspecifics. This could apply not only to solitary individuals but also to isolated herds. But perhaps in most instances a lone male is found in a herd of another species. This is a significant point, since in all the above species except zebra, the males are territorial and thus the most likely to be isolated from their own kind.

The mere fact that two or more species are frequently found in association is, then, no proof of any behavioral interaction between them. In fact, as a general rule associated herbivores fail to react to one another except in special circumstances. This is not really surprising because each species, after all, has its own language of displays, which is for the most part unintelligible to members of another species.

The exception to the rule is alarm signals, which seem to be understood by most associated species, even of different classes. Any ungulate that stands erect and stares in one direction, or suddenly breaks into flight, alerts all nearby game. Again, the alarm snorts given by most ungulates are basically alike and mutually alarming. Whether more specialized signals such as stiff-legged bounding by the gazelles are understood by other species is less certain, but not unlikely. To restate the rule to include the exceptions: interspecific response is limited to behavior of interspecific significance.

Turning now to relations between the gazelles, it is possible but by no means certain that there is more than the usual degree of mutual awareness and behavioral interaction between them. Thomson bucks show a slight tendency to herd *granti* does, and vice versa. The former may also adopt a mildly aggressive posture toward *granti* which enter their territories, and have occasionally been seen to threaten and even attack young males. On their part, *granti* females sometimes react to herding or threat behavior by chasing *thomsonii* males. This is about as far as their interaction goes. While it seems to argue a greater than usual degree of mutual awareness, perhaps even some tendency toward confused identity, it may also be explained as simply the result of lowered thresholds in the frequently isolated males for this kind of behavior, which is more readily released by another gazelle than by the other larger and less similar herbivores. The fact that territorial *thomsonii* also may assume threat postures in the presence of wildebeest and zebra supports this interpretation. Seemingly, the two gazelles do not respond to one another specifically.

Social System

The social system of the gazelles is typical of gregarious, territorial antelopes. Three classes based on sex and age may be distinguished: nursery herds of females and young, bachelor herds of largely immature males, and a relatively small class of territorial males. Since a nursery herd is usually presided over by a territorial male, it is often called a harem. But in gazelles and most other territorial species, a herd is not the exclusive property of any one male; rather it moves about within the territorial mosaic, despite each male's efforts to detain it. It is therefore more accurate to speak of a nursery herd.

Differences in the system between the gazelles may largely stem from differences in their habitat and food preferences; the simple fact that *granti* needs much more ground for feeding than does *thomsonii* may have a bearing. Thus, territorial *granti* are spaced about a half mile apart, on the average, compared to a spacing of 200–300 yards between *thomsonii* bucks. The spacing between nursery herds in the two species is on the same order. Not only are *thomsonii* nursery herds more closely spaced (more numerous), there also appear to be more animals in the average herd. Table 1, based on the aforementioned gazelle counts, shows the average numbers in nursery and bachelor herds. Student's t was used to test whether the indicated differences in mean herd size are significant; they were found to be significant for nursery herds ($t = 2.28$, 200 degrees of freedom, $P = 0.95–0.98$), and not significant for bachelor herds ($t = 0.596$, 39 df, $P = 0.60$).

Besides being smaller and more separated, *granti* nursery herds seem to be more stable in composition and to have a better-defined home range. A

TABLE 1.—*Comparison of herd size.*

	No. in sample	No. of herds	Average herd	±95% C.I. $\overline{X}$	SD
Nursery herds					
G. thomsonii	2287	100	22.9	5.47	27.60
G. granti	1640	100	16.4	2.81	14.20
Bachelor herds					
G. thomsonii	319	24	12.9	1.81	4.29
G. granti	154	15	10.3	2.66	4.84

marked female recorded 17 times in a year was always sighted within an area of 1½ square miles. Concentrations of more than 100 *granti* are rarely seen in the Crater, and represent merely temporary aggregations of several herds, whereas hundreds of *thomsonii* congregate for weeks at a time on preferred pasture. These concentrations appear to be broken up into separate herds purely through the activities of territorial males, and there is constant interchange between them of the females and young. However, the composition of known *granti* herds was also observed to be quite variable, and in fact neither species appears to have the sort of stable herd structure found in wildebeest, where small sedentary herds seemed to represent semipermanent and closed associations between adult females (Estes, 1966).

Reproduction and Care of Young

There may be a minor peak of gazelle births in January and February, between the short and long rains. These are months of peak calving for many East African ungulates. But relatively few are strictly seasonal breeders and observations of mating and young fawns of both species throughout the year indicate that the gazelles are no exception.

The two closely resemble one another in care of the young. Prior to parturition the females withdraw from their herds a short distance and drop their fawns in medium to long grass. In true short-grass steppe, the does may even drop their fawns on bare ground for lack of anything better (Fig. 1, C). Females with young fawns tend to associate, as Walther (1965) remarked concerning Grant's gazelles, and may remain apart from their herds for several days to a week or more following parturition. This tendency of new mothers to associate has an important bearing on the defense of the young against predators.

The same elaborate measures for effective concealment of the young are displayed by both species. During parturition, the female carefully examines the ground each time she changes position, and ingests all fluids and membranes. The fawn is cleaned with equal thoroughness. Once it has gained its feet and suckled, the fawn seeks a suitable hiding place, watched by the mother, who evidently memorizes its position by observations from several vantage points before moving away and beginning to graze. Walther (1965: 190) has given a detailed description of a birth in Grant's gazelle.

While considerable individual variability was observed in concealment discipline, ideally the fawn does not move until its mother returns and calls it to her, still from a distance. For the first few days, but not afterwards, the fawn will lie still even after discovery, to the point of allowing itself to be handled. Apparently the flight reaction is undeveloped at this stage, presumably because of the new fawn's inability to escape a predator if concealment fails; until its ability improves, an escape reaction would carry the danger of prematurely exposing its position. The strength of the concealment discipline was well demonstrated during a major grass fire in Ngorongoro in September, 1963, when several hidden *granti* fawns were literally plucked from the flames.

Finally, the risk of discovery by scent is reduced by still another behavior mechanism. During the concealment period, the fawn requires the active stimulus of licking by the mother before it can urinate or defecate, and the mother eats these waste products as they are produced (Walther, 1965). This behavior may be typical of antelopes that conceal their young.

Predators

In a study of predation on East African plains game, Wright (1960) concluded that *thomsonii* was the most widely-taken prey species, being preyed upon by nine of 11 identified predators. According to his evidence only two preyed on *granti*. But the evidence from Ngorongoro Crater indicates that they share the same six chief predators, though unequally: Asiatic and black-backed jackals (*Canis aureus* and *C. mesomelas*), wild dog (*Lycaon pictus*), cheetah (*Acinonyx jubatus*), leopard (*Panthera pardus*), and spotted hyena (*Crocuta crocuta*). This list is presented with the following reservations: 1) it includes only those species which are known to prey regularly on gazelles; and 2) it is not necessarily applicable throughout the gazelles' ranges at all times, since selection by predators is influenced by the availability of preferred prey species. In the Serengeti, for instance, lions would have to be added to the list, since certain nonmigratory prides subsist on *thomsonii* and other game when wildebeest and zebra, their preferred prey, migrate beyond their home range. In the Crater, lions very rarely hunt gazelles.

Jackals appear to be the most important predators on both gazelles, purely on the basis of fawn predation. Even healthy juvenile *thomsonii* are rather large for these fox-sized canids. The only other predator on the list numerous enough to be considered as a major cause of mortality is the spotted hyena, which also takes a certain number of fawns. But it does not appear to specialize in hunting gazelle fawns to nearly the same extent; rather it specializes in running down newborn wildebeest. In an anlysis of 188 hyena fecal samples from Ngorongoro, Kruuk (1966) found remains of wildebest in 83%, zebra in 46%, and Thomson's gazelle in only 16%. There were no traces of Grant's gazelles; the probable explanation is that there are relatively few of the latter on the floor of the Crater, where hyenas do most of their hunting.

Both Asiatic and black-backed jackals often employ the same efficient system of pair-hunting gazelle fawns. This is far more effective than hunting alone, for one can concentrate on finding and catching the fawn while the mother is occupied in chasing the other. Both gazelles defend their young vigorously against jackals (but not against larger predators), running out to attack as soon as one prowls anywhere near the hiding place. They use only their horns and do not strike with their feet, so far as I have seen. While the Asiatic jackal is seemingly more diurnal than the black-backed, one may readily see both species by day in Ngorongoro, singly or in pairs, quartering near gazelle herds in quest of hidden fawns. When a jackal comes upon a newborn fawn it immediately seizes and strangles it; older fawns bolt at the last instant, scuttling and twisting like rabbits until they drop in a new hiding place, while the mother tries to drive the jackal(s) away. Often it is aided by another doe (Fig. 1, E and F, in which two Grant's females were teamed against two black-backed jackals). In the instance figured, although the fawn was caught and partially throttled several times by a jackal, one of the does always managed to arrive in time to drive it off, and the jackals finally gave up the chase. The fawn escaped with no visible injuries.

Such cooperation purely on the basis of "altruism" would be extremely unlikely in an antelope, and indeed it is rare to see more than two does cooperating; it may be presumed that the helper also has a young fawn hidden nearby. The association of mothers with young fawns is therefore prerequisite to mutual defense. Yet the formation of such associations seemingly depends on the chance of females from the same or adjacent herds giving birth about the same time, for as often as not, mothers of young fawns (and only mothers of young fawns, among females) are found alone. Conceivably both team-hunting of fawns and the association-defense by mother gazelles are in the process of evolving.

Leopards prey heavily on gazelles where the opportunity arises, which is mainly in the vicinity of watercourses affording suitable cover adjacent to gazelle habitat. Thomson's gazelles made up seven of 15 leopard kills recorded by Wright (1960). In Ngorongoro, an adult female leopard that frequented a tree beside my camp brought in two adult Grant's males in a month, compared to 11 jackals, which it seemed to prefer to all other prey.

Of all carnivores that prey on the gazelles, the cheetah and wild dog specialize on them the most. In contrast to jackals, they generally select adults. If their numbers were at all comparable to the jackals, they might be considered the most important predators; but as both are sparsely distributed, it is hard to see how their predation could appreciably check the increase of any large gazelle population. Whether the cheetah has any preference between the two gazelles is not known. Wright (1960) recorded 12 cheetah kills, of which seven were *thomsonii*, three impala (*Aepyceros melampus*), one an immature wildebeest, and one a *granti*. Since *thomsonii* greatly outnumber *granti* in most areas, their representation here is not proof of a particular preference for them. But a clear preference was shown by a pack of wild dogs that was resident in Ngorongoro for more than a year. Of 50 kills, *thomsonii* represented 54% and *granti* only 8% (Estes and Goddard, 1967). Considering the relative abundance of the two species in the Crater, the high percentage of *thomsonii* is significant, and bears out information from other areas that it is the main prey of the wild dog in East African steppe-savanna.

Interestingly enough, 67% of the *thomsonii* killed were adult males. Most of them may have been territorial males. They are often the last to quit their place in the face of danger, perhaps partly due to inhibitions about trespassing on neighboring bucks, and when pursued tend to circle back instead of running in a straight line. The two factors together make them particularly vulnerable to wild dog predation. Females with young fawns behave in much the same way, and also seem more vulnerable than other members of the population. Thus wild dogs indirectly contribute to fawn mortality.

A differential adult sex ratio, often roughly 40 males to 60 females, has been reported for a number of territorial antelopes (Bourlière and Verschuren, 1960), whereas parity seems associated with nonterritorial systems (e.g., bovine species). The apparently greater risk of leading a territorial existence thus underscores the fact that only the holders of territories have the opportunity to reproduce in most, if not all, species with a territorial social system. Since in a normal population there always are more adult males than there are suitable territories, predator selection of territorial males would benefit the species by promoting a turnover, creating openings for vigorous young adults from the bachelor herds.

Territorial Behavior

The two gazelles differ from one another in territorial behavior more strikingly than in any other behavior. The much larger area defended by *granti* has already been mentioned; equally striking differences exist in the degree of territoriality, the means employed to mark the territory, and in the intimidation displays and fights which serve to maintain the territorial status quo. Thomson males are more vigorously territorial, both in their relations with rival males and in herding and chasing females. Grant bucks are comparatively languid in both respects, even to the extent of allowing other adult males to mingle with nursery herds and display to the does; however these males are generally members of bachelor herds and apparently not actual territorial rivals—even *thomsonii* males will sometimes ignore the intrusion of bachelors on their grounds.

According to Walther (1964: 883), one method of demarcating the territory—by linked urination-defecation (*Harn-Koten*)—is common to all the gazelles. The procedures and postures appear identical in *granti* and *thomsonii* bucks (Fig. 2, A, B, and Fig. 3, A, B). The buck first urinates, in a distinctive widespread stance, then brings the hindlegs ahead and squats to defecate on the same spot, which is often a regularly used dung heap on bare ground. The attitudes in territorial males are rather extreme, and Walther emphasized the possible display function of the second step by calling it "demonstrative defecation" (*demonstratives Koten*). But exaggeration of attitudes seems to me less significant than the linked performance of the two actions, for females and immature males practically never urinate and defecate in sequence on the same spot. Thus linked urination-defecation amounts to a display of territorial behavior. Female urination postures also are alike (Fig. 2, G; Fig. 3, F), although that of *thomsonii* looks more extreme.

Urination-defecation is the only method of scent-marking employed by *granti*, whereas *thomsonii* depend mainly on another system altogether. The bucks have large preorbital glands that stand out as prominent bulges on either side of the snout; these glands are small and not invariably functional in *granti*. As Brooks (1961) noted, territorial *thomsonii* mark decapitated grass stems, twigs, buds, etc., with a pitchlike substance secreted by the glands. They mark by carefully pursing the gland's orifice around the object (Fig. 3, D). Deposits built up over a long period may form a globule up to a half inch in diameter. An investigation of two territories yielded the following results: in one, 68 preorbital-gland deposits and 15 dung deposits were found in a sample of 14 100-ft quadrats, an average of four gland deposits per quadrat (the average distance between marked stems was 27 ft); in the other, 153 preorbital-gland deposits and 10 dunging places were found in 13 100-ft quadrats, an average of approximately 12 gland deposits per quadrat (the average distance between marked stems was 14 ft and the average height of the marked stems in this territory was 15 inches, compared to an overall grass height of about 8 inches). These data give some indication of how important preorbital-gland marking is in *thomsonii* territorial behavior. The arrangement and spacing of the deposits is such that invading males are certain to encounter one of these scent posts.

Since both types of marking are performed by bachelor-herd adults, Walther (1964: 874) contended that they are not necessarily linked to territorial behavior. But what could be more typical of ontogeny of territorial behavior than the appearance of such fixed action patterns before they begin to serve their intended functions?

The differences between the intimidation displays and fighting styles of the males of the two species are hardly less striking. In *granti* a special intimidation display largely substitutes for fights between adults, whereas in *thomsonii* fighting is the normal outcome of territorial conflicts.

Agonistic encounters between *granti* males begin with a ritualized display of neck development. From a distance of 5 to 10 yards two bucks display by drawing back their necks with chins tucked in and abruptly swinging their heads from side to side (Fig. 2, C). The ears are commonly held out horizontally, but may also be held back. Walther (1965: 182) considered that turning the head so that the chin and throat with its white gular patch were presented to the opponent was the main point of the display. But in what way this might be intimidating was not made clear. On the other hand, the way the neck is presented, with all muscles flexed, makes it appear much thicker than normal (compare Fig. 2, C with Fig. 2, F). Turning the head sideways causes an additional bulge of the opposite sternomastoideus muscle (just as in humans); the abrupt swinging of the head from one side to the other may be designed to make the muscles play, or simply serve to keep the bulge displayed in profile to the opponent as their relative positions change. The basis for the display lies in the importance of neck development in actual fighting, as described below.

The usual encounter goes no further than the neck display. But if two bucks happen to be so evenly matched in strength and motivation that progressively more intense neck display, during which they circle and draw closer, fails to intimidate either, then a fight is likely to develop. However, a whole series of further preliminaries may precede and follow actual fighting: repeated scratching of the neck and between the horns with the hindfeet; grooming of the flanks, shoulders, and legs; grazing; rhythmic swinging of the horns in the grass; and pawing followed by linked urination-defecation. All of these actions may also be seen in encounters between *thomsonii* males, and most are seen in other antelopes as well, including wildebeest. Scratching, grooming, and grazing under stress are probably agonistic displays derived from displacement activities; horning the grass may be threat or redirected aggression; pawing followed by excretion is accentuated territorial demarcation (*thomsonii* also frequently mark with the preorbital glands after fights).

In the final prelude to fighting, the bucks confront one another with their ears held out, duck their heads up and down (Fig. 2, D), and touch horns. Walther (1965: 177) fittingly termed this "taking of measure" (*Massnehmen*). Finally they engage horns firmly near the bases and push, heads almost flat to the ground as each strives to twist the other's neck and drive him sideways (Fig. 2, E). When mature males are invloved the fight generally ends after a brief joust; as a general rule, the more mature the contestants, the more likely they are to display rather than fight.

The most vigorous fights are to be seen between subadults in bachelor herds, particularly when a number of herds are intermingled, as often happens in the late afternoon. On these occasions the bucks may march and mill in a restless mob, some or all of the older animals displaying to one another.

There appears to be a definite age rank-hierarchy in bachelor herds with all subordinate to the master buck in whose territory they happen to be. Only bucks in the same stage of development are paired. Adults do not display to juveniles; they threaten them by abruptly ducking the head, or simply chase them. Fights between bachelors often are preceded by no more than the neck display, and may also develop from sparring matches. In any case the marking activities described above are commonly omitted. Engagements may last up to a minute and be renewed as often as the loser will turn and stand up to the pursuing winner. Not infrequently it happens that horns become locked together, whereupon the bucks spin in a circle, pulling and jerking away violently, so that the necks are often sharply twisted. But the animals invariably manage to pull free.

The intent to twist the opponent's neck may not appear in contests between equals, since often neither gains the advantage, but on rare occasions when poorly matched bucks were observed play-fighting, the bigger immediately twisted the neck of the smaller, which was clearly at a serious disadvantage. The risk of a twisted or broken neck—or more likely, a puncture by one of the opponent's long horns—is reason enough for avoiding encounters with stronger bucks, and no doubt every male learns to judge his own relative strength exactly through innumerable trials while maturing in a bachelor herd. The neck display is an effective and safe substitute for an actual trial of strength.

Male Thomson's gazelles have their own intimidation-challenge posture, which consists simply in approaching another male with neck held stiffly horizontal and chin in, with the horns presented vertically (Fig. 3, C). Yet most encounters between territorial males end, or begin, with fighting. In this species it seems that natural selection has operated on horn configuration and fighting style to produce a relatively harmless type of combat, rather than a substitute in the form of a display. The close-set horns are well designed for banging together, while their curvature and manipulation during fights largely rules out their employment for stabbing. Indeed the female's horns, feeble as they are, are better adapted for that purpose.

Instead of engaging horns, *thomsonii* bucks spring at each other (Fig. 3, C), bang horns and leap back, in a series of lightning thrusts and parries, with pauses during which they face one another and circle in the challenge position. Head-ducking, scratching, grooming, and grazing are frequently performed in the intervals, and a weaker or less confident buck gives ground by hopping backward. Fights rarely last more than a minute, but occasionally continue intermittently for 10 or more minutes. In some hundreds of fights, I have never seen an animal actually wounded. Yet one sees a surprising number of territorial males with more or less pronounced limps, which would most likely result from fighting; this could help account for their vulnerability to wild dogs.

Epigamic Behavior

Compared to the striking differences in their territorial behavior, the epigamic behavior of Thomson's and Grant's gazelles appears remarkably similar, differing only in emphasis and certain details. As Walther gave a full account of courtship (1964; 1965), the different stages and associated displays will be only briefly described here for comparative purposes.

The herding and pursuit of females is important in territorial behavior; perhaps mainly due to the smaller size of their territories, territorial *thomsonii* expend much more energy in these activities than do *granti* bucks. The characteristic "stalking" attitude and movements of a herding Thomson's buck (Fig. 3, E) strongly recall the behavior of a herding sheep dog. When chasing, and again, while closely following a female in courtship display, bucks of both species make a sputtering noise which sounds like a misfiring outboard motor.

During the actual courtship ceremony, the male follows persistently as the female continually, but at a gradually diminishing pace, moves away from him. However, the male only seriously begins to court if a female proves olfactorily "interesting." This is determined through the olfactory analysis known as *Flehmen* (though performed by nearly all ungulates there is no English term for this behavior; general use of the German term *Flehmen* dates from Schneider, 1930). Females respond to the male's persistent approach by stopping and urinating (Fig. 3, F). The buck sniffs the urine then performs *Flehmen*, standing with head erect, mouth open, and ears back (Fig. 1, A). The pose is identical in both gazelles. Having learned whether or not the female is nearing estrus, the buck accordingly continues, or breaks off, the courtship.

Perhaps the clearest species-specific differences in epigamic behavior are in the displays of the males during the courtship promenade. The behavior of displaying *thomsonii* males is the more varied and energetic (see Walther, 1964: 881), featuring periodic extreme raising of the nose (Fig. 3, G), kicking out with a foreleg if the female stops ("*Laufschlag*"), and a combined nose-lifting and goosestepping which Walther termed "drumroll" (*Trommelwirbel*). They do not raise their tails. A display similar to the last, given by the Uganda kob (*Adenota kob thomasi*), is termed "prancing" by Buechner and Schloeth (1965: 211). The *granti* display consists simply of walking behind the female with head and tail raised (Fig. 2, F), although the "goosestep" of *thomsonii* is sometimes faintly suggested.

The behavior of females during the courtship promenade seems nearly identical in the two species. Indeed their behavior consists of little more than moving away, starting at a run and ending in a walk. However, their state of receptiveness is further indicated by the position of the tail; when it is held out and up it is a sign that the female is almost or quite ready to copulate. It is also noteworthy that all females continue to hold out their tails for some time after urinating (Fig. 2, G).

The only difference in copulatory behavior that I have been able to detect is, again, in the way the males hold their tails. The tails of the females and of *granti* males are held out, but the *thomsonii* male's merely hangs. In mounting, the male stands erect, forelegs bent or occasionally held straight, but not touching the female (Fig. 3, H). Since she never stops moving, the male must accomplish intromission with a single ejaculatory thrust while walking on his hindlegs. If successful, he then loses interest for an hour or so, and meanwhile the female may leave him and copulate with other males.

The apparent similarity of epigamic behavior in the two gazelles may seem at first rather puzzling, especially since reproductive isolation is not achieved by other conventional mechanisms such as different times or places of breeding. On theoretical grounds one could therefore expect a marked divergence in the epigamic displays of two such closely related, sympatric species (Marler, 1957). As a matter of fact, a valuable comparison of courtship displays in gazelles by Walther (1964: 885) strongly suggests that divergence may be at least as great as, for instance, in the epigamic displays that serve to sexually isolate some of the Anatinae (Lorenz, 1951), and is similarly achieved through different emphasis, sequences, and combinations of the basic displays. Yet Walther himself finds it necessary to add three other possible isolating mechanisms, including differences in size, distribution (spatial separation), and territoriality, in order to account for the fact that *granti* and *thomsonii* do not interbreed. In my view these make unnecessary and unconvincing arguments, particularly the second. His idea of separation was based on observations made in Ngorongoro Crater in January, when the grass is high and the two species happen to overlap less than usual.

But in any case, until and unless some initial propensity to interbreed can be demonstrated, it hardly seems necessary to look beyond differences in recognition characters for an adequate species-specific isolating mechanism. In this connection, it may well be that olfactory differences are even more decisive than differences in markings. For instance, the fact that both male and female Thomson's gazelles have inguinal glands, the function of which remains to be demonstrated, and that *granti* do not have them, is highly suggestive. As Brooks pointed out (1961: 27), these glands emit a distinct, musky odor. Since even the human nose can detect a difference between the two species, the importance of this difference in species-specific recognition can hardly be doubted.

Warning Signals

The two gazelles have the same repertory of warning signals, with only minor qualitative and perhaps quantitative differences in their expression. The alarm snorts, already mentioned in discussing behavioral interaction, differ only in pitch, the smaller gazelle's being higher. Snorting is frequently accompanied by stamping with a foreleg, which may or may not have a signal function, but is almost certainly an expression of nervousness that is given by many different species (Walther, 1965: 179, thought it might mean threat). In addition, the gazelles have two visual warning signals: twitching of the flank skin and "stotting" (stiff-legged bounding). The former display is given, though not invariably, just as the animal begins running, when a predator or even a car comes within the minimum flight distance. Twitching of the flanks causes a rapid flickering of the side stripe and bordering white areas that has a semaphore effect. Since *granti* usually lack the side stripe, the display may retain little of whatever signal value it once had. Nonetheless, they continue to go through the motions, though perhaps less frequently than the other species.

Stotting (Fig. 1, D) is much the more striking and apparently the more effective of the visual warning signals. Its effectiveness is enhanced by the white rump patch, which has erectile hair that spreads fanlike during the display (and also curiously, during defecation). In fact the stotting gait and a white rump patch are so intimately linked that as a general rule probably only bovids with a conspicuous white rump perform this display.

The stotting display of *thomsonii*, in particular, is so closely identified with the appearance of wild dogs as to appear possibly specifically adapted to this predator's hunting behavior (Estes and Goddard, 1967). The wild dog, in fact, is the only African predator of healthy gazelles that simply runs down its prey without attempting a concealed stalk. As soon as a running pack comes into sight, the nearest gazelles immediately react by stotting and the display spreads wavelike in advance of the dogs. A broad path is cleared as the gazelles bound away from the disturbance. The peculiar thing is that the individual singled out by the pack continues to stot during the first stage of the chase, and appears thereby to lose precious ground until it finally settles to a full run. However, other observers (Brooks, 1961; Kühme, 1965) have argued that stotting is deceptively slow, and Percival (1928) felt that it might function to mislead the pursuer into a premature all-out effort. In any case, wild dogs usually succeed in running down either species within 1 to 2 miles.

Stotting is associated not only with danger but also with play between fawns and sometimes with intraspecific pursuit by members of either sex. In fact, it was only after a wild dog pack moved into Ngorongoro that the warning function of this display became clear to me, so seldom had I seen it performed in the context of danger. Walther, who apparently only saw it in the intraspecific context, has nonetheless advanced a hypothesis that would account for its performance in both situations, and even during actual pursuit by wild dogs (1964: 872). He considered it an expression of a high, but not the highest, level of excitation, most likely to appear in the transition from a high to a low level, and vice versa. Its evolution as a warning signal could thus have derived from the tendency of gazelles to stot in the situation of rising excitation occasioned by the approach of predators.

Acknowledgments

My research was supported primarily by a grant from the National Geographic Society, and by grants from the New York Explorers Club and the Tanzania Ministry of Lands, Forests and Wild Life. I am indebted to the Conservator of the Conservation Area for permission to live and work in Ngorongoro Crater; to F. Walther for assistance in investigating a Thomson's gazelle territory; to the staff and students of the Mweka College of African Wildlife Management for collaborating on one gazelle count and for helping investigate another *thomsonii* territory; and to Profs. W. Dilger, O. Hewitt, D. Thomson, P. Buckley, and Dr. F. R. Walther for criticisms of the manuscript.

Literature Cited

Andrew, R. J.   1965.   Intention movements of flight in certain passerines and their use in systematics. Behaviour, 10: 179–204.

Bourlière, F., and J. Verschuren.   1960.   L'Écologie des ongulés du Parc National Albert. Institut des Parcs Nationaux du Congo Belge, Brussels, 158 pp.

Brooks, A. C.   1961.   A study of the Thomson's gazelle in Tanganyika. Her Majesty's Stationery Office, London, 147 pp.

Buechner, H. K., and R. Schloeth.   1965.   Ceremonial mating behaviour in Uganda kob. Z. Tierpsych., 22: 209–225.

Estes, R. D.   1966.   Behaviour and life history of the wildebeest. Nature, 212: 999–1000.

Estes, R. D., and J. Goddard.   1967.   Prey selection and hunting behavior in the African wild dog. J. Wildlife Mgt., 31: 52–70.

Haltenorth, T.   1963.   Klassifikation der Säugetiere: Artiodactyla. Handb. Zoologie, 1 (18): 1–167.

Kruuk, H.   1966.   Clan-system and feeding habits of spotted hyenas. Nature, 209: 1257–1258.

Kühme, W.   1965.   Freilandstudien zur Soziologie des Hyänenhundes. Z. Tierpsych., 22: 495–541.

Lorenz, K.   1951.   Comparative studies on the behaviour of the Anatinae. Avic. Mag., 57: 157 *et seq*.

Marler, P.   1957.   Specific distinctiveness in the communication signals of birds. Behaviour, 11: 13–39.

Percival, A. B.   1928.   A game ranger on safari. London, 305 pp.

Schneider, K. M.   1930.   Das Flehmen. Der Zool. Garten, 3: 183–198.

Stewart, D. R. M., and L. M. Talbot.   1964.   Census of wildlife on the Serengeti, Mara, and Loita Plains. East African Agr. and Forest. J., 28: 58–60.

Talbot, L. M.   1962.   Food preferences of some East African wild ungulates. East African Agr. and Forest. J., 27: 131–138.

Vesey-FitzGerald, D. M.   1960.   Grazing succession among East African game animals. J. Mamm., 41: 161–172.

Walther, F. R.   1964.   Einige Verhaltensbeobachtungen an Thomsongazellen im Ngorongoro-Krater. Z. Tierpsych., 21: 871–890.

———.   1965.   Verhaltensstudien an der Grantgazelle im Ngorongoro-Krater. Z. Tierpsych., 22: 167–208.

Wright, B. S.   1960.   Predation on big game in East Africa. J. Wildlife Mgt., 24: 1–15.

Division of Biological Sciences, Cornell University, Ithaca, New York. Accepted 27 December 1966.

16

Reprinted from *Proc. U.S. Natl. Mus.*, 122(3597):1-9, 13-35 (1967)

A Comparative Study
In Rodent Ethology
With Emphasis
On Evolution of Social Behavior, I

―――――――

By John F. Eisenberg

Resident Scientist, National Zoological Park

―――――――

Introduction

The rodents exhibit, as an order, a complex series of morphological, physiological, behavioral, and ecological adaptations. During the long history of adaptive radiation within this order, many instances of convergent or parallel evolution have occurred, thus rendering the subordinal classification quite difficult (Simpson, 1959; Wood, 1955). In spite of such diversity, the rodents form a remarkably unified taxon, and such an example of ecological diversity superimposed on a common morphological theme has intrigued many ethologists concerned with the evolution of behavior. Among the more prominent workers, Dr. I. Eibl-Eibesfeldt has contributed much to our knowledge of the subject and is also responsible for the most recent review of rodent behavior (Eibl-Eibesfeldt, 1958).

Eight years ago the present author initiated a comparative study of rodent behavior, partial results of which have been published in several papers (Eisenberg, 1962a, 1963a, b, c) and abstracts (Eisenberg, 1962b, 1964). These earlier, separate papers do not adequately reflect the unity behind the undertaking, and in the present paper current results will be combined with previously unpublished data to establish several theoretical principles that have emerged in the course of the work.

Throughout the studies, the behavior patterns have been analyzed in order to determine which movements were stereotyped and, among these, which conformed to the concept of fixed action patterns. Thus, certain behavioral units meeting acceptable criteria could be treated in the same manner as morphological structures and compared from one species to the next in the classical manner developed by Lorenz (1957) and Tinbergen (1951). The ground rules for such comparative studies have been reviewed recently by Wickler (1961) and need not be reiterated here. In my studies, behavioral evolution was approached at the level of discrete behavioral units; such a description of discrete behavioral patterns allowed the formulation of descriptions of more complicated patterns consisting of predictable sequences of discrete units in time. Further analyses of the frequency of occurrence of different stereotyped units of behavior performed in a social context permitted a description of species-specific patterns of social structure. A knowledge of social structure permitted a partial answer to the following question: Do organisms that evolve independently in different geographical locations toward the same ecological niche also evolve similar social organizations? If the answer were yes, then it would be possible not only to predict a social system from a knowledge of niche requirements but also to determine from comparative studies what major environmental adaptations correlated with a given social system. Thus, the comparison of social systems demanded a description of the behavioral units and, in addition, the description of the different forms of social organization by means of some consistent experimental methodology (Eisenberg, 1964). Working in the laboratory with small rodents was possible only when adequate field data were available to correct for any inadvertent misinterpretation brought about by captivity. Finally, methods of quantification had to be developed that permitted a valid comparison among different species. After several years, it became obvious that simple criteria such as the presence or absence of a given behavior pattern would not suffice in determining relevant species differences. Indeed, it appeared that, under the same stimulus conditions, the different species were all capable of exhibiting almost the same discrete units of behavior and, unless major morphological differences interfered, only the different frequencies of occurrence served to separate species. Relative differences rather than absolute differences in behavior became the rule and led the present author to conclusions quite similar to those of Leyhausen (1965).

The study of social organizations ultimately led to an extension of the comparative method. A given type of social organization has a characteristic structure. The structure is the summation of the form and frequency of each type of interaction within an interacting group. Thus, the social structure reflects the mechanisms of interaction and the adaptive role of the interaction patterns determines the selective advantage of the system at the level of the individual member. On the other hand, a set of complementary emergent properties results when one considers that the social structure itself reflects an adaptation to a given set of environmental relationships that are reflected in the physiology or metabolism of the social group taken as a whole.

Future research on the adaptive nature of whole societies will necessarily have to concern itself with biologically viable societies, i.e., a social unit or series of social units that are maintaining their numbers over long periods of time. In essence, then, conclusions · concerning the adaptive nature of groups must be based on groups exhibiting a prevailing reproductive success.

The interrelationships of the various measurable phenomena exhibited at the level of the individual and the group are presented in figure 1.

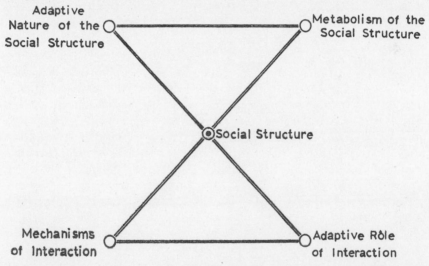

FIGURE 1.—Diagram indicating two possible levels of analysis commonly applied to social structures.

The present report thus demonstrates a method for the analysis of mammalian social structure, the application of the methods to several families showing varying degrees of convergence or parallelism, the reformulation of the term "species-specific behavior," and a discussion of social evolution within the Rodentia.

In a comparative study that spans 3 families and 34 species from three continents, the problems of acquisition and maintenance are of paramount importance. I should like to acknowledge the efforts of my laboratory assistant, Mr. David Williams, who lent his sensitive talents wholeheartedly to the work before us in Vancouver. Specimen donations were also an integral part of this study. I am indebted to Dr. D. Birkenholz of Illinois State University for 3 *Heteromys desmarestianus;* to Mrs. W. Downs of Yale University for 3 *Gerbillus gerbillus*, 2 *Jaculus jaculus*, and 2 *Jaculus orientalis;* to Dr. T. Reed, Director of the National Zoological Park, for 1 *Pachyuromys duprassi;* and to Mr. W. Preston of the University of British Columbia for 12 *Perognathus parvus*. In addition, I should like to cite Mr. Ralph Curtis of Miami, Fla., who was instrumental in arranging the importation of a series of rodents from Pakistan that formed the basis for our dipodid and gerbilline comparisons.

The studies reported on in this paper were financed in part by the following grants: E 350 and A 1371 from the National Research Council of Canada, and a general research board grant from the University of Maryland. The author is indebted to the Departments of Zoology at the University of California, Berkeley, the University of British Columbia, and the University of Maryland for the space and facilities utilized in the 7-year course of this study.

The Subjects

The choice of the specimens utilized in this study was influenced both by circumstances and design. Initially the behavior of one genus, *Peromyscus*, was selected for study. This North American group includes over 40 recognized species (Hall and Kelson, 1959) that have adapted to a myriad of habitats. As a genus it has lent itself well to comparative ecological, physiological, and morphological studies and offers a wealth of background studies that may be drawn on by ethologists (King, 1967). Four species were selected by the present author for behavioral studies, including two from the Transition life zone and two from the Sonoran life zone (see Eisenberg, 1962a, 1963a). The behavior of this group is the subject of a recent review (Eisenberg, 1967) and will not be redescribed here; however, relevant data will be introduced since this group of four species represents an example of the range of behavioral variation within one genus.

Several questions were raised in the course of the *Peromyscus* study concerning species-specific changes in social behavior resulting from adaptation to desert habitats. For this reason a complete rodent family, the Heteromyidae, was studied in order to correlate generic differences with niche requirements as I compared moist forest-adapted genera (e.g., *Heteromys* and *Liomys*) with desert or semidesert-adapted genera, *Perognathus, Microdipodops,* and *Dipodomys* (Eisenberg, 1963b). The results combined with the *Peromyscus* data suggested that any limiting factors a desert environment may impose concerning space requirements of a given individual rodent are not necessarily limiting with respect to the particular species-specific form of the social organization.

To confirm this hypothesis, two Old World taxa of desert-adapted rodents were studied, the Gerbillinae and the Dipodidae. Table 1 contains a list of the specimens included in this study with their geographical origins. Exact capture localities for specimens trapped by the author are included in previous publications (Eisenberg, 1962a, 1963a, b).

Adaptation to an arid environment involves the solution of several severe physical problems including (1) lack of water, (2) widely spaced food plants with concomitant loss of cover, (3) extremes of heat and cold. The behavioral, morphological, and physiological solutions to these problems are varied when the vertebrates are surveyed as a group (Schmidt-Nielsen, 1964), but all mechanisms are geared to the same ends: maintenance of a constant body temperature (at least during active periods), conservation of body fluids, and procurement of sufficient food. A number of small desert rodents have adapted in a similar fashion, providing typical instances of parallel physiological or behavioral evolution. For example, some species of the gerbilline, dipodid, and heteromyid rodents can recover metabolic water in the kidney and live on dry foods alone (Burns, 1956; Kirmiz, 1962; Schmidt-Nielsen, 1964). Furthermore, many desert rodents are nocturnal, construct burrows that are plugged during the day, and cache quantities of food. In several cases bipedal locomotion has evolved (e.g., within the Heteromyidae and the Dipodidae).

None of these behavioral attributes is confined to desert-dwelling species, but the total complex of attributes is unique to a certain type of desert-dwelling form (Eisenberg, 1963b). Within the genus *Peromyscus,* desert adaptation has taken place without the evolution of either bipedality or the ability to reuse metabolic water. The family Heteromyidae exhibits a range of morphological adaptations to a desert environment wherein only the genera *Microdipodops* and *Dipodomys* culminate their evolution with both an extremely specialized gross morphology and the ability to live on metabolic water alone (Eisenberg, 1963b).

Confronted with such a range of adaptations to the same set of problems in the desert, any correlation between behavior and ecology must take cognizance of the multiplicity of variables affecting the overall adaptation pattern. For this reason, tables 2 through 5 summarize the major similarities and divergences regarding reproduction, morphology, habitat, and ecology. Ecological data for the Gerbillinae and Dipodidae are adapted from Ognev (1963), Petter (1961), Zahavi and Wahrman (1957), and Kirchshofer (1958). Figure 2 compares the overall phylogeny for the three major families studied; it is derived from the work of Ognev (1963), Herold and Niethammer (1963), and Wood (1935).

The desert-adapted species show wide divergences when reproduction, maturation rates, morphology, and ecology are compared. The jerboas differ from the kangaroo rats in that some genera of jerboas hibernate and the maturation of the young is prolonged. On the other hand, the dipodids and the genus *Dipodomys* show a close parallel in their morphology and habitat requirements. Certainly the closest correspondence between rodent families occurs when some species of gerbil (i.e., *Gerbillus nanus*) are compared with the heteromyid species of the genus *Perognathus*. In any event, tables 2 through 5 emphasize the differences that may have a bearing on divergences in the form of social organizations (see p. 31).

Methodology

When not under direct observation, the subjects were held as individuals or pairs in glass-sided cages with wire tops and wooden floors. These holding cages were of two general sizes containing floor areas of 90 to 180 square inches. The animals were fed a mixture of sunflower seeds, commercial rat pellets, rolled oats, millet, and fresh lettuce (see also table 1).

The analysis of behavior began with a description of the discrete units that comprise the total repertoire of activity patterns displayed by the subject under study. In order to give the animal ample opportunity to display the full range of its behavior patterns, simulated natural habitat cages were employed. Such cages contained artificial burrows; they are fully described in a previous publication (Eisenberg, 1963b, p. 5). In the present study, the floor areas of the burrow cages included three sizes: 864, 550, and 360 square inches. Nocturnal forms were observed by a red light during their normal dark cycle of activity. Once the behavioral units had been described, the subjects were exposed to several testing situations that may be treated under two headings: (1) behavior patterns of the solitary animal and (2) patterns of social behavior.

The behavior patterns displayed by the isolated animal fall under the general categories of exploratory and maintenance behavior. In order to analyze the differences among species, standard tests were run by allowing an animal to move freely in a glass-fronted arena containing specific artifacts such as stones, hay, a tree branch, and a small amount of food. The arena sizes included floor areas of 748,

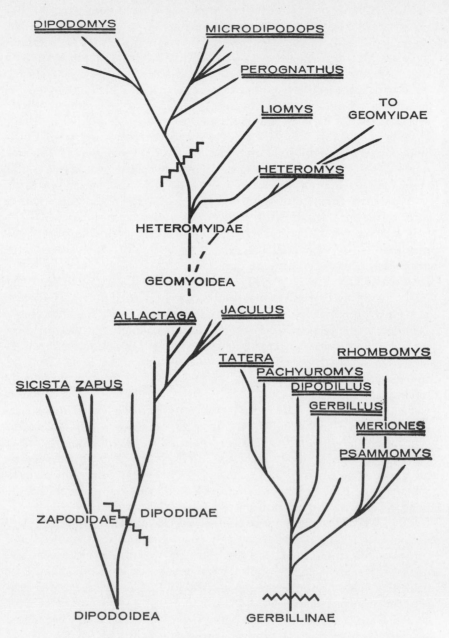

FIGURE 2.—Inter-relationships of major genera from three families of desert-adapted rodents (double underline=all genera included in present study; genera above jagged line are commonly found in arid or semi-arid habitats).

1598, and 2304 square inches. Employing a tape recorder, the behavior was described in a semicoded fashion by an observer concealed behind a screen. The verbal descriptions on tape were transcribed later, employing a constant 10-second time interval to allow further for a consideration of the temporal changes in the behavior patterns. An operations recorder coupled to a keyboard was also employed for recording behavioral units in time. Both data-recording methods are outlined in previous publications (Eisenberg and Kuehn, 1966; Eisenberg, 1963b). Such data were analyzed by various methods that permitted an accurate determination of the changes in the frequency of responses, the duration of specific activity patterns, and the frequency and durational changes over extended periods of time.

A problem in quantification arose in comparing acts that had a brief duration, coupled with a clear onset and cessation, with those acts having a variable duration of imperfectly definable onset and cessation. The more or less brief actions generally were counted and described in terms of straight numbers per unit time. On the other hand, such activities as driving, chasing, fighting, etc., were recorded either in terms of actual duration or were counted for a frequency analysis by employing a 10-second time criterion. Such acts having a duration of 10 seconds or less were treated as one unit; if the duration were greater than 10 seconds but less than 20, the action was treated as two units, and so on. Indication is made in the text whenever the 10-second criterion was applied for a frequency analysis of acts having a long duration.

Patterns of social interaction were also recorded by the general methods just described—by verbal means on tape or with an operations recorder. In addition, a third method involved a weekly check of cages containing freely growing populations. Pairing tendencies, incidence of wounding, litters born, and breeding condition of individuals were censused in this manner. The three approaches to the analysis of social interaction patterns are as follows:

(1) The encounter in a neutral arena: In this situation, two animals were placed simultaneously in a standard environment and the resultant interaction recorded. Such variables as the sex, the age, and the physiological state of the encountering animals could be controlled (see Eisenberg, 1962a, 1963b). The glass-fronted encounter arenas included 748, 1598, and 2304 square inches of floor area. Data from such standard encounter situations permitted species comparisons with respect to differences in the frequency and duration of various social behavior patterns.

(2) The territorial encounter: This was a variant of the simultaneous encounter technique and involved the use of a cage divided into two or three equal compartments as described previously (Eisen-

berg, 1963b, p. 4). The encounter boxes included 1110, 1728, and 1950 square inches of floor area. Utilizing this technique, some measure of the influence of the locus on the outcome of an encounter could be assessed.

(3) The formation of social groups: In this situation pairs or groups were allowed to remain together as a result of either a simultaneous or a territorial encounter. Several measurements were recorded, including the tendency to nest together, the incidence of wounding, the tolerance of the male by a female during parturition and rearing, the effect of the male on the female's estrous condition, and the effect of the adults on the survival, growth, and development of the young. Various cage sizes were employed in the study of the pairing tendencies and the social tolerance during the growth of confined populations. These included cages with floor areas of 180, 328, 550, 864, 1110, 1728, 1950, 2304, 5040, and 6480 square inches. Studies of this nature yielded information concerning species-specific differences in social tolerance. Pair tolerance and the growth of confined populations varied from species to species; the solitary species required a large space coupled with a complex environment in order to reproduce successfully.

[*Editor's Note:* Material has been omitted at this point.]

Behavior Patterns of the Solitary Animal

LOCOMOTION.—When a series of genera or species are compared, the presence or absence of a behavior pattern is often correlated with a corresponding presence or absence of some morphological feature. This is nowhere better illustrated than with bipedal locomotion. As discussed by Howell (1932), Hatt (1932), and Ognev (1959), bipedal, saltating locomotion occurs when a set of morphological features are present, including modifications in the musculature and vertebral column, reduction in the number of hind toes, lengthening of the hind feet, shortening of the forelimbs, fusion of the cervical vertebrae, and modifications of the tail. To be sure, most rodents possess the ability to stand on their hind legs and walk or even hop for a short distance, but sustained bipedal saltation does not occur as a normal behavior pattern with the absence of certain morphological and physiological correlates. In the group under study, full bipedal locomotion was exhibited only by *Microdipodops, Dipodomys, Allactaga,* and *Jaculus.* The latter two genera are more advanced in that the first and fifth toes are reduced (*Allactaga*) or absent (*Jaculus*) and in both forms the metatarsals are fused to form a "cannon bone."

COMFORT MOVEMENTS.—Sandbathing: Sandbathing is a complex movement pattern functional in dressing the pelage. It is composed of several distinct behavioral units, including side rubbing, ventrum rubbing, writhing from side to side, and rolling on the back. All these units are part of the total behavioral inventory of most rodents and other mammals; however, the organization of these units into a specific pelage-dressing pattern tends to be taxon-specific (Eisenberg, 1963c). The species-typical movement complex termed "sandbathing" is a normal part of the behavioral repertoire in most rodents that have adapted to arid habitats. It should be understood that sandbathing is not confined to desert rodents alone, but that the frequency of occurrence and stimuli necessary to elicit the pattern are characteristic of desert-adapted species. In the current study, *Gerbillus gerbillus, G. nanus, Tatera indica, Meriones unguiculatus, M. hurrianae, Pachyuromys duprassi, Allactaga elator, Jaculus jaculus, J. orientalis, Peromyscus crinitus,* and all species of *Microdipodops, Perognathus,* and *Dipodomys* exhibited functional sandbathing. Dressing the pelage by rolling or rubbing in dry soil may be accomplished potentially by any of the several discrete movements mentioned above. Typically the animals exhibit digging movements with the forepaws and the kicking back of accumulated earth with the hind limbs. Digging movements serve to dust the fur of the ventrum and often, while digging, the animal may exhibit slight forward extensions of the body as it presses its ventrum against the substrate. The components following digging and kick back are classified as a side rub, ventral rub, or rolling over, but this tripartite classification masks some subtle details. Since the side rub and ventral rub commonly occurred in the series of rodents studied, I include a brief description of the variations:

The side rub: This may involve lowering the head to the substrate and sliding forward on the side while first extending and then flexing the torso. The action may be swift with only a slight extension and flexion of the body, or it may involve a movement of both hind legs forward as the torso flexes. During the extension phase, the chin may rest flat on the substrate with the result that the chest is rubbed in the sand. The latter twisting is especially common in the dipodids studied.

The ventrum rub: In this study, the ventrum rub was defined arbitrarily as extension and flexion of the body axis with the ventrum pressed against the substrate. Typically the hind feet are thrust forward even with or beyond either side of the head during the flexion phase. The ventrum and perineal region are thoroughly scrubbed into the substrate. It can be seen from this description that digging and the side rub, plus the twisting of the body, will suffice to dust the entire pelage with soil. The ventrum rub is not always necessary as a pelage dressing component.

Rolling on the back while writhing from side to side was rarely shown by the species studied although it was observed occasionally with *Tatera indica, Meriones unguiculatus,* and *Pachyuromys duprassi.*

The frequency of side rubs to ventrum rubs varies in a species-specific manner. Table 6 indicates the trends for seven species from three families. The sandbathing Heteromyidae display a range of ventrum rubbing from 23 to 45 percent, whereas the dipodid *Allactaga* displayed only 8 percent ventrum rubs and the gerbils *Meriones* and *Gerbillus* displayed virtually no ventral rubbing as a component of sandbathing. The discrete components of sandbathing occur in sequences and each single event may be followed by a second event within an interval range of one-half to an unknown number of seconds exceeding 500. The frequency distributions for intervals separating sandbathing acts were plotted. Since the majority of intervals fell within 3½ seconds, this interval was selected arbitrarily as the limiting interval separating acts within the same sequence. For each species, all sequential acts were plotted in a Latin square in a manner described previously (Eisenberg, 1963c). Table 7 portrays the results, which give an indication of how the components of sandbathing are integrated. *Meriones* and *Gerbillus* are prone to alternate side rubs from one side to the other. *Allactaga* seems equally prone to alternate or rub the same side a second time. *Dipodomys deserti, Perognathus californicus,* and *P. parvus* tend to alternate sides whereas *D. nitratoides* shows a preference for rubbing the same side. All the Heteromyidae integrate ventrum rubs with side rubs into a functional sequence.

All species exhibit locus specificity in their sandbathing on consecutive days. There is also a pronounced tendency in some species to sandbathe at the locus of a partner during a social encounter (see p. 22). The occurrence of locus specificity and the exchange of loci during an encounter implies that chemical communication is taking place; thus, the functional classification of sandbathing really includes a marking function as well. In addition to the movements involved in dressing the pelage, these animals may have definite marking movements.

Marking: Marking is a functional term including several behavioral units. From a functional standpoint a marking pattern serves to spread some chemical substance at a specific locus. Such a chemical presumably has communicatory value. Many rodents utilize a marking pattern that involves a depression of the anal-genital region in order to bring these glandular areas into contact with the substrate as the animal moves about in its living space. This is termed the perineal drag. Urine and feces may serve as chemical markers in addition to glandular exudates. When special glandular areas are concentrated on other parts of the body such as the flanks or ventrum, side rubbing or ventrum rubbing may be expressed as a marking movement. The perineal drag is a movement common to all the rodent species studied. Ventrum rubbing is a ritualized marking movement shown by the genus *Meriones*. The latter genus is characterized also by a large gland-field in the ventral epidermis.

It should be noted that sandbathing can serve the dual function of dressing the pelage and marking (Eisenberg, 1963b, c); thus, in desert-adapted species the sandbathing behavior tends to be concentrated at a specific locus and can serve as a focus of activity for two or more interacting individuals.

Washing: Washing (autogrooming) is displayed in a typical myomorph fashion (Bürger, 1959). The chief modifications in movement pattern are the results of morphological modifications. Hence, the short forelimbs and short necks of the bipedal genera restrict the movement of the head in the vertical plane and necessitate a rotation of the head when it is washed with the forepaws.

Burrow construction.—The desert-adapted rodents of the families Cricetidae, Dipodidae, and Heteromyidae construct extensive burrow systems that employ basic digging patterns involving the forepaws and hind limbs (Eisenberg, 1963b). The teeth may be used to gnaw into a hard substrate; *Allactaga elator* is most prone to gnaw when constructing burrows. This behavioral trait is reflected in the protruding dentition, which enables *A. elator* to gnaw into a flat surface without a crack or crevice that would permit a starting point for gnawing.

Burrow walls are packed by a pushing and patting motion of the forepaws (Eisenberg, 1963b). This movement pattern is typical for many rodent species, and the nose and incisors may also be involved in packing loose soil. *Tatera* and *Meriones*, when packing, employ the forepaws and nose by jerking the body back and forth in the vertical plane while holding the forepaws rigid on either side of the nose. The forepaws and nose strike the soil and serve to tamp it firmly into place. *Allactaga* and *Jaculus* pack loose soil by raising and lowering the head in the vertical plane, thus repeatedly bringing the snout and incisors against the substrate. This packing method appears to be highly ritualized in the Dipodidae. The heteromyid rodents as well as *Gerbillus* and *Pachyuromys* appear to employ pushing and patting with the forepaws as the principal method for packing the tunnel walls.

ASSEMBLY OF FOODSTUFFS.—Studies of *Meriones persicus* (Eibl-Eibesfeldt, 1951) indicate a tendency to bite pieces of food (e.g., vegetable matter, stalks, roots, pods) into small pieces, which are then cached. This behavior pattern has been termed "Häckseln" and is here translated as chopping. *M. unguiculatus*, *M. hurrianae*, and *Tatera indica* all exhibited this trait.

The caching of foodstuffs either in the burrow or in discrete loci within the animal's home range is a behavioral trait shared by many species of rodents. The family Heteromyidae is characterized by a persistent tendency to gather and cache great quantities of grain, and this behavioral trait is correlated with the possession of capacious, externally opening, fur-lined cheekpouches (Eisenberg, 1963b). None of the other genera in the current study exhibited such persistent caching behavior, and the dipodid genera do not seem to cache very much at any time during their annual cycle. *Allactaga* and *Jaculus* will assemble dried grasses in their burrows, but this material is generally used in nest building rather than as food (see table 8).

DISCUSSION.—Behavioral and ecological convergences appear to be very close when the genera *Perognathus*, *Gerbillus*, and *Pachyuromys* are compared. Although the genera *Microdipodops*, *Dipodomys*, *Jaculus*, and *Allactaga* are ecologically similar and have evolved a similar morphology correlating with their bipedal form of locomotion, rather profound behavioral and physiological differences separate the bipedal Heteromyidae from the Dipodidae. The most basic correlate appears to involve the reduced caching tendency of *Jaculus* and *Allactaga* with a concomitant tendency to hibernate or exhibit periods of torpor (Skvortsov, 1955, 1964). *Dipodomys* caches seeds and is not known to hibernate.

Since the Heteromyidae are essentially solitary rodents with a very low threshold for the exhibition of agonistic behavior, this tendency toward asocial behavior may correlate with the fact that the genera *Dipodomys* and *Microdipodops* do cache and the fact that the selective advantage of caching is related to a dispersed or solitary social structure.

Sandbathing is a trait shared by all desert-adapted rodents. It would appear that increased sebaceous secretion is a necessary condition to reduce evaporative water loss through the epidermis. In addition, many desert rodents have a dense pelage with a concomitant increase in sebaceous glands as an adaptation to extremes of cold during the desert night (Sokolov, 1962). With the increase in sebaceous glands and secretion, one finds a corresponding necessity to dress the pelage. Since all species of rodents studied appear to dry their fur when it is moistened by means of either extending and flexing the body while lying on their side or ventrum or rolling over, the conclusion is unavoidable that ritualized sandbathing has evolved from the same set of basic movements in all rodent families. It is interesting to note that selection has favored a relatively stereotyped pattern that varies in a species-specific manner (see table 7). The higher taxonomic categories show less uniformity; however, the sandbathing Heteromyidae are remarkably uniform with their tendency to integrate side rubs and ventral rubs. This characteristic tends to set off the Heteromyidae from the Gerbillinae and Dipodidae.

Since sandbathing has the dual function of dressing the pelage and leaving a chemical trace of presumptive communicatory value, the evolutionary origins of sandbathing are inextricably tied to marking. Marking by means of the perineal drag probably had its origin in a common cleaning movement that consisted of wiping the anal-genital area on the substrate after urination or defecation. The stretch involving extension and flexion of the body also frequently accompanies elimination after the animal has awakened from prolonged sleep. Thus, selection could favor a combined ventral rub with perineal drag as a marking movement if the sebaceous secretions of the ventral epidermis had some inherent communicatory function that affected reproductive success or survival of the genotype. Such a ritualized marking pattern appears to have arisen as an independent element in *Meriones*, whereas in the Heteromyidae the ventral rubbing with its marking function has been combined with side rubbing in a functional sandbathing sequence (see table 9).

Patterns of Social Behavior

In a recent review of rodent social behavior (Eisenberg, 1966), I attempted to outline the origins and evolution of the various social systems to be found within the Rodentia. Social systems may be classified into two categories: solitary and communal. The communal systems have several subtypes including monogamous, polygamous, and family band groupings. For convenience, I will restrict the discussion in this paper to three categories: solitary, pair tolerance, and communal. The latter category corresponds to the family band as defined in the previously quoted monograph. It is the central thesis of this section that a given species has a typical social organization that falls into one of the three major categories.

In order to describe quantitative differences among species, three techniques have been employed: (1) the simultaneous or territorial encounter; (2) the maintenance of pairs through parturition; and (3) the study of groups derived from internal recruitment by births (see p. 26). The species-specific social organization is a result of species differences in the ability to tolerate contact with conspecifics. Thus, each species has a social tolerance that, when exceeded, will result in pathologies such as failure of the female estrus cycle, abandoning or destroying the litter by the mother, delayed maturation of the young, fighting to the point of wounding or death, and failure of the male gonadal development. The encounter allows one to make comparisons of the form and frequency of different postures from species to species. It also permits an assessment of the relative amount of agonistic behavior displayed by a given species. The pairing tests allow one to assess the effect of the male on the female throughout pregnancy and parturition. Again one can measure the relative social tolerance. Studies of groups allow one to observe the social tolerance throughout subsequent generations.

As discussed under Methods, most of the behavioral units described in this section involve an interchange of tactile or visual stimuli. Thus, in this investigation, consideration of presumptive auditory and chemical communication patterns has been minimized. It is understood that marking, naso-anal contact, naso-nasal contact, and, to an extent, grooming involve some chemical communication. Although auditory communication was investigated whenever the sounds were below 15 kc, ultrasonic sounds definitely were excluded in this study. Because of the incomplete nature of the sound recordings, a consideration of auditory communication will be deferred in this study.

Aspects of visual communication are difficult to evaluate among nocturnal rodents; however, the striking convergence in color patterns (especially of the tail) between the dipodid genera and the heteromyid genus *Dipodomys* deserves special comment. The kangaroo rats and jerboas have relatively large eyes, and in both groups ritualized upright postures are employed in fighting and during initial contact and courtship (see p. 23 and Eisenberg, 1963b). The white tip on the otherwise black terminal tuft of the tail appears to serve as an orientation point for a male when driving a female or when chasing another male. The white ventrum, displayed during a series of upright postures, produces a sharply contrasting reflecting surface that surely aids in orientation during sequences of mutual uprights and sparring. This latter characteristic is, however, common to all of the species included in this study and is not unique for the kangaroo rats and jerboas.

THE ENCOUNTER.—Encounters were run between males, females, and between males and females. In general, male to male encounters result in avoidance or in fighting, with the subsequent establishment of a dominant-subordinate relationship. Female to female encounters are less predictable and the most informative encounter type is that between a male and a female (Eisenberg, 1962a, 1963b). For the purpose of this paper, I will restrict the data to the male-female encounters under two spatial conditions and two physiological conditions.

The spatial conditions include (1) an encounter in a neutral arena and (2) an encounter in a territorial box (see p. 8). Since the dimensions of the encounter arena influence the outcome (Eisenberg, 1963b), the dimensions will be specified in all tabulations.

The two physiological conditions include (1) the estrous and (2) the anestrous states of the female. In all discussions of male-female encounters the male was judged to be in a sexually viable condition.

For tabulation purposes the behavior patterns are often classified into the following categories: contact promoting, sexual, and agonistic. The composition of each category corresponds to the classification presented on page 10.

Solitary Versus Tolerant Species: If we compare a series of encounters between males and anestrous females for different species, we find evidence for a distinct separation into two social types. One type either avoids contact or, if the arena space is small, engages in agonistic behavior. The second type initiates contact-promoting behaviors such as grooming. Table 10 lists the totals and average frequencies of bouts of the major behaviors displayed during encounters in a neutral arena or a territorial box. A tolerant species such as *Peromyscus maniculatus gambelii* engages in a grooming bout approximately once per encounter even in a territorial encounter, whereas a solitary species in a neutral encounter avoids contact aside from the naso-nasal. Tables 11 and 12 demonstrate a similar separation into contact-prone and avoidance-prone social types, although here the data are selected from a carefully controlled series that utilizes the neutral arena and the territorial box experiments. In the latter situation, contact-promoting behavior drops even further and agonistic aspects become more apparent.

An inspection of tables 11 and 12 suggests that *Meriones unguiculatus*, *Gerbillus gerbillus*, and *Allactaga elator* are more contact-prone, whereas *Dipodomys panamintinus*, *Perognathus californicus*, and *Gerbillus nanus* are more avoidance-prone when the female is in an anestrous condition.

The onset of agonistic behavior is not always immediate. In figure 3, the time course of an encounter between a male and female *Perognathus parvus* is treated in consecutive 100-second intervals. One may notice that on the day preceding estrus the male exhibits contact-promoting and sexual patterns during the first 100-second and third 100-second periods. During the second and fourth 100-second

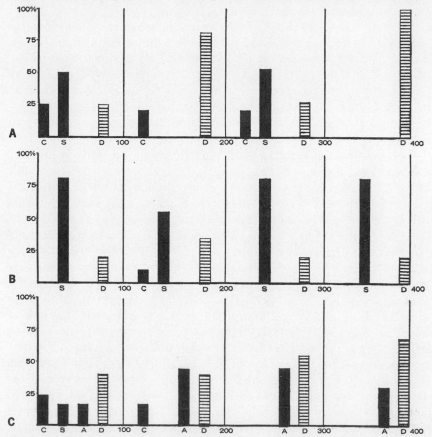

FIGURE 3.—Changing proportions of male *Perognathus parvus* behavior through the estrus cycle of his partner (A=day before estrus; B=day of estrus; C=day following estrus; ordinate=percentage of all acts for each 100-second interval; abscissa=1st, 2nd, 3rd, and 4th consecutive 100-second periods of encounter; s=all sexual acts; c=all contact-promoting acts; a=all agonistic acts; d=all individual behaviors).

periods, the male exhibits mostly individual behavior patterns that include digging and kick back, sandbathing, exploration, and sitting. On the day of estrus, sexual behaviors predominate throughout the 400 consecutive seconds. The day following estrus shows a gradual tendency for agonistic behavior to increase throughout the encounter. During the first 100 seconds, the male attempts contact and sexual behavior; some agonistic interaction also occurs. In the second period of 100 seconds, agonistic behavior increases and some contact-promoting behavior remains. In the third and fourth 100-second periods, the male alternates between agonistic behavior and individual acts of digging and sandbathing.

The reversion to agonistic patterns on the day following estrus was examined in detail by the successive encounter technique. Tables 13 through 15 indicate the reversion to agonistic behavior or avoidance on the day after estrus. The conclusion is evident: certain species of rodents such as the heteromyid *Liomys pictus, Perognathus californicus, P. parvus,* and *Dipodomys panamintinus* and the gerbillid *Gerbillus nanus* are prone to avoid or react in an agonistic fashion to one another except for the male-estrous female situations, whereas *Allactaga elator* and *Meriones unguiculatus* exhibit contact-promoting behaviors outside of the estrous condition of the female.

Whether solitary or social, chemical communication is implied in such postures as the naso-nasal or naso-anal configurations. In such semitolerant forms as *Meriones unguiculatus,* marking by the male plays an important role in an encounter. The desert-adapted species are prone to exhibit sandbathing during an encounter and in the intolerant species there is generally an exchange of sandbathing loci in order that the male may become familiar with the female's odor before achieving physical contact. The role of sandbathing in social integration has been discussed for the Heteromyidae in previous publications (Eisenberg, 1963b, c). Table 16 includes sandbathing data for three species: *Allactaga elator, Meriones unguiculatus,* and *Gerbillus nanus.* It is interesting to note that significant overlap between male and female sandbathing loci occurred only in the *G. nanus* encounters. This is in accord with the general theory since *G. nanus* is the only noncontact species of the three. The species conforms to the pattern of sandbathing during an encounter as outlined in previous publications for the intolerant Heteromyidae. There can be little doubt that chemical substances left in the course of sandbathing by *G. nanus* induce further sandbathing by an approaching conspecific. The exchange of chemical signals is thus a distinct possibility.

A further consideration of tables 12 through 14 points out some interesting aspects of behavior. During a male-female encounter on the days preceding or following estrus, the male exhibits more sandbathing and/or digging and kicking back than on the day of estrus. This is especially noticeable in the encounters with *P. parvus* and *D. panamintinus.* The digging and kick-back patterns exhibit all the characteristics of classical displacement activity and often follow a bout of preliminary sexual behavior or chasing. In both of the preceding examples the primary aggressive or sexual drives are not consummated, and the male switches suddenly to an apparently meaningless pattern of digging and kicking back, often while orientated toward the female.

Species-specific Patterns of Mating Behavior: In a previous publication (Eisenberg, 1963b), a series of heteromyid rodents was compared in order to demonstrate the differences between species and genera when the frequencies of the various mating behaviors were compared. It was found that the same basic components were present in almost all forms studied, but the relative frequencies were quite different. Table 17 presents a comparison of mating patterns for eight rodent species. All data were taken from behavioral records that had terminated in a successful series of mounts. Again the conclusion is unavoidable that, although the potential exists for expression of the same motor units by almost all species, the relative frequency of occurrence of any given unit exhibits unique characteristics for the species in question. Little information can be gained on taxonomic relationships from such an analysis since, in all probability, the quantified movements are not critical releasers as is the case with many birds and fishes. The role of chemical releasers that must act in part to promote sexual isolation in small rodents (Godfrey, 1958) remains to be investigated more thoroughly.

Although complete mating in the dipodids under study was not observed, certain characteristics of their precopulatory ritual render it unique. Males of *Allactaga elator* and *Jaculus orientalis* exhibit the following courtship patterns: As the female sits stationary, the male approaches and touches his nose to hers. The male may then groom the female on the head. If the female remains stationary, the male will hop to the rear and sniff the base of her tail or, if her rump is raised, he will perform an anal-genital sniff. If the female continues to remain stationary, he will straddle her tail and mount; however, she may initiate washing or sandbathing behavior or move away. In the former case, the male will pause and then hop around her to begin again with naso-nasal contact. On the other hand, if she moves away, he will follow and commence driving. While driving, the male *Allactaga* emits a buzzing sound followed by an audible squeak, whereas the male *Jaculus* emits only the buzz. This buzzing sound is unique among all the species studied. The mating behavior of *J. jaculus* has been described in part by Lewis (1965). It appears that the male *J. jaculus* utters a "chirping" call when driving the female. During the mount, which lasts less than a minute, the male employs a neck grip with his incisors. In the example cited, Lewis reports that the couple fall to one side during the terminal phases of the mount. (Compare with *Dipodomys* in Eisenberg, 1963b.)

Copulation involves the same basic movements for all species observed; however, the temporal patterning is variable when unrelated species are compared but uniform for a given species and frequently uniform for a genus. Figures 4, 5, and 6 summarize the copulatory patterns for species of *Meriones, Gerbillus, Perognathus,* and *Dipodomys.* *Dipodomys* is characterized by a long mount, whereas the other

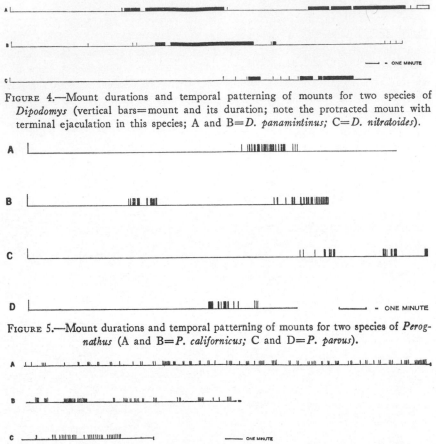

FIGURE 4.—Mount durations and temporal patterning of mounts for two species of *Dipodomys* (vertical bars=mount and its duration; note the protracted mount with terminal ejaculation in this species; A and B=*D. panamintinus;* C=*D. nitratoides*).

FIGURE 5.—Mount durations and temporal patterning of mounts for two species of *Perognathus* (A and B=*P. californicus;* C and D=*P. parvus*).

FIGURE 6.—Mount durations and temporal patterning of mounts for *Gerbillus* and *Meriones* (A and B=*M. unguiculatus;* C=*G. nanus*).

genera exhibit extremely short mounts. The *Perognathus* species and *Gerbillus nanus* generally have one bout of mounting with ejaculation on the terminal mount, whereas the more social *Meriones* exhibits a prolonged mount series with several apparent ejaculations. In this patterning, *Meriones* resembles the golden hamster, *Mesocricetus auratus* (Beach and Rabedaeu, 1959), and the Norway rat, *Rattus norvegicus* (Beach and Jordan, 1956).

At the time of ejaculation, the genus *Perognathus* exhibits a singular pattern. The male *P. californicus* rolls to one side while, in the case of *P. inornatus* and *P. parvus*, the female generally twists over on her side, frequently throwing the male over also.

This brief review should indicate that considerable uniformity in sexual behavior exists when species of the same genus are compared (e.g., *Perognathus* and *Dipodomys*). Table 18 summarizes the unique features of several rodent copulation patterns. Reference to this table further indicates that considerable variability can exist among the genera of a single family. At this time the adaptive significance of such differences is not apparent although one can surmise that *Dipodomys*, which exhibits such a sustained mount, surely must copulate underground in burrows.

PAIR TOLERANCE.—The data for this section were obtained by allowing a male to remain with a female throughout the parturition and rearing of the litter. Mere tolerance without fighting by a pair in the absence of reproduction was not accepted as evidence for social tolerance (see p. 3). Since the tolerance is in part a function of space, the dimensions of the cage are listed for all experiments. Further, the species ranged in size from 12 grams to 150 grams adult weight. To correct for this size bias, I have listed for each cage size two correction factors that express the area in terms of square inches per gram of animal and square inches per animal.

In general, even females of the most tolerant species will withdraw and nest separately at the time of parturition; however, tolerant species show a compatibility that permits sustained contact without extreme aggression throughout the first few hours after parturition and on through the rearing phase. Table 19 summarizes the pairing tendencies for several species prior to and through the female's estrous period. Those that exhibited solitary nesting tendencies were not paired beyond the mating; the more tolerant species were left together through parturition. This second class of data is summarized in table 20. Table 21 summarizes the data for those species that did not breed for various known and unknown reasons. The conclusions are as follows:

Regardless of the animal's size, given at least 70 to 90 square inches per animal, the normally tolerant species will conceive and remain compatible throughout parturition. The more intolerant species require greater space in order to avoid contact. One can breed the less tolerant species either by providing them with a larger space or by utilizing a series of staged encounters through a female's estrous period followed by total separation through the parturition and rearing phases (Eisenberg and Isaac, 1963).

GROUP FORMATION THROUGH INTERNAL RECRUITMENT.—By allowing
a pair to reproduce in a large cage, one can further evaluate social
tolerance. In general, reproduction will continue until social contacts reach such a peak that the female fails to cycle or deserts her
litter. If one adult male can keep other adults away and reduce the
level of contacts to which the sexually reproducing females are
subjected, then reproduction will continue. This is true for communal
as well as solitary species, but the latter rely on overt aggression more
often than do the former and also the solitary species are much more
sensitive to social interference than is the case with the more social,
communal forms. Thus, a communal species in a confined space
generally will exhibit an area in the cage where several sexually active
females are nesting and rearing litters. A single sexually active male
will have access to these females while the remainder of the population, adults and juveniles of both sexes, dwell communally and fail to
reproduce. (See also descriptions for *Mus musculus*, Crowcroft and
Rowe, 1963; and *Rattus norvegicus*, Calhoun, 1963a.) Eventually
reproduction may cease altogether. Figures 7 and 8 demonstrate the

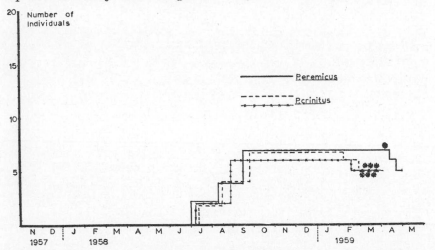

FIGURE 7.—Population growth for two species of *Peromyscus* (stars=number of individuals
showing wounds during the spring fighting).

population growth of several species of *Peromyscus*. Note in several
cases the tendency to form a stable plateau with a failure of reproduction. Note also, in the case of *P. crinitus*, the wounding caused by
male aggression. Although not illustrated in these particular graphs,
there is in some cases a slight trend to begin sexual activity at the
onset of spring after a fall and winter plateau; however, as shown
here the plateau holds. Table 22 includes the results of all population
growth experiments with eight rodent species. Note that the rela-

tively intolerant, solitary *Dipodomys nitratoides* ceased effective reproduction with the greatest number of square inches per animal and at approximately the same grams per unit area as did the less tolerant, semisolitary *Peromyscus crinitus.*

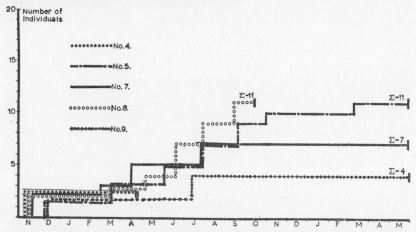

FIGURE 8.—Population growth for *Peromyscus californicus;* note the extreme stability.

An examination of the causes of population curtailment under confined conditions suggests four mechanisms: (1) During a plateau period, the females are generally anestrous with no evidence of normal estrous cycles; (2) the males born into the system show retarded gonadal maturation or even slowed growth; (3) males and even females may exhibit wounds as evidence of abnormal fighting, and such individuals may even die; (4) newborn litters may be neglected or destroyed. It would appear that different species have different thresholds of sensitivity for each of the four suggested pathologies. Thus, a solitary species, when forced to live as a pair in a small cage, may not exhibit reproduction because of a failure on the part of the female to cycle and, although they live compatibly, they are not exhibiting a natural tolerance. In a similar manner, the recruitment failure may be a result of male gonadal failure; agonistic behavior may be so intense that overt fighting and death result; and, if young are born, they may die from maternal neglect or cannibalism.

Lidicker (1965) has published the results of similar population studies utilizing *Peromyscus maniculatus, P. truei, Mus musculus,* and *Oryzomys palustris.* He concludes that species which normally do not experience high densities (or are dispersed in their native habitat) are more prone to exhibit overt fighting and cannibalism of young at high densities. My data concur in suggesting that tolerant species can form high densities with a minimum of agonistic behavior, but I believe the evidence shows that all systems of density regulation are latent in all species and only threshold differences correlate with a given social type (see p. 29).

In both Lidicker's study (1965) and my own, the species of *Peromyscus* show a tendency to increase their reproductive activity in the spring or autumn, following a cessation of reproduction after having achieved a density limit in the previous season. In one case a group of *P. eremicus* partially reared two litters, whereas two other *eremicus* groups showed increased male-male aggression in the spring but the females did not cycle. *P. crinitus* exhibited similar patterns with spring reproduction in one case and male-male antagonism in two cases (see table 22).

One may ask if a normally solitary species can be subjected to early experience that would be conducive to the production of more socially tolerant adult behavior. This may be the case with certain labile species that are adapted to exist at varying population densities. I have found, however, that strongly solitary species will not adapt to enforced proximity even as juveniles. It is true that juvenile groups can be maintained together for long periods of time, but normal reproduction does not take place. In figure 9 one can observe the reversion to agonistic behavior within eight days after a group of

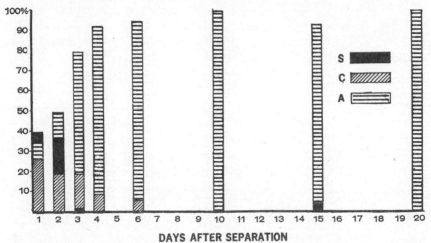

FIGURE 9.—Changing proportions of responses during *Perognathus parvus* littermate encounters after separation at 70 days of age (S=sexual behaviors; C=contact-promoting behaviors; A=agonistic behaviors).

Perognathus parvus littermates were separated. Not only is normal reproduction affected, but prolonged social experience often appears to have little permanent conditioning effect with this species and others. Littermate groups of typically solitary rodents that did not exhibit normal reproduction included *Perognathus parvus*, *Dipodomys nitratoides*, *Gerbillus nanus*, and *Liomys pictus*.

DISCUSSION.—As stated in the introduction to this section, rodent social systems may be classified as solitary, pair tolerance, and communal. The latter two represent in extreme form a closed social grouping, impermeable to outsiders. In reality there are rodent species that, as species, exhibit so-called loose social systems wherein the species may oscillate between a more dispersed solitary formation or a more tolerant, semicommunal system—depending upon the environmental conditions and the population density. The data from the current studies permit me to attempt several generalizations cast against the background of this social classification.

First, it appears that what separates a socially tolerant from a solitary species is the ability of the former to overcome an initially agonistic or avoidance interaction by means of special behavioral mechanisms to insure contact and familiarity without the danger of fighting. Among these mechanisms grooming appears to be important. Furthermore, a social species can achieve contact in the absence of primary sexual stimuli. A solitary species is either unable to achieve contact or, if it does, it is stressed to some extent; and, under conditions of prolonged contact, the physiological mechanisms governing the reproductive capacity break down. This difference between the extreme forms of solitary and social species is not viewed as absolute but rather as a result of differences in sensitivity to contact. Table 23 summarizes this concept whereby the same physiological pathologies are considered for each major social type.

A second generalization concerns the probable outcome of the various social tests if applied to a species of each major social type. Table 24 summarizes these predictions, which are drawn from the data in the previous section. In table 25 the species included in this study are ranked as to social tolerance according to the field and laboratory data.

General Discussion

The behavioral survey of the present study group has shown considerable uniformity in the orientation movements and fixed action patterns. Definite differences in fixed action patterns are usually correlated with corresponding differences in morphology as, for example, in bipedal locomotion. Even in such a markedly specialized behavior pattern, normally quadrupedal species have the capacity to assume a bipedal stance and even hop for a short distance. When we consider sandbathing as a fixed action pattern, we find that it is composed of two to three distinct movements that are part of the behavioral inventory of almost all mammals, and only a unique combination of these components results in a species- or taxon-specific behavior.

When we turn to social interactions, again it appears that the presence or absence of certain movements as an arbitrary criterion will not always suffice to delineate species or serve as a taxonomic tool. This results, no doubt, from the fact that small nocturnal rodents rely less on visual display in communication and, hence, less on visual releasers in sexual isolation with the result that spectacular differences in the fixed action patterns, observable in some avian and fish taxa, are not to be found. Certainly visual communication by movements, postures, and coat color are important in some mammalian taxa such as the primates and ungulates, but they are obviously less important in the small, nocturnal mammalian taxa. Again and again in this study, species-specific behavior patterns are found to be variations on universal neuromuscular patterns. Differences and similarities in the frequency of expression are more reliable indicators of nonrelationship or affinity than is the criterion of presence or absence of expression. In some cases even a frequency analysis fails to indicate affinity above the generic level, for example, in the temporal patterning of copulation. Thus, employing the presence or absence criterion, some behavior patterns will be family-specific, others genus-specific, and even more rarely a pattern will be species-specific; however, analysis of the discrete temporal patterning or relative frequency of occurrence almost always will demonstrate species differences although the adaptive significance of such differences may be unclear (consider again sandbathing). Since such an equipotentiality exists in the repertoires of these mammals—although frequency of occurrence serves to clearly delineate species—I can only conclude that differences in thresholds exist that account for the differential frequencies of expression. Thus, the behavior of a given species may be described in terms of the most probable sequence or set of patterns rather than in terms of its total potentiality for expression. In this sense, species-specific behavior describes the normal expression for a given environment and conforms to Leyhausen's "Verhaltenshäufigkeit" (Leyhausen, 1965).

By applying this definition of species-specific behavior to the social structure observed in different species, it is possible to characterize a given species as belonging to one of several categories of social behavior. For example, *Perognathus parvus* can be kept in groups with little overt fighting. This is especially true if the group is composed of littermates. Although they can tolerate one another, they do not reproduce. Utilizing reproductive success as the criterion for a natural social grouping, we can conclude that, although *P. parvus* has a range tolerance that permits group life without overt aggression leading to wounding, it does not have the ability to reproduce when subjected to social contacts above some minimum level. In the case of this species, the minimum appears to be only the pair association through estrus. Thus, as outlined in the previous section, there are probably several physiological mechanisms that are affected adversely in terms of reproductive success by social contacts. The relative thresholds for these mechanisms varies from species to species depending on their genetic makeup.

The adverse effects of high population densities on the growth and maturation of young rodents has been known for some time (Chitty, 1955), and the cessation of reproduction at high densities through a failure of ovarian and testicular function has been studied from a physiological standpoint by many workers (see Christian, 1963). Recently Calhoun (1963b) has developed an elaborate theory of social behavior based on the concepts of species-specific differences in social tolerance and the evolution of social groups. Wynne-Edwards (1962) has put forth an all-embracing theory on evolution of social groups and the adaptive value of self-regulatory mechanisms whereby populations are kept below maximum numbers by such reproductive failures as outlined previously. I do not wish to examine these various theories from a critical standpoint but only wish to point out that the evidence favors my interpretation of species differences in social tolerance and, further, that these differences have resulted from processes of natural selection to produce a social type adapted to its particular niche. I have attempted to outline a methodology whereby one can study social tolerance by means of several techniques and arrive at some quantitatively based conclusions regarding species differences in sociability.

There remains a further consideration. Given the demonstration that the rodents under study exhibit species-specific social tendencies, what adaptive correlates can one discern? The arid-adapted subfamily Gerbillinae spans the range from a relatively solitary form, *Gerbillus nanus*, to the more tolerant, communal *Tatera indica*. *Jaculus orientalis* and *Allactaga elator* are semitolerant; however, the family Heteromyidae, whether forest-adapted or arid-adapted, appears to be a solitary form. Within the genus *Peromyscus*, the desert species *P. crinitus* is intolerant whereas the equally xeric-adapted *P. eremicus* is more tolerant. Adaptation to xeric habitats with dispersed food supplies is not necessarily in and of itself conducive to selection for a dispersed, solitary existence. Equally important are other aspects of the species ecology including its mode of assembly of foodstuffs, its shelter construction, and its reproductive rate. At this point it seems safe to say that the Heteromyidae have retained the phylogenetically ancient trait of solitary existence because of the adaptive advantage accruing from its defense of cached food. It is not the case, however, that this social trait is always a concomitant of caching.

It is noteworthy that *Gerbillus (Dipodillus) nanus* resembles the silky pocket mice of the heteromyid genus *Perognathus* in several respects.   *G. nanus* is sensitive to crowding, and captive reproduction is possible only by means of a simultaneous encounter technique. During male-anestrous female encounters, the partners exhibit mutual avoidance or agonistic patterns of behavior.   Sandbathing as a means of chemical communication is displayed by this species and, after mating, the male and female return to behavior patterns expressing intolerance.   All the preceding attributes also characterize the behavior of *Perognathus parvus*, *P. inornatus*, and *P. longimembris* (see p. 22 and Eisenberg, 1963b).   Returning again to tables 2 through 5, there is a marked similarity between *G. nanus* and the silky pocket mice with respect to habitat, general ecology, size, and reproduction. This total convergence may well be indicative of a special adaptation syndrome that correlates with the attribute of social intolerance; however, until further comparisons are made on the energy needs and recruitment rates of different desert-adapted species, the selective advantage of the various social systems in all probability will remain obscured.

Summary

Selected species from three rodent families have been studied, including the following genera: Cricetinae: *Peromyscus;* Gerbillinae: *Tatera, Gerbillus, Dipodillus, Pachyuromys,* and *Meriones;* Dipodidae: *Jaculus, Allactaga;* Heteromyidae: *Heteromys, Liomys, Perognathus,* and *Dipodomys.*   Species have been selected in order to give a series of forms that have evolved in a convergent or parallel fashion in adapting to desert environments.

The discrete behavior patterns exhibit a profound similarity, but species and generic differences can be discerned.   Sandbathing has been selected for intensive study and it has been found that functional sandbathing has evolved independently in all xeritic-adapted forms from the same basic movement patterns.   Differences in the frequency of occurrence rather than in the form of the movement have proved to be the most effective criterion for delineating taxon-specific differences.

Social behavior has been investigated with a standard methodology including staged encounters, pair tolerance, and group formation tests. Using reproductive success as the criterion for describing a functional social group, it has been found that each species can be characterized into one of three general social types.   The different social types appear to represent different degrees of tolerance for the presence of conspecifics.   All social types will exhibit similar social pathologies under social stress, but the more social species have higher thresholds of tolerance.   It is proposed that species-specific social tolerance represents an adaptation to average densities that are of selective advantage to the species under consideration.   The adaptive correlates of the differing social systems remain, in part, unknown.

Literature Cited

BEACH, F. A., and JORDAN, L.
  1956. Sexual exhaustion and recovery in the male rat. Quart. Journ.
        Exper. Psychol., vol. 8, pp. 121–133.
BEACH, F. A., and RABEDAEU, R. G.
  1959. Sexual exhaustion and recovery in the male hamster. Journ.
        Comp. and Physiol. Psychol., vol. 52, pp. 56–61.
BÜRGER, M.
  1959. Eine vergleichende Untersuchung über Putzbewegungen bei Lago-
        morpha und Rodentia. Zool. Gart. (Leipzig), vol. 23, pp. 434–506.
BURNS, T. W.
  1956. Endocrine factors in the water metabolism of the desert mammal *G.
        gerbillis*. Endocrinology, vol. 58, pp. 243–254.
CALHOUN, J. B.
  1963a. The ecology and sociology of the Norway rat. U.S. Dept. Health,
        Education and Welfare, Public Health Service, viii+288 pp.
  1963b. The social use of space. Chapt. 1 *in* vol. 1 *of* Mayer and Van Gelder
        (eds.), Physiological mammalogy. New York: Academic Press.
CHITTY, D.
  1955. Adverse effects of population density upon the viability of later
        generations. Pp. 1157–1167 *in* The numbers of man and animals.
        Edinburgh: Oliver and Boyd.
CHRISTIAN, J. J.
  1963. Endocrine adaptive mechanisms and the physiologic regulation of
        population growth. Pp. 189–353 *in* vol. 1 *of* Mayer and Van Gelder
        (eds.), Physiological mammalogy. New York: Academic Press.
CROWCROFT, P., and ROWE, F. P.
  1963. Social organization and territorial behavior in the wild house mouse
        (*Mus musculus* L.). Proc. Zool. Soc. London, vol. 140, pp. 517–532.
EIBL-EIBESFELDT, I.
  1951. Gefangenshaftsbeobachtungen an der persichen Wüstenmaus (*Meri-
        ones persicus persicus* Blanford). Zeitschr. Tierpsychol., vol. 8,
        pp. 400–423.
  1958. Das Verhalten der Nagetiere. *In* Handbuch der Zoologie, vol. VIII,
        pt. 10, sect. 13, pp. 1–88.
EISENBERG, J. F.
  1962a. Studies on the behavior of *Peromyscus maniculatus gambelii* and *P.
        californicus parasiticus*. Behaviour, vol. 19, pp. 177–207.
  1962b. Behavioral evolution in the rodentia. American Zool., vol. 2, no. 4,
        abstr. 22.
  1963a. The intraspecific social behavior of some cricetine rodents of the
        genus *Peromyscus*. American Midl. Nat., vol. 69, pp. 240–246.
  1963b. The behavior of Heteromyid rodents. Univ. California Publ. Zool.,
        vol. 69, pp. 1–100.
  1963c. A comparative study of sandbathing behavior in Heteromyid rodents.
        Behaviour, vol. 22, pp. 16–23.
  1964. Experimental techniques in the analysis of grouping tendencies in
        insectivores and rodents. American Zool., vol. 4, pt. 3, abstr. 13.

EISENBERG, J. F.
   1966.   The social organizations of mammals. *In* Handbuch der Zoologie, vol. VIII, pt. 10, sect. 7, pp. 1–92.
   1967.   The behavior of *Peromyscus*. Chapt. *in* King (ed.), The biology of *Peromyscus*.

EISENBERG, J. F., and ISAAC, D. E.
   1963.   The reproduction of Heteromyid rodents in captivity. Journ. Mammal., vol. 44, pp. 61–67.

EISENBERG, J. F., and KUEHN, R. E.
   1966.   The behavior of *Ateles geoffroyi* and related species. Smithsonian Misc. Coll., vol. 151, no. 8, publ. 4683.

GODFREY, J.
   1958.   The origin of sexual isolation between bank voles. Proc. Roy. Physical Soc. Edinburgh, vol. 27, pp. 47–55.

HALL, E. R., and KELSON, K. R.
   1959.   Mammals of North America. New York: Ronald Press, 2 vols., xxx+1083 pp.

HATT, R. T.
   1932.   The vertebral columns of ricochetal rodents. Bull. American Mus. Nat. Hist., vol. 63, no. VI.

HEROLD, W., and NIETHAMMER, J.
   1963.   Zur systematische Stellung des Südafrikanischen *Gerbillus paeba* auf Grund seines Alveolenmusters. Säugetierk. Mitteil., vol. XI, pt. 2, pp. 49–58.

HOWELL, A. B.
   1932.   The saltatorial rodent *Dipodomys:* The functional and comparative anatomy of its muscular and osseous systems. Proc. American Acad. Arts Sci., vol. 57, pp. 377–536.

KING, J. A. (ed.)
   1967.   The biology of *Peromyscus*. [In press.]

KIRCHSHOFER, R.
   1958.   Freiland- und Gefangenschaftsbeobachtungen an der Nordafrikanischen Rennmaus. Zeitschr. Säugetierk., vol. 23, pp. 33–49.

KIRMIZ, J. P.
   1962.   Adaptation to desert environment: A study on the jerboa, rat, and man. London: Butterworth, xi+168 pp.

LATASTE, F.
   1887.   Documents pour l'ethologie des mammeriferes. Bordeaux, 659 pp. [Reprinted from Actes Soc. Linn. de Bordeaux, vols. 40, 41, 43.]

LEWIS, R. E., and LEWIS, J. H.
   1965.   On a collection of mammals from northern Saudi Arabia. Proc. Zool. Soc. London, vol. 144, pp. 61–74.

LEYHAUSEN, P.
   1965.   Über die Funktion der Relativen Stimmungshierarchie. Zeitschr. Tierpsychol., vol. 22, pp. 412–494.

LIDICKER, W. Z., JR.
   1965.   Comparative study of density regulation in confined populations of four species of rodents. Journ. Res. Popul. Ecol., vol. VII, pt. 2, pp. 57–72.

LORENZ, K. Z.
   1957.   Methoden der Verhaltensforschung. *In* Handbuch der Zoologie, vol. VIII, pt. 10, sect. 2, pp. 1–22.

MacMillen, R. E.
    1965. Aestivation in the cactus mouse, *Peromyscus eremicus*. Comp. Biochem. Physiol., vol. 16, pp. 227–248.
Ognev, S. I.
    1959. Säugetiere und ihre Welt. Berlin: Akademic Verlag, 362 pp.
    1963. Rodents, 508 pp. Vol. 6 *in* Mammals of the U.S.S.R. and adjacent countries. Israel Program for Scientific Translations.
Petter, F.
    1961. Repartition geographique et ecologie des rongeurs desertique. Mammalia, vol. 25, pp. 1–219.
Rauch, H. G.
    1957. Zum Verhalten von *Meriones tamariscinus*. Pallas Zeitschr. Säugetierk., vol. 22, pp. 218–240.
Rood, J. P.
    1963. Observations on the behavior of the spiny rat *Heteromys melanoleucus* in Venezuela. Mammalia, vol. 27, pp. 186–192.
Schmidt-Nielsen, K. S.
    1964. Desert animals: Physiological problems of heat and water. London: Oxford Press, xvii+277 pp.
Simpson, G. G.
    1959. The nature and origin of supra-specific taxa. Vol. 24 *in* Cold Spring Harbor Symposia on Quantitative Biology, pp. 255–272.
Skvortzov, G.
    1955. On the conditions of hibernation of *Allactaga jaculus* in Turkmenia. Vol. 4, pp. 39–51, *in* Rodents and the fight against rodents. [In Russian.]
    1964. Hibernation in *Dipus* and *Alactagulus*. Zool. Zhurnal, vol. xlii, pp. 1848–1854.
Sokolov, W.
    1962. Skin adaptations of some rodents to life in the desert. Nature, vol. 193, pp. 823–825.
Tinbergen, N.
    1951. The study of instinct. London: Oxford Press, 228 pp.
Wickler, W.
    1961. Ökologie und Stammesgeschichte von Verhaltensweisen. Pp. 303–365 *in* vol. 13 *of* Fortschritte der Zoologie. Stuttgart: G. Fischer.
Wood, A. E.
    1935. Evolution and relationship of the heteromyid rodents with new forms from the tertiary of western North America. Ann. Carnegie Mus., vol. 24, pp. 73–262.
    1955. A revised classification of the rodents. Journ. Mammal., vol. 36, pp. 165–187.
Wynne-Edwards, V. C.
    1962. Animal dispersion in relation to social behavior. Edinburgh: Oliver and Boyd, 653 pp.
Zahavi, A., and Wahrman, J.
    1957. The cytotaxonomy ecology and evolution of the gerbils and jirds of Israel (Rodentia: Gerbillinae). Mammalia, vol. 21, pp. 341–380.

Reprinted from *A Handbook of Social Psychology,* Carl Murchison, ed.,
Clark University Press, Worcester, Mass., 1935, pp. 185–203

THE BEHAVIOR OF MAMMALIAN HERDS AND PACKS

FRIEDRICH ALVERDES

Zoological Institute, Marburg

INTRODUCTION

Among most mammalian species every male, as a rule, exhibits a temporary or lasting relationship to one or several females; such a relationship is spoken of as either monogamy or polygyny, as opposed to promiscuity. Promiscuous sexual intercourse, though it may be observed occasionally in these species, never appears as the normal form of sexual relation but only as an accessory phenomenon. The monogamous or polygynous mateship may be either seasonal, i.e., ending after every mating-period, or permanent, i.e., outlasting even the annually recurring periods of sexual rest. It may be either solitary, as when the male forms a separate group with his female or females, or it may take the form of mateship within a herd, as when several (monogamous or polygynous) mated animals join to form a herd. Through combinations of these various possible relationships there result eight different forms of mateship which are characteristic of mammals. Unfortunately many facts are not yet known, and therefore, in regard to many species, we do not know whether they live in monogamous or polygynous relationships or in seasonal or permanent mateships.

PROMISCUITY

Genuine promiscuity seems to exist among bats which live in herds. Cases have been observed in which several males covered the same female one after another, the surplus males showing complete indifference. Promiscuity is also said to exist among the North American bison, but the data on which this statement is based have not been verified. This relationship has likewise been attributed to rabbits, although it is possible that mateship exists in this case, too, but has been overlooked by observers because the males do not keep the mating bond strictly. For a long time foxes were said to be promiscuous, but it has recently been shown that they live monogamously. The fact that house dogs pair promiscuously is the result of domestication, inasmuch as their characteristic way of living with men makes it impossible for them to form a regular mateship. It has not been ascertained whether wild boars form polygynous mateships during the mating period or whether promiscuity exists among them. During the mating-season individual males or packs of males unite with packs of females. These females may be intermixed with younger animals which are not yet capable of reproduction.   Combats which last over long periods of time develop among the males and lead either to the expulsion of rivals by one male or to the mutual tolerance of equally strong male individuals in the same pack.   In the latter case it is possible that the pairing is promiscuous (it might be permissible to call this a kind of "group-mateship"), but it is also possible that the rivals settle their rights of possession over individual females.

Monogamous Solitary Seasonal Mateship

A monogamous solitary mateship of the seasonal type does not last longer than a reproduction period, and at the beginning of each such period every male and every female looks for a new partner.   There are no relations between a partner in this mateship and individuals outside of this union.   A few beasts of prey live in a monogamous relationship, but none continue so throughout life.   With many species of cats and martens the two sexual partners remain together during and after the mating-season in order to protect and feed the young.   Lions choose a new partner in each mating-season, but throughout one season they live in a monogamous union.   Male and female jaguars form such a mateship for four or five weeks; during the rest of the year they are solitary. Foxes are also monogamous and remain together after the young are born; the fathers, contrary to earlier opinions, defend their young and bring them food.   Wolves live in couples during the spring, but form packs during the winter.   Wolves, foxes, and bears live in pairs for a short while even after the mating-period.

In the case of many mammals it has not been determined whether seasonal or permanent monogamy is the customary relationship.   They are seen in pairs during the whole year, part of the time accompanied by their young.   Certain species of dolphins, sirens, pigs, stags, antelopes, and blue whales belong to this group.   Many of them, especially the small antelopes, probably live in permanent monogamy.   Our lack of knowledge in regard to the duration of mateship is in part due to the fact that we do not have exact information concerning the oestrous and parturition seasons of many tropical wild animals.   According to Berger (1922), the theory that many African animals may bear young during the whole year is incorrect.   The duration of pregnancy differs according to the species; the oestrous period likewise varies from species to species. Schuster, contrary to Berger's opinion, believes that at least antelopes show no periodical oestrus and that, accordingly, the parturition season extends over the whole year.   Unfortunately, animals in captivity give us no reliable information about such matters because the environment is so totally changed.

Monogamous Solitary Permanent Mateships

In the case of many mammalian species which are found in couples during the entire year it has been impossible to determine whether they live in seasonal or permanent mateships.   However, it seems that at least the various species of rhinoceros form solitary permanent mateships. It is possible that the permanency of these mateships is related to the fact that, according to Schuster (1923), the oestrus of this animal is not interrupted during the whole year. Deeg (1922), however, reports a periodic oestrus for these animals. However this may be, we must not neglect the importance of the urge for mateship which operates together with the propagative instinct.

Monogamous Seasonal Mateships within a Herd

The very nature of such mateships makes it very difficult to determine whether monogamous animals which live in groups during the entire year form seasonal or permanent unions. To obtain satisfactory information it would be necessary to mark paired animals in some harmless way and observe them year after year. It is known that many rodents which build their dens in colonies and, within a stationary community of this nature, keep up distinct relations with one another are found in couples in their structures during the reproductive period; this is true for instance of the marmots (*Arctomys*), rabbits, ground-squirrels (*Spermophilus*), prairie-dogs (*Cynomys*), chinchillas, beavers, and rats. Since rabbits are said to live in mateship for several years, we may expect the same to be true of some of the other species mentioned above. *Oryx* and many species of whales are monogamous in wandering herds. The whales live in pairs within their herds and each pair is accompanied by one young animal. The sea-cows (*Rhytina stelleri*), which are now extinct but formerly lived in the Bering Sea, were also monogamous within a herd. Steller, who discovered them, reported that a male came to the shore for two days to a female which had been killed. The otters (*Latax*) are likewise monogamous within the herd; a male, a female, and a young animal keep together within the larger group. In all these cases it is not certain, as we have said, whether the mateship is seasonal or permanent; if it is the latter, then the species in question does not belong to this section but to the next.

Monogamous Permanent Mateship within a Herd

This form of mateship is an especially important one because it is the characteristic relationship in many races of men. If we analyze a herd of a species in which this type of mateship exists we may observe that it is not an unorganized group of individuals but consists of monogamous couples. Domesticated guinea pigs exhibit this type of relationship. The oestrus of this species is continuous throughout the year. There may be combats between unpaired males for the possession of an unpaired female, but one of the males will eventually pair with the female in question and the others will respect his right from then on. A mated male and female of the maras (*Dolichotis*), a species related to the guinea pigs and, like them, living in herds, form such a close association that they find one another again even after months of separation, during which time each one has been locked up with an individual of the opposite sex (Brehm,

1914-16). Wild rabbits build their dens in colonies, and within a colony the tunnels which serve as dwellings are interconnected. These animals are said to be monogamous over a period of many years; the males, however, seem to adopt accessory promiscuity inasmuch as they occasionally have intercourse with unpaired females. Further investigation of rodents which live in colonies may reveal other species which are monogamous over a period of years.

POLYGYNOUS SOLITARY SEASONAL MATESHIP

Many mammals observe this form of mateship, and the following behavior is typical of many species. Outside of the mating-period males are not interested in females and live in packs or alone. When the mating-period begins they look for the females, join their packs, or drive together as many females as possible. They combat each rival and only young males which are not yet capable of reproduction are tolerated. The Indian arnibuffaloes separate from the herd and form small groups at the beginning of the mating-period, each group of females being joined by a bull. Such packs of females, watched by a single male and followed eventually by young animals, are called "harems." It is characteristic of a seasonal harem that at the end of every oestrus it is deserted by the possessor of the harem. The harem scatters then (e.g., as in the elks) or stays together as a "mother's herd." As to the organization of solitary seasonal harems it may be said that in many cases the male assumes the leadership; frequently, however, the function of leadership is retained by an old experienced female that has been at the head of the pack when it was without a male. Strong stags drive away all younger males and rivals from the packs of six to twelve females which they have chosen. At the head of the pack there is always a female who determines the behavior of all the other females. This is also true during the oestrus. During the lively combats between grown-up stags young males sometimes succeed in approaching the pack and covering the females; they also appear and cover the females when the grown-up males are sexually exhausted. (This phenomenon should be considered as accessory promiscuity.) Other species of deer and many species of antelope, wild sheep, and wild goats behave similarly during the oestrus. The harems of Indian and African elephants, of mufflons, and of waterbucks (*Kobus*) also have female leaders. The male belonging to the harem is only a beneficial companion of the females. In earlier times such an elephant harem consisted of from thirty to fifty grown-up females, but due to the destruction which they have suffered during the last few decades we now find some harems that are much smaller. It has been observed that most deer, many species of antelopes, wild sheep, and wild goats form solitary harems, but it has not been possible to determine in every case whether a male or a female leads the group. The data existing are in part contradictory. In the case of gaurs, an Indian species of wild buffaloes, every herd of from eight to ten individuals has two bulls during the time when the females are not in oestrus. During the mating-period, however, a bull, which at other times lives solitarily but which proves himself to be superior to the bulls already in the herd, joins the herd and becomes absolute ruler. The two other bulls remain in the herd in spite of his presence.

Polygynous Solitary Permanent Mateships

In the case of guanacos and vicugnas every pack consists of a number of females and only one male except for the duration of the mating-period. Only young males not yet capable of reproduction are tolerated by the old male who leads the herd (the harem), sees to its security, and covers its retreat. If this leading bull is killed, the female vicugnas roam about aimlessly.

Among Asiatic aboriginal wild horses every herd consists of a leading stallion and a harem of from five to fifteen mares. The same organization may be observed in the herds of zebras; if the leading stallion dies, the mares find refuge in another harem (Schillings, 1906). Several zebra herds, each consisting of a leading stallion and several mares, may unite temporarily to form a larger herd, but such unions are loose and separate easily into their constituent elements, the individual polygynous mateships. During such a temporary union of herds the strong leading stallions watch their harems zealously. Zebras have no definite period of oestrus and parturition. Deeg (1922) has observed copulation and newly born animals in every month of the year. In countries where domesticated animals live in the half-wild state, as for instance in South America, every stallion is given from twelve to eighteen mares which he will keep together. Each group lives separately, and, after an incidental jumble, such as may occur when they are driven together, the individual groups immediately separate again. The superfluous stallions are castrated; these castrates live in groups by themselves. All stallions, even those who were living in stables until the time of observation, show the instinct of driving together a number of mares and of taking possession of them (Brehm), and the latter submit to it; evidently stallions as well as mares are directed by an instinct for polygynous mateship. The leading animal of Kulan herds, an Asiatic species of wild ass, is always a stallion; the older he gets the more females he keeps together and the number so kept may vary from three to fifty. If the leading stallion is killed the herd breaks up. The leaders of Nubian wild asses, herds of from ten to fifteen mares, on the other hand, are said to be old mares (Brehm), but each herd is accompanied by one male, who defends his rights stubbornly. Every kangaroo herd keeps one or several definite pasture-grounds that are connected by well-trodden paths. One herd always remains together and never mixes with another. Every herd is led by an old male and follows him blindly when looking for food or when fleeing. At the beginning of the mating-period the leading male demands the females of the herd for himself but has to establish his supremacy by combats with males that have grown up since the previous mating-season. Sometimes a herd divides, forming several herds that keep separate and are each led by a male.

Polygynous Seasonal Mateships within a Herd

If individuals of a certain species live in herds during the whole year, then the structure of the herds may differ greatly according to the season. During the period of sexual rest the individuals may separate into a male and a female herd, or the males and females may keep separate in two groups within one herd. The awakening of the sexual urge, however, brings about a complete change, inasmuch as the herd then consists only of harems. The fur-seals (*Callorhinus*) represent an especially clear example of this type. Outside of the mating-period the bulls live in separate herds in the open sea, and so do the "bachelors," i.e., those males who have had no chance to pair; some of the bachelors, however, seem to join the females with their young ones. At the beginning of the mating-period the old bulls come to the shore every year and occupy there the same rock they have occupied before. Multitudes of females appear later, and the males begin at once to form harems of from fifteen to twenty-five, at most forty. The females either are far advanced in pregnancy, or, if only one year old, are still virgin. The males enter violent combats, until gradually equilibrium, mutual recognition, and a division of interests are established. The old defeated and expelled bulls and the bachelors who were excluded from reproduction by the owners of the harems form two separate groups. When all females are covered the harems break up, and all animals move out into the open sea. The mating of other species of seals takes place in a similar manner.

Polygynous Permanent Mateships within a Herd

This form of mateship has been established by many human races. In regard to other mammals the following facts are known. The tarpans, extinct European aboriginal horses, lived in herds of many hundred animals. Each of these herds separated into a number of families, each one consisting of a leading stallion and several mares; no stallion would tolerate any younger stallion or older rivals capable of begetting in his harem. The stallions frequently drew away domesticated mares, and it was for that reason that the South Russian peasants pursued them. In the case of zebras, Kulans, and South American horses which have become wild, several stallions may unite with their harems and form more or less loose unions, but each one keeps his females for himself.

Families

From mateship a family may originate, but that is not necessarily so. Parental, paternal, and maternal families may be distinguished, depending upon whether both parents, the father, or the mother remains with the offspring. If the offspring are deserted by both parents, but keep together for a while, they form what is called a "children's family."

Among many mammals not only the mother, but both the father and the mother guard and train the young; it is probably due to the nature of the subject in question that monogamous fathers take more care of their young ones than do polygynous fathers. In the case of mammals which have a long period of development two or more litters may remain with their parents. In the case of many beasts of prey the fathers remain with their mates and children for a certain time. Male lions are said to gather food and to protect their mates and young ones. Many beasts of prey learn the methods of hunting under the supervision of their parents. Foxes, too (in contrast to earlier opinions), defend their females and young ones. On the other hand, it is uncertain whether male wolves take part in supplying their young ones with food. The males of many monogamous ungulates (peccaries, *Phacochoerus,* certain deer, antelope, gazelles, etc.) also defend and lead the young ones. In the case of wild rabbits one reads that males protect their young ones and that the females hide them from other males. It is possible that their behavior varies according to regions. In the case of ungulates living in polygynous permanent mateship the leading stallions always protect their whole harem and their young. With catlike beasts of prey the mothers drag in small living and half-living animals for their young, but sometimes the fathers take a share in the training. With most beasts of prey, however, the mothers have to keep the males from the young ones, or else they will eat them. The mothers release their prey in the presence of their young ones, who thereby get practice for their future occupation of killing such animals. Later the mothers take their young along with them when they hunt. Such instruction may take months. Only when they are fully trained do the young separate from their mothers. While the prey-seeking instinct is inherited, its development depends upon instruction. Female foxes, like the females of catlike animals mentioned, train their young with living prey. Female ice-bears and seals carry their young ones into the water and teach them how to swim.

Solitary maternal families exist among many ungulates (e.g., the elks). There, where mammals which live in large herds are often pursued by their enemies, herds sometimes separate to form maternal families. Among elephants Schillings frequently observed small herds consisting of a mother and six to seven younger animals of varying size; in all probability these young animals were young of the elephant mother born during the past twenty to fifty years. When many mothers and their offspring unite without the fathers, the group is called a "mother herd." There are also children families among mammals. Rather young individuals of sperm-whale (*Catodon*) keep together in groups up to sexual maturity. Young deer remain with their mothers until the next mating-period and then are driven away by their mothers and frequently form independent groups. Young male and female reindeer which are not capable of reproduction unite to form strong packs led by an old virgin female.

HERDS

Among many mammalian species living in herds one can find old solitary males, the *"Einzelgänger,"* who probably are unable to be members of a herd because their sexual urge has weakened. It is quite possible that the other members of a herd drive away the individuals that are getting old. Solitary individuals can be found among elephants, buffaloes, deer, antelopes, giraffes, hippopotami, kangaroos, rodents, etc. It has been observed that strong male gnus drive older males away from the herd and keep them away permanently. Those driven away remain as if they were sentinels at a distance of a few hundred meters from the herd and warn it by their behavior in case of danger. Very old bulls later separate entirely from the herd and pass the rest of their lives solitarily. It has been reported that old bulls of the bantengs (*Bos banteng*) are driven away by the young ones. Old male individuals either keep strictly isolated or look for their equals of their own species or of others. Thus old gnu bulls and male elephants unite to form groups of two or three. Schillings observed two old elephant bulls in company with an old giraffe bull for weeks. As a matter of course there are always a large number of males among polygynous animals that are excluded from reproduction either temporarily or even for their whole lives. In the case of fur-seals and related species, where every male gathers a large harem, the excluded males unite to form two groups: the old expelled or wounded bulls make up one group, and the younger males, the "bachelors," who have not yet been permitted to pair, form another group. Young mature males of the guanacos, driven away from the herd by the leading stallion, unite with their equals and with young immature females. Young male vicugnas remain with their mothers until they are grown up, but then the entire pack of females drives away the young males which are capable of reproduction by biting and beating them. These males unite to form herds of twenty to thirty head, admitting a few individual defeated males. In such groups there are no leaders, and there is continual quarreling, especially during the mating-period. Such bachelor packs are to be found among deer, antelope, gazelles, buffaloes, and the castrate horses in South America.

Males and females of many mammalian species form common herds outside of the mating-season. Wolves, jackals, and hyena-dogs (*Lycaon*) live in packs and hunt together. It has been observed during wars that in periods of stress domestic dogs instinctively form hunting packs. When a certain section has been deserted by its inhabitants, the farms burned down, and the cattle and most of the dogs have perished, the remaining dogs frequently form packs; these groups hunt on the fields in a formation that looks like a line of skirmishers. A group may be made up of dogs of different races, colors, and sizes. Hunting in long arrays is no tradition among dogs; they unite to form such hunting companies in times of stress, driven by an old instinct. Related wild species of the domestic dog hunt in a similar way. Wolves live in packs during the winter and the members of a pack go in single file; in snow one animal treads in another's footsteps. When hunting together they divide into two packs; one pursues the prey, while the other cuts off its path. The African hyena-dogs (*Lycaon*) hunt large prey in packs. The rear animals of the pack cut off circuitous paths and relieve those animals that were leading before, and this is repeated until the prey is exhausted. In countries rich in wild animals, lions hunt together, sometimes accompanied by their young ones; some drive and come catch the prey; roaring seems to play an important part in their driving.

After the mating-season the two sexes of many animals live in separate herds, but in the case of pronghorns (*Antilocapra americana*) and saiga-antelopes (*Antilopa-saiga*), males and females form a common herd throughout the year. Also males and females of mammals living in colonies remain together. Many rodents and coneys (*Procavia*) live similarly in herds. Within such colonies individuals, either alone (ground-squirrels) or in pairs (wild rabbits), build separate dens and structures, which may be connected to one another, or several individuals build a common dwelling. Common dormice (*Muscardinus*) build skilfully constructed nests in bushes. In the case of jumping-rats (*Dipodinae*) and related species, marmots, wombats (*Phascolomys*), and kangaroo-rats (*Bettongia*), several individuals live together in dens which they dig. Viscachas (*Lagostomus*) and beavers live in colonies with the members in close contact with one another.

In a number of species living in herds the two sexes live separately except for the duration of the mating-period. It is not unusual that the males unite to form one herd, and the females another. When young ones join female herds, one calls them maternal herds. We can see, therefore, that not only with men, but also with animals, individuals of the same sex may unite for purposes other than sexual. According to Schillings many species of antelope form herds divided according to sex during seasons other than the mating-period. Wild sheep, steinbocks (*Ibex*), and other ungulates also live in packs separated according to sex outside of the mating-period. African and Indian elephants form separate packs of males and females. The female herds have a female leader and are accompanied by the young ones. In earlier times it was possible to see herds of several hundred head, but they were really temporary associations of several herds. In recent times the greatly reduced herds of Africa no longer observe the same strict division as before when there were a great many elephants. When occasionally a male and a female herd unite, the separation of the two sexes remains unchanged; this can be clearly observed during rest periods of their marches. The two sexes enter into a close relation only during the oestrus when a male takes possession of a female pack. The millions of bison that were living in North America did not form large disorganized herds, but every herd, when carefully observed, showed a subdivision into individual groups, bull herds with six to sixteen head each and cow herds with thirty or more head each. The female groups were led by a young male, and every male group had a leader as well. The male groups formed the edge of the entire herd, while the female groups were in the center. While marching and eating, the bison were generally in file. It has been observed that the animals did not form pairs or harems during the oestrus, and yet the order as described before must have been disturbed considerably for some time. During other times than the mating-period one can find the two sexes of the seals separated into different herds, the females being accompanied by young ones and sometimes by "bachelors." The old males of mule deer (*Caria-cus*) unite to form packs, but young males join packs of females. The females of many species live with their young ones in maternal herds, the young males capable of reproduction join to form male packs, and the old males live solitarily and care for female packs only during the mating-period. Wild boars, stags, and fallow bucks belong to this group. The males of bats live alone after the mating-period, but they live in groups at other times. The females form special groups; up to the present time no male has ever been found in these female herds. The female herds of the European bison, now completely extinct as a wild animal, consisted of ten to twenty head and were always led by an old cow and accompanied by two to three bulls. Young bulls lived in packs of fifteen to twenty head, but old ones solitarily. Female European bison rarely were solitary, and, if so, it probably was due to the destruction of this species. During oestrus the old bulls joined the packs of females, and violent combats developed among the rivals. Similar facts have been reported regarding other buffalo species. The males and young animals of the elk unite to form packs of fifty individuals at most, while the females and their young ones frequently remain apart. Typical maternal herds have been observed among many whale species. Sperm whales appear in schools consisting of twenty to thirty animals, all of which are females and their young ones; the leading animal of each school is an old male. Sometimes several schools and their leaders unite. Schools of gray whales (*Rhachianectes*) have been observed to have female leaders and they have no males at all. In the case of otters (*Lutra*) living solitarily, several females with their young ones not infrequently join to form a maternal herd. Sometimes two or three lionesses with their young are observed to form a group.

In one herd there may be individuals of different species. Dolphins of different species join and form troops that are led by individuals; sometimes whales follow a ship, propelled by such an "escort instinct." Antelopes of different species sometimes mix, but water-antelopes (*Redunca*) offer an exception, for they always remain apart. Peruvian brockets (*Hippocamelus*) and elks join herds of horned cattle. Wild zebras follow domestic horses and graze among them. Gazelles mix with herds of cattle. Sometimes one can see Kulans, antelopes of different species, yaks (*Poëphagus*), and scattered horses joining together. Wild buffaloes join elephants. Zebras chum with various species of gazelles, with ostriches and balearicans. Such cases have been considered as "mutual insurance," for the long-legged birds just mentioned are "eye-animals," while the hoofed mammals in question are "nose-animals"; that which the eyes cannot see the nose may detect, and vice versa. We do not know yet to what degree experience and tradition play a part in this form of association. In a similar way antelopes and ostriches sometimes stay among herds of baboons for hours.

One may distinguish between closed and open groups. New members are not accepted without resistance in groups of the former type, and individuals that have been accepted separate only under special circumstances; there frequently exists a distinct hierarchy among the members of a group, but members of open groups join and separate without special difficulties. Solitary monogamous mateships, families, and solitary harems are closed troops, excluding other individuals. Parents always know their young ones and many parents bite or kill strange young animals. Herds occupying certain sections and defending them against intruders belong to the type of closed groups. Different animals put into one cage may form a closed group in the course of time. Mutual recognition and acceptance of individuals in cases as mentioned above depend upon the fact that one individual knows another. But with the increasing number of individuals, closed groups come to be open ones. The half-wild pariah-dogs of the Orient, according to Brehm, quarrel at once with their equals who have not grown up among them. And every alley of an Oriental city has a group of half-wild dogs of its own who never forsake their headquarters. If a dog enters a strange alley all the dogs living there attack and kill him, unless he saves himself by fleeing instantly.

Frequently a group has a leader, which may be a male or a female according to the species. Frequently it is the strongest and most experienced individual that occupies this rank. But we must nevertheless not assume that experience is the only factor, but that there are specific "qualities of a leader" playing a part. When an animal tries to enter a closed group, it is usually received in a very unfriendly manner. Schulz added a third rhinoceros to a couple of other ones already accustomed to one another; the two attacked the newcomer at once and only later became friendly. Similar attitudes can be observed when a new animal is brought into a cage of a zoological garden. Beasts of prey, and also other species, as for instance prairie marmots (*Cynomys*), attack a newcomer of their own species or genus at once, and maltreat him or bite him to death. If the newcomer survives this reception, the animals frequently make peace; the meaning of these combats probably is the establishment of a hierarchical order.

It is not always possible to draw a sharp line between closed and open groups. Marmots build structures underground and even during seasons other than the mating-period live together in large numbers. But when these animals appear on the surface they place sentinels. The several viscachas connect their underground dwellings. Such a settlement is called a "viscachera." An old male seems to play a leading rôle in every one of them. Neighboring viscacheras are connected by paths that are used by the animals for visiting one another. When a viscachera is demolished the members of a neighboring settlement dig out the inmates. The few beavers still living in Europe usually live in couples, but beavers in rather quiet countries, as for instance in Canada, may live in small or large groups. In thickly populated countries they usually live in simple tunnels underground and do not build houses (Burgen), but in sparsely settled countries they may build houses.

The houses are hills which look like baking-ovens and have thick walls; they are built from peeled pieces of wood, branches, earth, clay, and sand; they contain a dwelling and a storeroom for food. In places where animals can work without being disturbed a great number of individuals usually work together in the building of one structure. The females are the builders, the males rather the carriers and hands. When the level of the water changes during the course of a year, or if the water is not of the proper depth, the beavers build dams across the brook. The dams are made out of the same material as the houses and have an almost vertical wall on the side of the water, while on the opposite side they have a slope. Holes in the dam are discovered and filled in. Canadian beavers sometimes reinforce the first dam by building other dams below the first one. It is said that the dams of European and Canadian beavers become hundreds and thousands of years old. Innumerable generations cooperate in building them and caring for them.

In captivity individuals of species that when free have no relation with one another develop friendly relations, being impelled by an urge for company. Animals in captivity usually become friends of the person who takes care of them. Schillings was able to improve the degree of comfort of a young rhinoceros which was in captivity for a short time by giving him a grown-up goat as a companion. They soon were such good friends that the goat frequently rested on the rhinoceros.

The herds and groups of mammals that are formed during other seasons than the mating-period frequently are open groups. One individual is the leader, and some serve as sentinels during the stops of a trip. Gnus and buffalo herds not infrequently rest on hills which permit a view of the surrounding territory. Giraffe and cervine antelope (*Bubalis*) herds are led by a male or a female. Cervine antelopes frequently post their sentinels on termite hills. The saiga-antelopes live in large herds and the old bucks remain with the herds at all seasons. There are always some animals watching; no watching animal rests without disturbing another animal. Only when the other animal gets up and watches does the first lie down to rest. In case of danger these antelopes gather and then flee in the order of a long line, with usually a male, but sometimes a female, leading. With other species the leader is either always a female or always a male. The female leader of a reindeer herd always stands when watching, even if all the other individuals of her herd rest; when she goes to rest, another female gets up at once. In case of danger, the leading reindeer sometimes pushes some individuals of the herd with her antlers to stir them up. It is well known that animals living in herds sometimes follow their leaders blindly, even down an abyss if the leader falls into it. If a leader is killed a herd may be headless for a time. A herd of bottle-nosed whales has an old male as leader whom they follow blindly. Brehm says that a bottle-nosed whale (*Globiocephalus*) who escaped a massacre committed by the inhabitants of a certain shore returned again and again to his dying and dead comrades. In the case of cows the herdsman places the largest bell on the leader. The cows know the bells of their herd very well and individuals that have gone astray find their way back again with the help of the ringing bells. The leading cow shows a certain pride regarding her rank and does not permit any other animal to walk ahead of her. In South America where large herds of llamas are used for the transport of foods a male richly decorated with blankets serves as leader. In the same way leading animals of South American caravans of mules are richly decorated. A leading sheep carrying a bell can keep together from three to four thousand individuals. If the bell is missing, the herd separates into groups of six to twelve head, each one under the leadership of one individual. By marking one of the leading animals so that it may be seen from afar, man enables this one to lead several thousand.

General Attitudes of Mammals Living within Herds

The subject of mutual help in the realm of animals, and the opposite, mutual injury, have frequently been discussed in the field of animal sociology (see Kropotkin). Cases of mutual help are not rare; on the other hand one should not generalize such cases too much and look at them in a sentimental way, for mutual injury is also fairly frequent in the realm of animals. This occurs even among individuals of the same species or group. Mutual help is shown in the case of an animal that rushes to the aid of another animal, setting up cries of warning and distress, and helps without being attacked itself. Schulz reported that when one of his two young rhinoceros was attacked, the other one rushed to the spot and took part in resisting the attack. Prairie marmots (*Cynomys*) and viscachas pull wounded comrades into their dens. According to Berger no species but the apes have as strong an instinct to help their sick comrades as the elephants have. If an elephant is wounded by shooting, other animals support him and, if he falls, his comrades kneel beside him and push their tusks under his body, while others put their trunks around his neck and try to put him on his legs. Mutual help of animals can be observed when they pair and care for their nestlings. The males defend their females; one or both parents protect their young ones. Young animals that have lost their mothers are sometimes adopted by another mother without any fussing. This is true of elephants and wild sows. It is well known that one can foist young ones of very different species, such as young rats, mice, rabbits, wild rabbits, dogs, foxes, squirrels, etc., upon a female cat, on account of her well-developed nursing instinct.

Many sick or wounded animals are thrown out of groups of their former class, or even killed. Before moving into their winter quarters the marmots join, and, according to Girtanner, it sometimes happens that several individuals attack an old thin animal and bite it to death. The biological significance of such a phenomenon is clear, for when an individual dies during hibernation the whole troop is endangered by its dead body. It has been observed that beavers sometimes drive an old, or sometimes even a young, animal out of their colony by biting it. Such animals live in a special house from that time on. There has been much talk about so-called "executions" among animals. The following facts are verified with regard to mammals. "Executions" occur occasionally among many species of beasts of prey and even among cattle. Groos believes that they cannot always be traced to an instinct, useful to the species in question, that urges the animals to remove an individual that has become dangerous to the entire group, for it would be sufficient to expel that individual. The phenomenon rather manifests an inborn instinct of destruction. Cannibalism, i.e., the killing and eating of animals belonging to the same species, occurs frequently in the realm of animals. A condition for such an act is, of course, that one individual is able to overpower another, i.e., the defeated animal must either be weaker than the other or must be hindered in its defense. Individuals of solitary species, e.g., shrews (*Sorex*), and beasts of prey are relentless against other individuals of their species; even male hamsters (*Cricetus*) kill females when they meet them at times other than the mating-period, without, however, eating the dead bodies. The individuals of many species eat their young ones; the parent who does not participate in the care of the young attacks the offspring. Many beasts of prey do this. Animals in captivity leave each other alone as long as they are of equal strength, but if one gets sick it may be attacked and sometimes eaten. This may even happen among brothers and sisters. When a muskrat (*Fiber zibethicus*), or a brown rat (*Mus decumanus*), or a water rat (*Arvicola amphibius*) is caught in a trap, it is torn to pieces while still living and eaten. Sometimes, if one pinches the tail of a rat in captivity so that it shrieks, the rest of the animals in the cage attack it and bite it to death.

If many individuals are together, the entire herd may join a few animals in action. Herds of horses and domestic pigs fight successfully against a wolf, while an individual horse or pig is frequently lost. Similarly wolves organized as a pack are much bolder than individuals hunting alone. Animals living in herds, e.g., half-wild horses of South America, are frequently seized with a panic. Hundreds or thousands of horses gallop along as if mad, running against rocks and into abysses. They carry away with them draught animals and mounts of travelers who may happen to rest nearby (Brehm). Sheep run like mad away from wolves and, in Australia, from dingos. Many ordinary activities are intensified by group practice. This is true for eating; farmers are familiar with the fact that many animals, e.g., pigs and horses, especially when young, eat better when with companions than when alone.

Many species of mammals show characteristic phenomena of courtship and mating. Males usually are stronger, more conspicuous, and more aggressive than females. They fight for and strive to win the females, while the latter remain passive. In the case of some species the urge to fight has developed to excess. The combat of stags appears to be a goal in itself during the mating-period. Stags not infrequently forsake their harems in order to meet their rivals whose call they hear. Combats are jousts, though they may sometimes turn out to be serious. We may say in a general way that many mammals with antlers and horns enter this kind of combat during the oestrus. European bison, intoxicated with the sex drive, tear up medium-sized trees by the roots; not infrequently slain bisons, bulls and cows, can be found during the oestrus.

All animals may be said to play together, young ones especially, but old ones as well. Not only individuals of the same species play together, but also those of different species, especially in captivity. We must be careful, however, in interpreting these activities not to confuse real hunting, or pairing, or similar activities with playing. In games an element of rivalry can frequently be observed, as for instance in sham fights and hunting games. According to Groos hunting games may be classified as follows: those with living prey (a cat with a mouse); those with living mock prey (one dog hunting another one); and those with lifeless mock prey (playing with a ball). According to Pfungst young wolves and dogs arrange jousts and hunting games with living and with lifeless mock prey, in which young wolves without any kind of training show some elements of fetching and carrying. According to Schmidt the impulse to play may come from the young ones, or from the mother. The latter brings a half-dead mouse or something similar, and the young ones practice with it. Dogs (more social animals) take up games more easily than cats. Dogs, cats, and other mammals play more and also more intensively with youthful persons than with grown-up people and submit to rougher treatment when playing with children. They seem to have a certain amount of understanding for the awkwardness of children. If the playful activity of young mammals turns to earnest, their mothers sometimes separate the young fighters. Brehm reported a certain activity of chamois that can be interpreted only to be play. These animals, completely undisturbed, would glide in a crouching position, one by one, over a slope of snow. The distance would sometimes amount to 150 meters. Then they would climb back to the starting-point and glide down again.

Animals now and then produce sounds in common. These sounds may be interpreted as imitation pure and simple, as rivalry in the case of males, and as shouts of greeting. The latter are expressions of excitement and are heard when two or more animals meet. When a lion, free or in captivity, roars, other lions nearby join at once in roaring. A donkey may cause all other donkeys of his neighborhood to bray. In a similar way entire herds may roar. Tamias have the habit of calling one another with chirping sounds for hours. Sometimes one can recognize a certain rhythm. This habit probably has no relation to sex. Two bats that meet call one another.

Many mammals claim a certain territory for their residence; when two dogs meet they may fight one another. Every kangaroo tries to keep a certain territory or several territories connected with each other.

Every herd of the North American pronghorn (*Antilocapra*) occupies a certain area which it claims for itself. They undertake long trips into various sections of it. Many animals keep a supply of food and consider it more or less strictly as property of their own. According to Schillings leopards have the habit of hanging the flesh of an animal which they have killed on the branches of trees and bushes, after having eaten its heart and liver and buried the intestines. Now and then they climb up high to do this. Hamsters, living solitarily, build one or several storerooms in their houses; moles, also solitary, gather earthworms, biting off their front ends to prevent them from escape; squirrels (*Sciurus*) hide supplies of food.

Imitation plays an important part in animals living in herds. It has been frequently observed that, if a trained beast of prey attacks a tamer, all other animals of its group join in the attack. According to Sokolowsky animals of a herd do not stop a common action, but obey some kind of "agreement," and observe the motions of the animal that started the attack. In a similar way many animals living in herds imitate one another when fleeing or attacking, and the majority are not aware of a cause of the change. A horse, hearing the sounds of galloping caused by another one, reacts very distinctly. Many animals have a characteristic warning call: chamois and marmots whistle, wild rabbits beat the ground with their hind legs, and kangaroos show alarm in a similar manner. Many solitary and gregarious species understand each others' calls of warning (Schmid). When reedbucks (*Redunca*) warn with a whistling sound, not only animals of the same species, but also other mammals and many birds flee.

Von Uexküll studied carefully the manner in which dogs make known their home place to other dogs. The spots they distinguish by means of their "marks of odor" are identical with those that men use to remember places by sight. If two dogs take a walk with their master, they start a regular contest in urinating. A high-spirited dog is inclined to provide the next outstanding object with his "mark of odor" whenever he meets another dog. Also he will try to cover "flags of odor" of other dogs whenever he enters their home territory. A dog without temperament, however, will timidly pass the "marks of odor" when he enters the territory of another dog and will not reveal his presence by placing marks of his own. The tall bears of North America also mark their territory. Bears, standing erect, rub with their backs and mouths the bark of trees that stand alone and are visible from considerable distance. These marked trees serve as warnings to other bears to avoid these signals and the territory marked by them. In the case of many mammals, odors and excretions of special glands serve the individuals to establish and to maintain a mutual relationship. The white spots on the posteriors of many deer also have been said to be marks by which individual animals are enabled to follow their leading animal and their pack. In the case of deer the white spots are distinct in winter only; this is said to be due to the fact that the females are alone with their young ones during summer. The white spots of wild sika (*Pseudaxis*) are invisible while the animals rest, but very distinct while they flee. According to Berger those species that have no white spots usually have tails which are light-colored on the lower side. The beating movements of the tails are supposed to be signals to show the way. According to Schillings giraffes wave their tails vigorously right and left when they flee or become suspicious.

According to Pfungst wolves have various sounds at their disposal, about half of which express anger; even young wolves kept in complete isolation develop real barking when they are menaced. This is sufficient to prove that wolves do not learn barking by imitation from dogs or from the voice of men. Domestic dogs evidently bark more frequently because they are without the restraint of wild animals forced to be silent; and therefore they learn to make use of barking as an expression of various emotions, e.g., of joy, while originally it expressed only anger. Female wolves throw themselves on their backs to express helplessness.

According to Darwin the scratching of dogs and wolves after completed defecation is a rudiment of an urge to hide the excrements. Pfungst believes, however, that it is impelled by the animals' instinct to spread their scent. The fact that many animals leave their excrements on elevated spots or stones is said to verify this theory. If we accept von Uexküll's opinion with regard to the meaning of urinating of dogs, we may conclude that excrements also mark the territory. Free hamsters, badgers, martens, and rabbits have definite dung places. All llama species and various species of antelopes pile up their dung. Guanacos start a new pile of dung when one has reached a certain size. Cervine antelopes (*Cape hartebeests*) deposit their dung on certain places that have the shape of a circle and are made even by treading. The center of such a place is a small ant-hill covered with dung. These level places come into existence because a pack of animals gather on or beside an ant-hill, use it as a playground, and trample it down. Schillings reports that African rhinoceros show a certain preference in leaving their excrements at fixed places and spreading it by scratching with their hind legs. According to Schillings these dung places serve the widely scattered individuals as "posts" and as means of orientation in finding one another. The hippopotami display a peculiar behavior which is said to have the same meaning. They hurl their dung high up on bushes by means of their tails, which have the appearance of brushes and are provided with stiff bristles. It is possible, however, that they, too, wish to mark their territory in this manner.

Only animals which are capable of adaptation, such as horses and donkeys, could be domesticated. Zebras, for instance, are very difficult to tame; in training them for shows it is necessary that ponies be used as companions and "instructors." Indian elephants, even if captured in old age, will establish a friendly relation to men in a few weeks and will willingly perform work they are taught to do. African elephants qualify much more poorly for such a purpose. Schulz, who has lassoed zebras and antelopes from horseback in East Africa, has given us an illustration which shows how well animals may learn to understand their masters by means of suggestion. A horse rapidly learns to recognize the goal of his rider, flies into a very ardent hunting passion, and sometimes even bites the captured zebra. Many antelopes turn sharply at full speed when they are pursued and a horse will make the same turn. Hunting dogs behave in a similar way. Due to their versatility, dogs are also used for the watching of cattle and sheep. They change their behavior properly according to the situation, being much more gentle in dealing with lambs than with sheep. Only animals who offer a special advantage to man have become domesticated in the proper sense of the word.

Man has not brought about anything new by breeding domestic animals, but has only transmitted certain variations of quality especially agreeable to him from one generation to later generations of domestic animals by means of selection and elimination. It is impossible to create something entirely new, but the breeder must have the ability to pick out those morphological, physiological, and psychological variations of quality which are favorable. Pfungst thinks dogs originate from wolves and jackals, not from foxes, and makes the statement that the domestication of dogs did not produce anything new. He mentions the fact that young wolves at play show elements of fetching and carrying without any training. By consistent selection man has made near-automats out of some of his domesticated animals, who perfom his work willingly and sometimes go to the limit of their ability. That which training can add every generation has to learn again (hunting in dogs, training of horses, etc.). Habits acquired by learning are not passed on by inheritance, but dispositions for learning such habits are. By associating with man many domesticated animals have completely lost their original inclination toward certain sociological structures of mateship, family, and herds. Accordingly, in most cases, it is not permissible to draw conclusions regarding the behavior of the original species ("das Tier" überhaupt) from the behavior of domesticated animals.

BIBLIOGRAPHY

ALVERDES, F. 1927. Social life in the animal world. London: Kegan Paul; New York: Harcourt, Brace. Pp. ix+216.
————. 1932. The psychology of animals in relation to human psychology. London: Kegan Paul; New York: Harcourt, Brace. Pp. viii+156.
BERGER, A. 1922. In Afrikas Wildkammern als Forscher und Jaeger. (2nd ed.) Berlin: Parey. Pp. xvi+327.
BREHM, A. 1914-16. Tierleben: Säugetiere. Vols. 1-4. (4th ed.) (Ed. by O. zur Strassen.) Leipzig & Vienna: Bibliog. Instit.
DEEG, J. 1922. Ueber Brunst- und Setzzeit in der Tropen. *Wild u. Hund,* 28.
DEEGENER, P. 1918. Die Formen der Vergesellschaftung im Tierreiche. Leipzig & Berlin: Verein. wiss. Verleger. Pp. xii+420.

DOFLEIN, F. 1914. Das Tier als Glied des Naturganzen. Vol. 2 of *Tierbau und Tierleben,* by R. Hesse and F. Dolflein. Leipzig & Berlin: Tuebner. Pp. xv+960.

ESPINAS, A. 1879. Die thierischen Gesellschaften. (Trans. by W. Schlosser.) Braunschweig: Vieweg. Pp. xiii+561.

GIRTANNER, A. 1903. Aus dem Leben des Alpen-Murmeltiers. *Zool. Garten,* **44.**

GROOS, K. 1930. Die Spiele der Tiere. (3rd ed.) Jena: Fischer. Pp. ix+348.

KROPOTKIN, P. 1908. Gegenseitige Hilfe in der Tier- und Menschenwelt. (Trans. by G. Landauer.) Leipzig: Thomas. Pp. xvi+294.

PFUNGST, O. 1914. Versuche und Beobachtungen an jungen Wölfen. *Ber ü. d. VI. Kong. f. exper. Psychol.,* 127-132.

SCHILLINGS, C. G. 1905. Mit Blitzlicht und Büchse. Leipzig: Voigtländer. Pp. xvi+558.

————. 1906. Der Zauber des Elelescho. Leipzig: Voigtländer. Pp. xiv+496.

SCHMID, B. 1921. Von den Aufgaben der Tierpsychologie. (*Abhandl. z. theoret. Biol.,* No. 8.) Berlin: Gebr. Borntraeger. Pp. iv+43.

SCHULZ, C. 1922. Auf Grosstierfang für Hagenbeck. (3rd ed.) Dresden: Verlag dtsch. Buchwerkstätten. Pp. 184.

SCHUSTER, L. 1923. Brunst- und Setzzeit des Wildes in den Tropen. *Wild u. Hund,* **28.**

UEXKÜLL, J. v., & KRISZAT, G. 1934. Streifzüge durch die Umwelten von Tieren und Menschen. Berlin: Springer. Pp. viii+101.

ZIEGLER, H. E. 1913. Tierstaaten und Tiergesellschaften. In Vol. 9 of *Handwörterbuch der Naturwissenschaften,* ed. by E. Korschelt, G. Linck, *et al.* Jena: Fischer. Pp. 1204-1220.

18

Reprinted from *Folia primat.*, 11:80–101, 105–118 (1969)

CORRELATES
OF ECOLOGY AND SOCIAL ORGANIZATION
AMONG AFRICAN CERCOPITHECINES

By T. T. STRUHSAKER

Introduction

The relation between environment and social organization is a problem receiving considerable attention in recent field studies of primate behavior. Notable among the publications contributing to an understanding of this problem are those of DeVore [1963], Hall [1965a, b; 1966], Rowell [1966], Crook and Gartlan [1966] and Chalmers [1968a, b]. Unfortunately, most of the reliable information in these studies deal only with savanna or arid-country cercopithecines in Africa. Rowell's [1966] study of baboons living in the gallery forest and savanna of Ishasha, SW. Uganda is a partial exception to this. Chalmers' [1968a, b] is the only comprehensive study on an African rain-forest cercopithecine. Needless to say, the formulation of meaningful hypotheses on the relation between environment and social organization must consider the many species of rain forest cercopithecines which vastly outnumber the non-forest species.

There are a number of factors that can affect the social organization of a species, including a variety of environmental parameters such as habitat, food, predators, competitors, population density, etc. In addition to these immediate or proximate variables, one must also consider the evolutionary history of the species; namely, its phylogeny. To what extent is the social organization of a species a function of its ecology? How do species with radically different phylogenies respond to similar environments? How do species with similar or closely related phylogenies respond to different environments?

339

It is the intent of this paper to present data collected between November 1966 and May 1968 on the group size and, to a lesser extent, on group composition of rain-forest cercopithecines living in Cameroon, West Africa. It is further intended to discuss the relevance of these data to the current hypotheses relating ecology and social structure of African cercopithecines and to consider some of the aforementioned questions. A general paucity of detailed information on the ecology and social organization of most of the species mentioned in this paper necessitates that the comparisons be made at a very general and gross level. This is unsatisfactory, but will demonstrate those relations that can be explained at a 'gross' level and those that require more refined study.

General ecology of species considered in this paper

Cercopithecus nictitans and *C. pogonias* are rain-forest species of West and Central Africa, being apparently more successful in mature forest than in younger successional stages and are seen most commonly in the upper heights of the forest. *Cercopithecus erythrotis camerunensis* and *C. cephus* are also monkeys of the West African rain forest, having similar ecologies but differing from the former two in apparently preferring the lower strata and younger or more immature kinds of forest. For most of their range these latter two species are allopatric and seem to be ecological counterparts. *Cercopithecus mona* is found in a variety of forest habitats in West Africa, but is apparently most abundant in mangrove swamps. When in lowland rain forest *mona* can be seen at all heights, but seems to have a propensity for lower levels. *Cercopithecus l'hoesti* occurs in two disjunct populations; one in West Cameroon and one in Eastern Congo Kinshasha and Western Uganda. In Cameroon it seemed most abundant in the montane forests, but did occur in a variety of forests including mature lowland rain forest. This species has a distinct preference for lower levels in the forest and is commonly seen on the ground in contrast to the other aforementioned *Cercopithecus* species. All of the preceding species seem to be opportunistic omnivors. *Cercopithecus aethiops* shares this habit, but differs from the other *Cercopithecus* species in being a savanna and arid-country species and in having a much wider distribution than the others. It occurs in nearly all of the savanna south of the Sahara and is definitely more terrestrial than the other *Cercopithecus* spp. *Erythrocebus patas*, considered to have strong phylogenetic affinities with *Cercopithecus* species, is also a savanna and

arid-country species, but seems even more independent of trees than
C. aethiops. It is also omnivorous, but unlike *C. aethiops* does not
occur south of northern Tanzania. Both species of *Cercocebus* are
typified as omnivorous, lowland rain-forest species. There is some
evidence indicating that *Cercocebus torquatus* is more terrestrial and
more common in mangrove and seasonally inundated forests than is
C. albigena (JONES and SABATER P. [1968]). Drills *(Mandrillus leucophaeus)* live in the lowland rain forests of a very restricted geographical region of West-Central Africa. They are most abundant in
West Cameroon and their western and northern distribution termin-
ates somewhere in eastern Nigeria, while their southern and eastern
limits, although unknown, probably do not extend far beyond the
borders of East Cameroon. Drills are omnivors who spend considerable time both on the ground and in the trees. Most populations of
Papio spp. are living in savanna habitats. Some spend a large proportion of their time in gallery forests and on the edge of more extensive high forests but, as far as is known, always have some contact
with savanna. This is in contrast to drills who are strictly a forest
species. Most populations of *Papio* spp. can be considered terrestrial
omnivors. *Papio hamadryas* and *Theropithecus gelada* are restricted
to the arid high country of Ethiopia, *gelada* apparently preferring
higher altitudes than *hamadryas*. Both species are typically terrestrial,
but rely heavily on steep cliffs as refuges and sleeping roosts. *Gelada*
may be more of a herbivor than *hamadryas*. Gross ecological comparisons of these species are summarized in Table I.

Social group – An operational definition

The term troop or group is widely used in contemporary field
studies of primates. However, I have been unable to find an operational definition of social group in any of the recent publications, e.g.
DEVORE [1965], ALTMANN [1967]. One probable reason for this is that
most of the monkey species that have been studied live in relatively
open country and their spatial relations are so obvious that a definition seems pedantic. Due to the inherent difficulties of observation
in rain forests, these spatial relations are not so evident and one is
forced to ask the basic question: what is a social group ? Reading the
publications on the savanna and arid-country monkeys of Africa gives
one a good indication of what the authors mean when they refer to
groups or troops. The following features would appear to be the main
ones distinguishing monkey groups in these various studies and also

represent my concept of a monkey social group. Membership in the same social network would seem to be the major feature distinguishing group members from extra-group conspecifics. Individuals of the same social network, by definition, have the majority (at least 80%) of their non-aggressive social interactions within this social network. A given individual does not necessarily interact with all members of its social network, but is associated with all at least indirectly. For example, if monkey A equally divides its non-aggressive social acts between monkeys B and C, but monkeys B and C don't interact with one another they are, none-the-less, members of the same social group; namely, the group consisting of monkeys A, B, and C. Temporal stability is another key character in describing a social group. Although most field studies of old world monkeys have been for only one or two years, it is generally assumed that group members remain together for longer periods. Subadult and adult males in several species seem more mobile than females and juveniles in changing from one group to another. However, even in most of these cases the males remain with a given group for about five months (STRUHSAKER [1967d]). Seasonal variation in group size and composition has not yet been demonstrated among old world monkeys. However, seasonal changes do occur in the frequency and size of temporary aggregations of two or more social groups, e.g. *gelada* (CROOK [1966]); *patas* (STRUHSAKER, pers. observ.). Spatial proximity is often implied in descriptions of intragroup relations, but this seems to have limited value as it is very dependent on the environmental conditions and their effects on the dispersal of the animals and on the observer's ability to measure group dispersal. This factor is discussed more fully in the section on intragroup dispersal. But, one can say that group members occupy the same home range. This also has limited value in defining social group, because in many species the groups have considerable overlap in their home ranges (DEVORE and HALL [1965]). Distinct social roles seem a key feature among group members, e.g. there are adult males who perform the majority of copulations (HALL and DEVORE [1965]; STRUHSAKER [1967d]) and adult females who lead the majority of group progressions (STRUHSAKER [1967d]). Synchronized progressions also delineate members of the same group. Such progressions are especially prominent when the monkeys are fleeing from a predator or are moving from one feeding place to another. Group members exhibit different behavior toward non-group conspecifics. This different behavior may be in the form of unique patterns, e.g. intergroup aggressive calls of vervets (STRUHSAKER [1967a]), or may consist of avoidance or a difference in the frequency with which certain behavior patterns are performed, e.g. aggression may be greater among non-group than group conspecifics. Social groups are relatively closed to new members. Exceptions to this have been noted for vervets and baboons, but these cases are relatively infrequent and don't invalidate the concept that social groups are relatively closed. Data presented throughout this paper support the conclusion that the rain-forest monkeys of Cameroon, West Africa live in social groups having these general characteristics and differ only in some details of their social organization from the African savanna and arid-country monkeys previously studied (also see CHALMERS [1968a, b]).

Methods

During an 18-month field study in the rain forests of Cameroon, West Africa, attempts were made to count all monkey groups that were encountered. Due to the inherent difficulties of observing monkeys in the thick vegetation of a rain forest combined with the heavy hunting pressure exerted by humans on monkeys in Cameroon, observations and accurate and complete counts of monkey groups were exceedingly difficult to obtain. As a consequence, relatively few reliable counts were made. Often the counts were incomplete, but the amount of noise made by the group permitted what was to be considered a reliable estimate. Sometimes these estimates were confirmed by accurate and complete counts made of the same group at a later time. The most accurate counts were made as the monkeys climbed along the same arboreal route in synchronized progression. During the last year of this study I was accompanied by Mr. FERDINAND NAMATA, a local African who proved to be extremely perceptive and reliable in his observations. The counts made by Mr. NAMATA were compared with those made by me as a test of observer reliability. A high degree of consistency was attained in our counts and estimates and thus further increased my confidence in their accuracy. In addition to a general survey of many areas and a variety of rain-forest types, one particular area of about 0.41 square miles was mapped in detail.[1] This area was marked with a grid of trails and numbered aluminium plates were placed at regular intervals along these trails permitting very accurate location of the monkey group. The localization of monkey

---

[1] A minimal estimate of this area, based on aerial photographs, was 0.35 square miles. It did not include the considerable area on cliff faces and slopes that could not be measured accurately from aerial photographs. A very crude estimate of this additional vertical or sloping surface area is 0.057 square miles. This is based on a rough measure of the linear distance of cliff face multiplied by a rough estimate of the average cliff height, i.e. 5.330 yards times 100 ft. Consequently, a more accurate estimate of this area is 0.41 square miles.

groups allowed what are considered to be reliable inferences on group stability, i.e., groups of similar size and composition were often seen at the same specific locality. Every attempt was made to remain with a group or association of monkeys for as long as possible. All social events were described and tabulated.

Compared to the majority of field studies on the behavior and ecology of non-forest African cercopithecines, these data are relatively scant and unsophisticated and may seem unsatisfactory to some readers. However, in the absence of other more reliable and complete data on the species I studied, the data presented in this paper are valuable for our development of theories on the parameters most influential to primate social structure.

Results and Discussion

Counts made in the major study area that was mapped in detail support the hypothesis that groups of *Cercopithecus nictitans martini* were stable in size and composition. Usually groups of the same size and composition were seen in the same region of this mapped area. These data not only indicate that groups are temporally stable, but also that they are relatively closed. Most of the data for the other species listed in Table I were collected outside of the mapped area and therefore do not necessarily permit this same conclusion. However, prolonged observations of some groups of these other species implied that they too lived in stable social groups. CHALMERS [1968a] studied two groups of *Cercocebus albigena* and found that their membership remained relatively constant over an 11-month study. This finding is also consistent with studies on the majority of nonforest cercopithecines. Even in *gelada* and *hamadryas* baboons who form large and unstable aggregations there are small and relatively stable social units (CROOK [1966]; KUMMER [1968]). BUXTON [1952] found that group size of *Cercopithecus ascanius*, a rain-forest species, fluctuated with the time of day. His results are rejected on technical grounds that are fully discussed in the appendix.

Social Network

Membership in the same social network and thus of the same social group is most readily demonstrated by tabulating the frequency with which particular individuals interact socially. To do so, one must be able to identify all participants of many social interactions. In my study of rain-forest cercopithecines, recognition of individual monkeys was virtually unaccomplished. Individuals with natural marks were seen, but rarely more than once. Indirect evidence of membership in the same social network is provided by observation of non-aggressive social·interactions between monkeys in the same aggregation. As stated above, census results in a mapped area indicate that aggregations of *nictitans* were stable over time and relatively closed socially. The occurrence of non-agonistic social acts within such aggregations lends support to the hypothesis that in fact these are social groups as defined above.

Social grooming was seen in the following rain-forest species: *Cercopithecus nictitans n.*, *C. n. martini*, *C. mona*, *C. pogonias grayi*, *C. erythrotis*, *C. cephus*, *Miopithecus talapoins*, *Cercocebus albigena*, *Mandrillus leucophaeus*, and *Colobus satanas*. Frequency data seem of little value in analyzing this behavior pattern among rain-forest monkeys because of the inherent problems of sampling their behavior. Visibility is greatly restricted and, even when visible, a given monkey is seldom in view for any great period of time. Consequently, it is difficult to understand if grooming is socially and functionally as important in rain-forest species as it is among savanna species. Although several species of biting fly are common in the Cameroonian rain forest, ectoparasites that are kept in control by grooming are not, such as ticks. This is in contrast to the African savanna where an ungroomed monkey soon becomes covered with ticks (Struhsaker [1967c]). This may be a major factor accounting for the relative infrequency with which grooming was observed. The scant data permit few generalizations on grooming, but, as with most old world monkeys, all age-sex classes seem to participate.

Heterosexual behavior was observed in groups of *C. nictitans n.*, *C. nictitans martini*, *C. erythrotis* or *C. erythrotis-cephus* intermediates, and *C. mona*. Ejaculate was once seen on the perivaginum of an adult female drill who had no perineal swelling typical of estrous females. Similar to grooming, frequency data have little value in understanding heterosexual behavior among rain-forest cercopithecines. Mounting was observed once in *C. mona*, once in *C. erythrotis* or *C. erythrotis-cephus* intermediates, five times among four different pairs of *C. nictitans martini*, and three times by one pair of *C. nictitans n*. One of the *C. nictitans martini* mountings involved two juveniles about two-thirds adult size. All other mounts were among adults or older subadults. A distinct call was given by one of the *mona* during the mount. It would appear that in *nictitans* the ejaculatory mount may be preceded by a series of incomplete or non-ejaculatory mounts.

Play behavior was seen in *C. nictitans martini* (7), *C. pogonias p.* (1), *C. pogonias grayi* (1), *C. erythrotis* (3), *Mandrillus leucophaeus* (7), and *Colobus satanas* (1). The numbers represent frequencies of play encounters and are probably not too meaningful because of the inherent difficulties of sampling behavior under rain-forest conditions. On only two occasions did adult *nictitans* play with a juvenile. Once an adult played with a juvenile and, on another occasion, two adults played together and with a juvenile who was less involved. All other cases of play involved only juveniles. Among the *nictitans* two play sessions had two participants, four had three and one had four. The drills had five encounters of two members and two with three. Grouping the data for the other species revealed that five of the play sessions had two members and one had three.

One of the most outstanding features of groups of rain-forest cercopithecines is the high frequency of vocalizations that seem to be contact calls. They may be functional in maintaining group cohesion. Several individuals give these calls and in a typical case one gains the definite impression that they are 'exchanging' calls. The first vocalization seems to evoke similar calls from other monkeys, which in turn stimulate the initial caller to call again. These contact calls are quite distinct for each species and have been given descriptive names such as bleats, grunts, chirps, staccato-barks, crowing, and panting. Frequency data of such behavior is more meaningful than for non-audible behavior because they are independent of actual observations of the monkeys. An indication of the importance of these calls can be seen from selected examples. On March 28, 1968 I maintained contact with a group of about 20 drills for three hours and forty minutes. In this period 22 bursts of crowing were heard; about 0.1 crows per minute. These crows were given by only one caller at any one time. In one exceptional case two drills crowed simultaneously. Nineteen inter-crow intervals were available for measurements. Fourteen of these were between 30 s and 2 min, three were between 4 and 5 min, and two were 20 and 30 min respectively. The mean interval was 4 min 15 s. If these calls were uniformly distributed in time, one would expect a crow to be given about every eleven minutes. The fact that most of them are given with shorter intervals would lend support to the hypothesis that a crowing drill evokes crows from other drills and that the call functions to facilitate group cohesion.

Further support is given this hypothesis by an observation made on January 22, 1968. Two sections of a large group of drills (estimated at 65) were separated by 100–150 yards. Several bouts of exceptionally loud and shrill crowing were given by two sources, one in each section. There seemed little doubt that they were calling in response to one another in what might be described as an exchange of calls. These two sections eventually joined.

Crowing is not the only call drills use for maintaining auditory contact. On October 10, 1967 we inadvertently disturbed a large group of drills, evoking alarm calls and mild flight of a relatively short distance. As we stood between a subadult male and the rest of the group, an exchange of panting calls was given by the subadult male and several sources in the main body of the group. This occurred at least three times and after about one hour the subadult male climbed to the ground and fled out of view, presumably rejoining the others.

A very clear case of call exchange and auditory contact among *C. nictitans martini* was seen on April 22, 1968. An adult female was sitting about 40 ft up. Each time a grunt was given by another monkey about 10–15 yards from her, she looked in its direction and replied with a similar grunt. This exchange of grunts was observed five times in a period of about two minutes.

There were other kinds of social behavior whose significance was less clear, but were clearly of a non-aggressive nature. Participants in these miscellaneous social behaviors were most likely members of the same social network. Some of these patterns seemed to be greeting behavior. For example, on November 27, 1967, a juvenile *C. nictitans martini* about two-thirds adult size approached an adult female, very briefly made muzzle-muzzle contact with her, and then climbed away from her. Other cases of muzzle-muzzle seemed to involve food investigation. On July 28, 1967, a subadult *nictitans martini* approached an adult who was feeding. The subadult made muzzle-muzzle contact with the adult. The adult turned its head away, but the subadult made muzzle-muzzle contact again and then it too fed alongside the adult. Similar observations have been made of group members of *C. aethiops* (STRUHSAKER [1967 d]).

Other kinds of miscellaneous social behavior had elements suggestive both of greeting and exploratory behavior. For example, on September 23, 1967 adult female A *nictitans martini* with an infant clinging to her venter approached another adult female B who sat and had an infant of similar size nearby. This latter infant climbed about, but remained in contact with adult female B. A grabbed and elevated the tail of this infant and then muzzled its perineum, released the infant and climbed away with her own infant still clinging to her venter. B was in contact with the recipient infant throughout this encounter, but showed no response. This encounter was virtually identical to patterns observed in vervets and baboons. Another interesting social interaction was observed between two juvenile *nictitans martini* on May 17, 1967. A slightly smaller juvenile approached a larger juvenile who was sitting. They embraced and then sat maintaining the embrace for about thirty seconds when the larger juvenile climbed away with the smaller juvenile clinging ventrally to it.

One case demonstrating both 'aunty' and protective behavior was observed on December 13, 1967. An adult *nictitans n.* saw me and immediately ran toward a juvenile (about three-fourths adult size) thirty feet away who was holding an infant. The adult grabbed the infant who clung to her venter and then fled. A social encounter of rather ambiguous significance was seen between two *C. erythrotis camerunensis* on April 24, 1968. An adult male was climbing and closely followed by an adult female. They sat and foraged. The female climbed toward the male who sat facing her. She stopped about one foot from him and while looking into his face lowered her hindquarters. She then turned and climbed away. About thirty seconds later, he followed. In some ways this was reminiscent of behavior between consort pairs of some macaques.

In all cercopithecines so far studied the social interactions between members of the same social network are not restricted to grooming, heterosexual behavior, play, contact behavior, etc. but also involve agonistic behavior. In some species this is considerable. Rainforest cercopithecines in Africa seem to be no exception. Intragroup aggression was seen in the following species: *C. nictitans martini*, *C. pogonias p.*, *C. erythrotis*, *Cercocebus albigena*, and *Mandrillus leucophaeus*. Few of these encounters were clearly observed. In a typical case one became aware of an aggressive encounter by the screams, screeches, chutters, and other calls associated with agonism and by the sight of monkeys leaping about rapidly, chasing and counterchasing 60 to 100 ft overhead. Those cases that were most clearly observed involved only two participants. For example, the one case of aggression among *C. pogonias p.* occurred on May 17, 1967 when one monkey ran toward and grabbed another monkey who immediately fled. Screaming was heard but the vocalizer was not determined. The entire encounter was very brief. A mild case of non-vocal aggression was seen on August 1, 1967 when a smaller *C. erythrotis* approached a larger *erythrotis*. The smaller extended its muzzle toward the larger's. The larger slapped with one hand toward the smaller, but did not make contact. The smaller fled and the encounter was terminated. During a group progression on January 5, 1968 two *C. nictitans martini* had a mild agonistic encounter. They were leaping across a gap from one branch to another. Monkey B was following monkey A and, as B leapt onto the same branch with A, A lunged toward B. B immediately leapt to another branch. Chutters were heard during this brief aggression, but the caller was not determined. A rather prolonged agonism occurred between an adult male *C. albigena* and a conspecific that was about two-thirds to three-fourths his size. This took place while the group was fleeing from us on February 14, 1968. The adult male chased the smaller *albigena* for about 40 yards at heights ranging from about 20 ft up in the trees to 70 ft up. The chase lasted about 30 s and no contact was made. The adult male was always about 10–15 ft behind the chasee who screamed and screeched almost continuously. During the chase the other *albigena* in the group seemed greatly agitated and ran about in the trees. Their excitement could have been a combined response of alarm toward us and a response to the chase. After the adult male ceased chasing, he leapt about the tree branches 80–90 ft up, making great bounds and moving in a circular manner in the top of one tree in what was obviously a branch-shaking display lasting 15–20 s. Five minutes after the rest of the group had left this area and our proximity, the chasee followed them (also see Chalmers [1968b]): The final dyadic or possibly tripartite case of aggression concerns drills. This occurred on March 28, 1968. Two drills about the size of adult females, but whose gender was not determined were grooming 18 ft up in a 60 ft secondary forest. After two minutes this grooming was terminated when an adult male approached them and one of the groomers climbed away. The other stood quadrupedally and turned its hindquarters toward the adult male. The male touched the right hip of this other drill with his right hand. The recipient then moved away from the adult male and out of my vision. The two drills of the preceding grooming encounter were now together again and less than 10 ft from the adult male. The adult male stared and bobbed his head once in the direction they had gone. One shriek was heard from the vicinity of these two smaller drills. A few two-phase grunts were heard from the adult male. These grunts are usually considered to facilitate group cohesion. The two smaller drills then climbed up to 35 ft and were followed by the adult male.

All were climbing slowly and in a non-panicked manner indicating that the aggression had terminated.

The multipartite agonistic encounters were less clearly observed, possibly because with an increased number of participants the chances are less of observing all behavior of all the monkeys. Consequently, these cases were usually very confusing. An extreme example of this was observed on September 20, 1967 among *C. erythrotis*. This was a long encounter lasting from 09.16 to 09.55 h and involved at least 17 monkeys including adults, clinging infants, and juveniles. Chutters, grunts, growls, screeches, and chirps were common. There was much leaping, chasing, and counter-chasing. This occurred at heights of 10–15 ft in a 100 ft secondary forest. There seemed to be two clusters of monkeys and combined with the long duration and relatively high intensity was reminiscent of intergroup conflicts among *nictitans* and *aethiops*. However, when they saw me at 09.55 h they fled. Initially in several directions, but eventually they all fled in the same direction indicating that they were members of the same social group.

Frequency data of vocal agonistic encounters are much more meaningful than similar data for non-vocal encounters, because of the problems of observation discussed earlier. Interpretation of frequency data is further compounded by the fact that the frequency of most social behavior is dependent on the size and composition of the social group being observed. The size and composition of the groups being observed were usually not known. Earlier data were presented on the frequency of crowing by drills. These were collected on March 28, 1968 during a three hour and forty minute contact period with a group of about 20 drills. Frequency data on shrieks, screams, and screeches were also collected during this period. Eighteen bouts of such calling were heard during this contact period. Sixteen of these involved only one vocalizer and were relatively brief. On one occasion there were at least three sources and on another there were several sources. Sixteen inter-call intervals were available for measurement. If these calls and therefore the vocal agonisms were uniformly distributed in time, one would expect a bout of calls every thirteen minutes and forty-five seconds. Although the distribution was not exactly like this, it did approach it and was radically different from the temporal pattern of crowing bouts in the same observation period, as presented earlier. The mean interval between bouts of agonistic calls was ten minutes. Three of the intervals were between thirty seconds and two minutes in length, three between 2.1 and 5 min, four were between 5.1 and

10 min, four between 10.1 and 20 min, and two were greater than 20 min. Intraspecific comparisons are compounded by the fact that most contact periods with other species were of shorter duration than for drills. Most contact periods for *C. nictitans martini* were less than one hour. Only ten intragroup vocal agonistic encounters were tabulated for *C. nictitans martini* during the entire study period of eighteen months. This was in spite of the fact that it was one of the most common species encountered. One of the longest contact periods with *C. nictitans martini* lasted three hours and fifteen minutes and was with a group of about sixteen to nineteen individuals on January 16, 1968. In terms of contact time and group size, it is comparable to the preceding drill data. In this period only two vocal agonisms occurred. They were separated by one hour. It seems reasonable to conclude that *C. nictitans martini* engages in far fewer vocal agonistic encounters than does *Mandrillus leucophaeus*.

Social Roles

Limited evidence suggests that in *C. nictitans martini* warning or alarm calls are only given by adult females and juveniles. For example, on April 19, 1968 a group of *nictitans* was encountered and almost immediately they detected us. Two juvenile or adult females looked toward us, gave alarm chirps and fled. A large *nictitans* in this same group sat looking about. It had a very contrasting white chest and no visible nipples and was therefore probably an adult male. It suddenly saw us and fled without vocalizing. On this same date, but in the afternoon and with a different group of *nictitans*, similar observations were made. An adult female with a clinging infant saw us, gave many alarm chirps, and then fled. An adult male in the same group saw us, fled into dense foliage and out of view without vocalizing. A related observation was made on December 28, 1967. We were cutting a trail through an area of very dense underbush when we came upon a group of *C. nictitans martini*. One adult female sat 35 ft up in a 60 foot secondary forest. She was perched on the edge of an old chimpanzee nest in a very exposed and prominent position. While sitting thus, she looked about in an obvious scanning manner and gave alarm chirps. It seemed quite apparent that she had heard us but not yet seen us. This appeared not only as alarm behavior but had aspects of vigilance as well. These observations further imply that adult males of *C. nictitans* play no vocal role in warning the group of the proximity of potential predators, whereas females and juveniles do.

Some of the most prominent features of several cercopithecine species in the rain forests of Cameroon are the loud calls given by adult males. The two-phase booms and hacks of *C. mona*, the single-phase booms and staccato-hacks of *C. pogonias*, the pows and hacks of *C. nictitans*, and the whoop and rolling calls and staccato-barks of *Cercocebus albigena* and *C. torquatus* are the most commonly heard. These calls are given at all times of the day and under a variety of circumstances. It is postulated that these calls serve as a point of central focus for maintaining intragroup cohesion during occasions when the group is likely to become dispersed, e.g., during intergroup agonistic encounters, after loud noises such as those made by thunder or a falling tree, after the attack by a crowned-hawk eagle, and prior to a group progression. These loud calls may have other functions that could be performed simultaneously. For example, they may also serve in intergroup spacing. The loud call of any one of these species conveys the basic information that an adult male of that species is present. In all of the stimulus situations evoking loud calls, it would seem advantageous to have this basic information conveyed whether to a predator or another group of conspecifics or a solitary male. If these loud calls do enhance group cohesion, it means the group's movements and activities are, to the same extent, centered around the adult male of the group. This latter point remains to be demonstrated. However, in any situation in which the group becomes scattered, such as after an attack by a predator or flight evoked by the loud crash of a falling tree, it would seem advantageous to have some reunion point for all group members. In a rain forest where visibility is considerably limited it would be further advantageous to have this point a vocal one. This might be provided by the adult male and his loud call without further involving him as a center of the group's other activities and movements, such as progressions under non-alarm situations. Thus, the adult male might be the central focus of the group only in such situations and even then only temporarily.

These conclusions are based on observations of the following nature. On February 19, 1968 at 18.45 h and very near sunset a group of *C. mona* was clearly observed in a tall *Ceiba* tree towering above the surrounding cultivation. There was only one adult male in this group and he gave two two-phase booms and hacks. After calling, he followed nine other *mona* (including three clinging infants) from the

352

Ceiba into another tree along one horizontal branch and in an obvious and coherent group progression. Only one other *mona* followed the adult male. Thus, although the male was clearly the vocal center, he was not leading the progression.

On April 16, 1968 a group of *C. pogonias p.* was located. They saw us almost immediately, gave staccato-chirps and began to flee. Three were fleeing together toward the east until typical adult male *pogonias* loud calls were heard from a source to the northwest of them. They immediately reversed their course and headed back along the same tree branch toward the source of loud calls.

On July 28, 1967 observations were being made on a mixed association of *nictitans* and *mona*. They had not detected us. Adult male *mona* loud calls were heard from one source. The stimulus was not obvious, but another *mona* immediately climbed toward the caller.

C. nictitans martini were clearly seen to orient toward adult male loud calls on two occasions. On March 1, 1968 a group of *nictitans* had seen us and six of them were giving chirps toward us. From deeper in the river gorge another *nictitans* began giving loud calls. Some, but not all of the preceding six moved toward this source. On March 26, 1968 while I was watching *nictitans* of adult female size, loud calls of this species were heard from one source in this group. The stimulus was not apparent and they had not seen us, but, while the loud calls were being given, the monkey under observation looked about with sharp lateral jerks of its head. After thirty seconds of this scanning it rapidly climbed toward the source of loud calls and was followed by a juvenile *nictitans*.

As postulated above, loud calls may be given in response to any stimulus that is potentially disruptive to the group's cohesion. An example of this was apparent on January 9, 1968 when a brief and vocal agonistic encounter occurred in a group of *C. pogonias p.* Immediately after this agonism an adult male *pogonias* responded with typical loud calls. He sat and intermittently fed while calling.

A case reminiscent of this occurred on February 8, 1968 at the unusual time of 22.53 h. There was a loud outburst of *mona* barks and hacks similar to the response given by monkeys to a predatory attack. This outburst was immediately followed by *mona* loud calls from two different sources and probably two different but proximal groups.

Two cases were observed in which the vocalizer was apparently solitary. On December 16, 1967 between 07.51 and 08.00 h a large and presumably adult male *C. nictitans n.* gave loud calls during this nine-minute period. No other monkeys were seen or heard in the vicinity and it was assumed to be solitary. On December 1, 1967 a large, adult male *C. nictitans martini* was observed for twenty-one minutes between 16.09 and 16.30 h. During this period he gave loud calls almost continuously as he climbed 75 ft up in a 125 ft mature forest. He was climbing very rapidly and in so doing covered an actual distance of at least 350 yards. He ran in a large oval, returning to the place where he was first encountered and where he eventually disappeared. There was no indication of any other monkeys in this vicinity throughout the entire sequence. If in fact these monkeys were solitary, then the preceding hypothesis requires expansion. It should allow that calling by solitary males is functional. It is possible they can compete vocally with harem males and that it may be a manner in which new groups are formed or a way in which new harem males are established. It is quite conceivable that solitary males giving loud calls would evoke similar calls from any harem males that were nearby and would thus convey information about the locality of a heterosexual group and the presence of an adult male in that group. This would allow ready replacement of harem males that die or otherwise leave their groups. Alternatively, these 'solitary' males may have become temporarily separated from their groups and the calling may have been relevant to reunion with their group.

Adult males of *C. l'hoesti* also give loud calls, but they differ from other *Cercopithecus* species in that the calls are given only between 17.50 and 18.40 h[2] and that they seem much more contagious. After the first oop-uuh call is heard in the evening, many others soon follow, giving the impression of a spreading wave of calls. One gains the impression that each vocalizer calls only once in an evening. Only twice did it seem that the same vocalizer called twice. Nine evenings were spent listening for these calls in areas known to have dense populations of *l'hoesti*. Calls were heard on all of these evenings. The maximum number heard in any one evening was eight, the minimum was one, and the mean was 5.27. The period of time during which all calls were given during any one evening ranged from one minute thirty seconds (five calls were given) to thirty-two minutes (eight calls were given). The mean period length was eleven minutes and twenty seconds. In all cases the vocalizer was presumed to be an adult male *l'hoesti*. This seems a reasonable presumption in view of the fact that

---

[2] Sunset is between 18.30 and 18.45 h in the rain forests of Cameroon.

an adult male *l'hoesti* was once seen in the exact position from which an oop-uuh call had been given a few minutes before and that in three of the areas where these calls were most commonly heard *l'hoesti* was the only monkey species living there. The function of this call is not clearly understood, but its contagious effect and marked diurnal periodicity would imply that it plays some role in conveying information about the number and distribution of adult males and may in turn play some role in social spacing. Regardless, it would appear that the adult males of *l'hoesti* play a special social role in this respect.

The staccato-barks of *C. albigena* and *C. torquatus* are apparently given by all age-sex classes and are thus distinctly different from the loud calls of *Cercopithecus* species. However, the staccato-barks of adult males are lower pitched and distinct from those of other group members. Because there are often more than one adult male in *albigena* groups, it is unlikely that their staccato-barks function as a central focus. It seems more likely that the staccato-bark functions as a cohesion call and that the adult males may be more important in maintaining group cohesion through vocalizations than other members of the group. For example, on December 15, 1967 a large adult male *albigena* was climbing about 20 ft up in a 100 ft secondary forest. He gave staccato-barks almost continuously and he was obviously in front of a group progression. Later, a large adult male, probably the same as preceding, gave staccato-barks almost continuously for five minutes while he occasionally fed. Staccato-barks were heard from other sources, but were very sporadic and brief. There was no obvious stimulus evoking barks from this male, except it seemed apparent that he barked more frequently when other *albigena* were also barking. He also responded with staccato-barks to the whoop and rolling call given by another *albigena* in the group.

The whoop and rolling call is apparently given only by adult and possibly subadult male *albigena* and *torquatus*. The data are few and primarily from *albigena*. There may be some diurnal periodicity, with whoop and rolling calls being more commonly given in the early morning. They also seem to evoke similar calls from other and often distant sources. This implies that the whoop and rolling call function in spacing, possibly both inter- and intragroup. For example, on February 14, 1968 an *albigena* rolling call was heard in the very far distance at 06.33 h and then at 06.33.5 h a similar call was heard from a different source. Later in the day and at a different location two series of *albigena* whoop and rolling calls were heard from one distant source at 11.07.5 h. At 11.15 h similar calls were given by a different and nearer source and these were immediately followed by more whoop and rolling calls from another near source.

Inter- as well as intraspecific stimulation can also occur. On February 18, 1968 one *torquatus* whoop and rolling call was immediately followed by *torquatus* staccato-barks, *pogonias p.* loud calls and *nictitans martini* loud calls each from one source and all from the same association.

Sexual Dimorphism

DeVore [1963] has advanced the hypothesis that sexual dimorphism, especially regarding body size and canine size, is an adaptation by terrestrial primates against predation. He states: "... morphological adaptations for fighting and defense are clearly correlated with adaptation to the ground." And again: "The trend toward increased fighting ability in the male of terrestrial species is primarily an adaptation for defense of the group." This hypothesis is not supported by observations on forest cercopithecines. Drills demonstrate an extreme form of sexual dimorphism. The males are considerably larger than females in both body (about twice) and canine size and they have distinct colouring. There is no evidence that adult males play any role in the defense of the group. Man is the major predator of drills in Cameroon. When they detected us, the drills either fled immediately or the adult females and immature drills approached, looked toward, and gave pant-barks toward us before eventually fleeing. Adult males never did this and a subadult male only once. On January 2, 1968 while watching a fairly large group of drills I was given mild threat gestures by a subadult male. He sat about sixty feet below a cliff edge and below me. While sitting with his hands on the ground, he jerked his forequarters forward, briefly held this forward position, and then returned to the upright and original sitting posture. This jerking was not repeated rapidly and thus differed from the more rapid, but similar pattern in vervets and baboons. He also jerked his forequarters sidewards in a lateral movement while retaining his sitting position. Sometimes he went from a sitting to quadrupedal position in a rapid and jerky manner and then oriented his longitudinal axis perpendicular to my line of vision. He then stared toward me. After a few seconds, he would sit again. This entire sequence was interpreted as a defensive threat toward me. Usually adult and subadult males were not seen after the group had detected us and given 'alarm' calls. It has been argued that perhaps the best defense drills can exert against human predators is flight. However, if this were so, then one would expect to see a similar response to humans by the adult females and immature drills. As stated earlier, drills spend a considerable amount of time on the ground. Although not so marked as among drills, *Cercocebus albigena* is obviously a sexually dimorphic species. On only one occasion did the larger male demonstrate defensive or protective behavior. This occurred on February 15, 1968 when I was observing a group of at least twelve or thirteen *C. albigena* who were feeding 60–80 ft above. Upon seeing me, six fled in one direction and two or three in another. They were all larger than one-third adult size. Remaining behind in the food tree were three juveniles (about one-third adult male size) and the only adult male in the group. For the next ten minutes, after the others fled, the adult male gave staccato-barks almost continuously, eating fruits intermittently between barks. The three small juveniles remained close to the adult male throughout this period. After this ten-minute period, the three juveniles fled in the same direction as did six other *albigena*. The adult male was the last to leave, following the three juveniles. The forest and more arboreal *Cercopithecus* demonstrate as much sexual dimorphism in body and canine size as do the more terrestrial *C. l'hoesti* and the non-forest and more terrestrial *C. aethiops*. The majority of evidence indicates that adult males of forest *Cercopithecus* perform no special role in the defense of the group. Only once did an adult male *C. mona* demonstrate threat or defensive behavior toward me. This occurred on April 18, 1968 when a group of at least six *mona* were encountered. They all fled immediately except for an adult male who remained momentarily, gave low-pitched sneeze calls, and bobbed his head and forequarters toward me. Sometimes his hands left the substrate during the upward portion of the bob. The rate of bobbing seemed more rapid and his body seemed more hunched than a similar pattern in *C. aethiops*, but was clearly a threat. Among the more terrestrial and savanna-dwelling *C. aethiops*, an adult male was seen only once to make threatening lunges toward a potential predator, the Martial eagle *(Polemaetus bellicosus)*. The rarity of defensive behavior by adult males does not conform with DEVORE's [1963] hypothesis and I am inclined to agree with CROOK and GARTLAN [1966] in thinking that this sexual dimorphism in body and canine size is not primarily adapted to group defense. It seems reasonable that it is basically a result of sexual selection, either intra and/or inter, and with some species and under certain circumstances has become secondarily adapted to protection of the group, e.g., DeVore's savanna baboons in the Nairobi Park. The basis of this argument is weakened by the relative paucity of positive data and by its heavy dependence on negative data, i.e., few cases of aggression between adult males and defense of the group by adult males have been observed. However, it should be emphasized that intra- and intersexual selection usually involve displays that are often very subtle and may rely on nothing more than relative size and not involve physical aggression at all. For example, dominant male vervets assert and presumably reinforce their dominance by performing the Red, White, and Blue display to subordinant males. These same dominant males copulate significantly more than the subordinants and thus their genotypes are selected for with a minimum of physical aggression (Struhsaker [1967b]). Furthermore, the data supporting the hypothesis that sexual dimorphism is primarily an adaptation for group defense are also few. Cases of adult males chasing or threatening potential predators from the group have only been reported by DeVore [1963] for *Papio anubis* and by Stevenson-Hamilton [1947] and Bolwig [1959] for *Papio ursinus*. In contrast to these observations, Rowell [1966] reports that adult males of *Papio anubis* groups in southwest Uganda did not demonstrate a protective role against predators. She states: "... a stronger stimulus produced precipitate flight, with the big males well to the front and the last animals usually the females carrying heavier babies." Earlier I mentioned a single case of an adult male *C. aethiops* threatening a Martial eagle. In the Waza Reserve of Cameroon I once observed several adult male *patas* monkeys chasing a jackal *(Canis adjustus* or *C. aureus)* who had caught an infant *patas* that it eventually released as it fled from the adult males. However, I have also seen adult male *patas* threatening and chasing peripheral adult and subadult male *patas* away from his group, which seem best interpreted as cases of intrasexual selection. Thus, data supporting both hypotheses are available. Hall [1965b] was of the opinion that the primary function of the extreme sexual dimorphism among *patas* monkeys in Uganda was in the performance of a diversionary display. Apparently, the adult male would leap about noisily on bushes and trees as he ran away from his group and presumably diverted the attention of the potential predator.

Group Size

What is the relation between group size and the degree to which a species is terrestrial and the degree to which it is a forest dweller? Within the area of detailed study there were five groups of *C. nictitans martini* of whose identity I was quite certain. The average counts of these groups and the mean of the estimates made of these groups (in brackets) are as follows: 7 (7.3); 7.2 (10.3); 17.0 (19.0); 14.2 (16.1); and 9.0 (9.5). The average (mean) count and estimate of these five groups are 10.9 and 12.3 respectively and are probably not significantly different from the means for all counts of this species in Table I. Most of the other forest *Cercopithecus* species lived in heterosexual groups of similar size. The data for *C. cephus* are few and not too reliable. *C. l'hoesti preussi* is an outstanding exception, having a mean group size about half that of the other forest *Cercopithecus* species. It is noteworthy that *C. l'hoesti* is also the most terrestrial of the forest *Cercopithecus* monkeys in this sample. It invariably ran along the ground when fleeing from man and was often seen foraging on the ground. None of the other forest *Cercopithecus* fled on the ground and were very rarely seen there.[3] This conflicts with DEVORE's (1963) conclusion that "... a trend toward larger groups in the terrestrial species is apparent...". In contrast to this and consistent with DEVORE's [1963] hypothesis, drills *(Mandrillus leucophaeus)*, who spent a considerable amount of time feeding, progressing, and fleeing on the ground, have mean group sizes larger than any of the *Cercopithecus* forest monkeys observed, who are primarily arboreal. How-

---

[3] It has been argued that because *C. l'hoesti* is more terrestrial than the other forest *Cercopithecus* spp. it may be exposed to greater hunting and trapping pressure by man than the other species and that this might account for its smaller group sizes. If this were so, one would expect to see different sizes of *l'hoesti* groups in areas where different amounts of trapping occur. Not only would it be difficult to measure the effective trapping and hunting pressure being exerted on a given population, but my sample of *l'hoesti* is not adequate for statistical analysis. However, the data on *l'hoesti* group size came from several widely separated areas in West Cameroon and in none of these areas was the impression gained that *l'hoesti* groups differed in size (Tab. I). However, the trapping and hunting pressure varied considerably in these same areas, as evidenced by the number of traps encountered and the number of gun shots heard. These impressions would support the conclusion that, if trapping and hunting of *l'hoesti* have any population effect, they result in a reduction of the number of groups and not a reduction in group size. Furthermore, additional support of this conclusion is given by the fact that other terrestrial forest species such as drills and mangabeys live in large groups with no apparent reduction in group size in areas of different trapping and hunting pressure.

ever, the drills apparently spent more time in the trees than non-forest baboons and yet the group sizes of these species who occupy such radically different habitats do not appear significantly different. This also conflicts with DeVore's hypothesis. Hall [1965b], in his study of *Erythrocebus patas*, was one of the first to demonstrate the inadequacy of DeVore's preceding hypothesis; "... patas, which are as terrestrial as the baboons in the same habitat region, have a mean group size of about half that of the Murchison Falls Park baboons...".

The drill data on group size would also appear to contradict the conclusion of Crook and Gartlan [1966] that "Outside the forest the social units are generally larger and this appears to result from open country conditions of predation and food supply affecting ground dwelling populations." A comparison of *C. l'hoesti* group size with *C. aethiops* would support this hypothesis (Tab. I). However, excluding *C. cephus*, the other forest *Cercopithecus* seem to have group sizes of the same size as *C. aethiops* on Lolui Island (Hall and Gartlan [1965]) and in Murchison Falls Park (Hall [1965a]), but of a smaller size than *C. aethiops* in the Amboseli Reserve (Struhsaker [1967b]). The habitat of Lolui Island is a mosaic of moist semi-deciduous forest, fringing forest, thickets, grasslands and swamps, whereas the habitat at Amboseli is best described as semi-arid savanna with very sparse and restricted tree groves in the water hole areas. The vegetation of Murchison Falls Park is rather intermediate to these two extremes and the relation between group size and ecology among *C. aethiops* is worthy of further investigation. Comparison of different populations of *Papio anubis* does not support Crook and Gartlan's hypothesis, for the more forest-dwelling baboons of Ishasha (Rowell [1966]) lived in groups of the same size or slightly larger than those living in the savanna near Nairobi (DeVore and Hall [1965]) and in the arid savanna of Waza, Cameroon (Tab. I). In contrast and in support of Crook and Gartlan are the data on group size on *Papio cynocephalus* in the semi-arid savanna of Amboseli where the mean group size is eighty and significantly larger than the *Papio anubis*, *P. ursinus*, and drill groups. It is of interest that the largest *C. aethiops* groups are also found at Amboseli.

Group Composition

Comparison of group compositions is hindred by the difficulty of making reliable observations of forest *Cercopithecus*. Adult females

[*Editor's Note:* Table 1 has been omitted because of space limitations.]

could often be distinguished by the presence af pendulant nipples. Sometimes the scrotum of adult males could be seen. The slightly larger size of adult males was most apparent when they were near adult females. Sex determination of immature *Cercopithecus* was not possible under the prevailing observation conditions. Age and sex determination of drills and mangabeys was easier, but never was the complete age-sex composition of a group determined. The vocalizations of adult males were distinct from all other age-sex classes in the forest *Cercopithecus*, drills, and probably, but less certainly, in the mangabeys. This sexual difference in behavior provided indirect clues on group composition.

Among the forest *Cercopithecus* in this sample, never more than one adult male was seen in each heterosexual group. With only four exceptions, there was one source of adult male loud calls in each group. In the four exceptions, two sources of these loud calls were heard once each in *C. nictitans n.* and *C. l'hoesti preussi* and twice in *C. pogonias grayi*. On one occasion two very large adult and presumably male *C. nictitans martini* were seen running away from a group. This encounter appeared to be a chase. Only one of these, which proved to be an adult male, returned. Solitary monkeys were commonly observed in forest *Cercopithecus*. With one exception these solitaries were adult and when their sex was verified, proved to be adult males. The following data are the frequencies with which solitaries were seen and the number of these that were definitely verified as adult males: *C. nictitans martini* (17, 1A♂); *C. nictitans n.* (5); *C. pogonias p.* (1, A♂); *C. pogonias grayi* (1, not A♂); *C. mona* (2, 1A♂); *C. cephus* (1); *C. erythrotis* (5, 2A♂); and *C. l'hoesti preussi* (1). Many of those listed whose sex was not determined were probably adult males judging from their large size, lack of nipples, and more contrasting pelage coloration that is typical of adult males. These observations suggest that heterosexual groups of forest *Cercopithecus* are characterized by the presence of only one adult male. The other adult and/or subadult males are solitary or very peripheral to these heterosexual groups. This structure would appear to be maintained by the aggressiveness of the adult male in the heterosexual group.

HADDOW [1952] collected fifteen solitary *Cercopithecus ascanius* which is also a forest species. These fifteen included: ten adult males, three old adult males, one subadult male, and one adult female.

A different system prevails among drills, *Cercocebus albigena*, and probably, though less certainly, *Cercocebus torquatus*. Even though solitary or extremely peripheral adult and subadult males are moderately common, the heterosexual groups usually contain more than one adult male (Tab. I). Verified solitaries were seen with the following frequency: drills (2 A or SA♂, 3 A♂, 1 SA♂) and *Cercocebus albigena* (1 SA♂). Adult male drills often give a low-pitched two-phase grunt that carries for a considerable distance and presumably facilitates group cohesion. The number of sources of this call in each group provides indirect, though less reliable, information on the number of adult males present. Such information indicates that drill groups of about twenty or less have only one adult male, whereas those greater then twenty usually have more than one adult male. The few reliable counts would support this (Tab. I). There is no evidence for all-male groups in any of the forest monkeys in this sample.

Apparently no relation exists between the number of adult males in heterosexual groups and the extent to which the species is terrestrial or arboreal among the forest cercopithecines. *C. l'hoesti* is considerably more terrestrial than the other forest *Cercopithecus* and yet they are similar in having only one adult male per group. Drills are more terrestrial than *Cercocebus albigena* and yet both species have several adult males per group. The one relationship which does seem consistent is that solitary adult and subadult males are common among forest cercopithecines but not among non-forest species. Although the ecology of the forest *Cercopithecus* corresponds to CROOK and GARTLAN's 'Grade III' their social structure does not[4]. Drills have a social organization comparable to their 'Grade IV', but an ecology between

---

[4] CROOK and GARTLAN [1966] have attempted to categorize primate species into a series of five grades. These grades represent levels of adaptation in forest, tree savanna, grassland, and arid environments. They are based on the assumption '… that patterns of social organization determining population dispersion are intimately linked to species ecology…'. The features and members of their grades are summarized in their Table I. The major features (habitat, diet, diurnal activity, size of groups, reproductive units) of each grade are as follows: Grade I: forest, mostly insectivorous, nocturnal, usually solitary, pairs where known; Grade II: forest, fruit or leaves, crepuscular or diurnal, very small groups, small family parties based on single male; Grade III: forest-forest fringe, fruit or fruit and leaves, stems, etc., diurnal, small to occasionally large parties, multi-male groups; Grade IV: forest fringe, tree savanna, vegetarian-omnivore, occasionally carnivorous, diurnal, medium to large groups, multi-male groups; and Grade V: grassland or arid savanna, vegetarian-omnivore, occasionally carnivorous, diurnal, medium to large groups, variable size in *gelada* and *hamadryas*, one-male groups.

'Grade II' and 'Grade III'. Revision of their categories seems necessary in this regard.

In contrast to the forest cercopithecines most of the non-forest species have more than one adult male in each heterosexual group (Tab. I). The most notable exception is *Erythrocebus patas* in which one adult male heterosexual groups are the rule. No exceptions have yet been observed. It seems relevant to mention that most taxonomists consider *patas* to be closely related to the *Cercopithecus* monkeys. JOLLY [1966] has subdivided the subfamiliy Cercopithecinae into three tribes. One of these, Cercopithecini, consists only of the genera *Cercopithecus* and *Erythrocebus*. Although *gelada* and *hamadryas* have one-adult-male heterosexual units, there are usually several of these units which move together and thus there are in effect several adult males present in the bands and herds at any given time. Solitary animals are rare among non-forest cercopithecines. KUMMER [1968] has observed only one solitary *hamadryas*; an adult male. Similarly, among baboons of South Africa and *patas* monkeys of Uganda, HALL [1965a] has only observed an occasional isolated adult male. In Cameroon I observed one solitary adult male *patas* and two adult male *patas* who were each very peripheral to heterosexual groups. It seems justifiable to conclude that selection has favored solitary males in forest cercopithecine species and not in non-forest species. Presumably, this is related to lower pressures of predation and greater concealment from predators in the forest. It seems significant that among the non-forest cercopithecines those species having one-adult-male heterosexual groups are also the only cercopithecines having all-male groups. HALL [1965b] observed one such group among *patas* monkeys in Uganda. This consisted of "... one full-grown adult male, two near-adult males whose fur colouration was similar, but their size was somewhat smaller and one young adult male". I observed three all-male *patas* groups in the Waza Reserve of Cameroon. In 1967 one consisted of three adult males and three subadult males and the other consisted of four subadult males. In 1968 what was presumed to be a third and different group had two adult males and four subadult males. CROOK [1966] mentions that all-male groups occur in the herds of *gelada*. Young and subadult male *hamadryas* move on the periphery of the one-adult-male units (KUMMER [1968]). These all-male groupings and the peripheralness of males support the hypothesis that the solitary mode of existence is selected against in the non-forest cercopithecines.

Intergroup Spacing

Without individually recognizable monkeys and because of the inherent problems of making observations in a rain forest, few data are available on the distance between groups that were not engaged in intergroup conflicts. It is estimated that as many as five groups of *C. nictitans martini* used the mapped area of 0.41 square miles. These data do not permit accurate estimates of population or group density. Assuming that these monkeys spent their entire life within this 0.41 square miles, which is almost certainly not the case, the maximum density would be 12.5 groups of *nictitans* per square mile or about 155 *nictitans* per square mile. Usually considerable time lapsed between the time when one group was left and another one located. Consequently, it is virtually impossible to present exact data on the distances between any two groups at any given moment in time. The only exceptions to this occurred when two groups were fighting. A few examples are given of distances between *nictitans martini* groups that were not engaged in aggressive encounters. On January 17, 1968 I left one group at 15.00 h and encountered another group 200 yards away at 15.20 h. On February 27, 1968 a group was left at 12.25 h and another one located 800 yards away was seen at 13.32 h. This group was left at 14.11 h and a solitary male was found 700 yards away at 17.22 h. He was left at 17.30 h and still another group was located 430 yards from him at 17.57 h. April 1, 1968 a group of *nictitans* was left at 11.35 h and another group encountered at 13.25 h and 1,033 yards from it. This group was left at 14.10 h and a third group met at 16.34 h and 733 yards away. April 5, 1968 a group was left at 12.58 h and a second group met at 14.07 h and 1,066 yards from it. This second group was left at 14.56 h and a third encountered 666 yards away at 15.58 h. The third group was left at

---

Fig. 1. Four examples of intergroup dispersion of *Cercopithecus nictitans martini* within the major study area at Idenau, West Cameroon. The encircled numbers refer only to the sequential position in which the group was seen. They do not refer to a particular group and, in fact, any given number may refer to a different group in these four cases. The encircled S refers to a solitary monkey. The other numbers refer to the hour at which the group was first contacted and at which it was left. All four examples were within the major study area, but not necessarily within the same specific area. The circumscribed areas for each group indicate the space they occupied during the period they were observed.

16.36 h and still another group was found at 17.00 h and 366 yards away. This group was left six minutes later and a solitary male *nictitans* was seen 300 yards from it at 17.16 h. The final example occurred on April 10, 1968. The first group was left at 11.02 h and a second group met at 13.20 h and 933 yards away. This group was left at 13.38 h and the next group encountered was 433 yards away at 15.00 h. I left this one at 16.03 h and saw a fourth group 800 yards away at 17.54 h. These latter four cases are summarized in Figure 1.

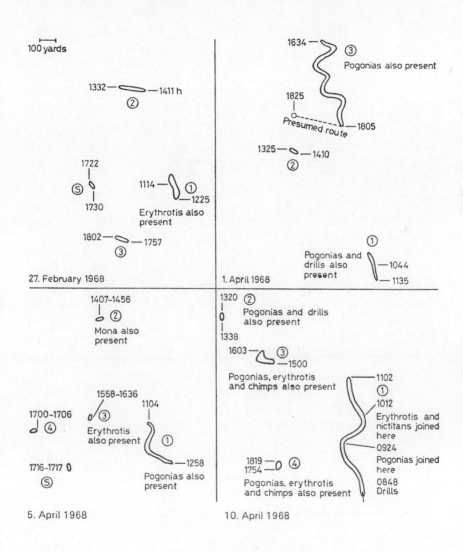

Intergroup Relations

Five intense intergroup agonistic encounters were observed among *C. nictitans*. At least one of these involved physical contact and biting. With so few data it is not possible to assert whether these encounters were defense of territories or not. No aggression was verified between groups of other forest cercopithecines. *Cercopithecus aethiops* is markedly territorial in at least two of the three well-studied populations, but territoriality apparently does not exist in other non-forest African cercopithecines. Limited data suggest that *Erythrocebus patas* is territorial, but this is no more certain than the data for *C. nictitans*. It would seem likely, therefore, that territoriality is independent of the degree to which a species is either terrestrial or forest-dwelling. This is in agreement with CROOK and GARTLAN [1966] who have placed territorial species in three of their Grades (II, III, IV) ranging in ecology from forest to tree savanna. Because intergroup conflicts among rain-forest cercopithecines are rare events, details of the two most clearly observed cases are presented here. The first case occurred near Bifos in the Dja Reserve of East Cameroon on December 10, 1967. Observations of the monkeys involved extended from 10.13 to 10.52 h. At 10.13 h we located an association of *C. nictitans n.* The following occurred between 10.19 and 10.40 h. Many agonistic chutters and screams were heard from several sources. These were followed by *nictitans* loud calls (pows and hacks) from two sources that were near each other. These loud calls are given only by adult males. My guide and I were about fifteen feet from the monkeys and could therefore observe them very well. Most of the agonism occurred about twenty feet up in a dense liana thicket within a 100 ft tall forest of a distinctly secondary nature. Throughout this encounter there was much shacking of foliage, jumping and climbing back and forth. Chirps and long-duration, low-frequency growls were common. At 10.40 h some of the *nictitans* saw us, but, instead of fleeing away, fled toward us and overhead. They were followed by members of the other group. While overhead, some intense fighting occurred about six feet up from the ground. This clearly involved physical contact including grabbing and apparently slapping and biting. One pair of combatants was in aggressive contact for about seven continuous seconds. Their heads were together and they may have been biting one another about the face, neck, and shoulders. This contact was terminated when one of them fell or leapt to the ground and fled. Screaming, chutters, chirps, and growls from several sources were very common now. *Nictitans* were leaping back and forth in all directions. Many fell, but only two fell to the ground. The others grabbed lianas before reaching the ground. The level of aggression seemed comparable to some of the most intense intergroup aggression that I have observed among *C. aethiops*. At about 10.46 h the fight seemed to have terminated as the two groups of *nictitans* moved apart. Accurate counts of these two groups weren't possible, but nine (estimated at about fifteen) moved in one direction and six (estimated at about ten) moved in another direction. Apparently, all members of both groups participated, including adults and juveniles, all of whom were at least two-thirds adult size. At 10.46 h adult male loud calls were heard from one source in the smaller group. At 10.52 h no more were seen and we left at 11.15 h.

The second case occurred in my major study area at Idenau, West Cameroon on January 1, 1968 between 10.58 and 11.14 h. A group of *C. nictitans martini* was located at 09.18 h. They were feeding and at 10.25 h began progressing upstream. About 150 yards from the preceding place and at 10.58 h they apparently met another group, for suddenly there was a loud crash, screeches, chutters, barks and a few chirps. At 10.58.5 h adult male loud calls (pows and hacks) were heard from one source in this group. Very suddenly at 11.00 h thirteen *nictitans* fled back to their previous location of 09.18 h and then turned south and continued fleeing. Growls, chirps, chutters, and trills were frequently heard during this flight. There was also much crashing of branches and leaping about. At 11.10 h adult male loud calls were heard from one source in this group. The intensitiy of the agonistic encounter subsided at 11.12 h and only a few grunts and growls were heard until 11.13 h when there was a brief burst of chutters. The encounter apparently terminated at 11.14 h. Subsequent counts and measurements revealed that a group of eighteen *nictitans* (including only one adult male) had been chased about 220 yards by another group of nine to twelve *nictitans*. The chase terminated at a particular tree where the group of eighteen had been seen many times in the past, and the smaller group then headed east and back in the same general direction it had come from.

The other three intergroup agonistic encounters were less clearly observed. Two of these occurred in the Idenau forest. *Nic-*

titans martini were actively involved in both cases and, although present, it was not apparent that the *C. pogonias p.* fought or were in any other way actively participating. Another less clearly observed encounter occurred in the Campo Reserve, East Cameroon. Two groups of *C. nictitans n.* and at least one group of *C. cephus* were involved here. One group of *nictitans* and *cephus* reversed direction and fled together, but direct conflict with the other group of *nictitans* was not seen.

Intragroup Dispersal

Limited data indicate that group or troop members of cercopithecine species living in rain forests are not more widely dispersed than are those of non-forest species. DeVore and Washburn [1963] describe baboon troops as sometimes being dispersed over a distance of a quarter of a mile. They also report that some of the larger troops divided, although "Such a division lasts for only part of the day and the troop reunites before nightfall". Hall [1965a] substantiates this with the observation that baboon groups in Uganda were widely spaced while feeding in the savanna. *Erythrocebus patas* groups in the savannas of Uganda are sometimes dispersed over distances at least as great as 500 m (Hall [1965b]). Members of vervet groups are also sometimes widely spaced. In the semi-arid savanna of Amboseli, Kenya, I have observed a group of sixteen vervets spread over a linear distance of 300 yards. One night this same group was spread over 750 yards. Rain-forest *Cercopithecus* monkeys were never so widely dispersed. In fact, most groups of these species seemed more compact and concentrated than non-forest cercopithecines. In contrast, groups of drills and mangabeys seemed to have dispersal patterns similar to baboons, *patas*, and vervet monkeys. Some indication of these differences in intragroup dispersal pattern can be seen from a few cases of what were considered exceptionally wide dispersions. On one occasion a group of *C. nictitans martini* and again a group of *C. nictitans n.* were spread over a linear distance of 50–60 yards. A case of whose validity I am less certain involved one individual *C. nictitans martini* who was at least 250 yards from a social group of conspecifics. This individual may have been a peripheral male or only temporarily separated after the group fled from us. A large group of drills, estimated to have about sixty-five individuals was once observed to have a dispersal of at least 150 yards. At this time the group was segregated into two sections that were more than 100 yards apart. Eventually, these two sections joined and became relatively compact. A relatively large group of *Cercocebus albigena*, of at least twenty but for whom neither an accurate group count or estimate was made, were once seen dispersed over a linear distance of at least 200 yards.

Conclusions

The data presented in this paper on rain forest primates in Cameroon call for revision of several hypotheses that correlated certain aspects of social organization and ecology. 1) The direct relation between group size and the degree of terrestriality was not supported by these data. 2) The hypothesis that larger groups occur in non-forest habitats was contradicted by some of these data. 3) Several combinations of ecological and social structure variables were described that did not fit into the classification scheme of CROOK and GARTLAN [1966]. This is not to say that ecological factors cannot have an important effect on social organization. Evidence was presented that supports a hypothesis directly relating the occurrence of solitary males with a forest habitat and their absence or rarity in non-forest areas. The fact that all-male groups have been seen only in open, non-forest habitats further supports the idea that the amount of vegetational 'cover' can have an effect on social structure. It would, thus, appear that certain ecological factors and certain aspects of social organization can be correlated, but obviously not all. HALL [1965 a] has stated "... social adaptations shown in the varying patterns of dominance, from the conspicuous adult male hierarchies of baboons and rhesus to the lack of any such demonstrable relationships in other species, can readily be understood only within the ecological framework with all its complexity of food-getting, sheltering and resting, avoidance of predators, and the changing seasonal internal pattern of births and mating." This statement has several attributes and in some respects is applicable to many other aspects of social organization. However, in considering the relation between ecology and society it must be emphasized that each species brings a different phylogenetic heritage into a particular ecological scene. Consequently, one must consider not only ecology but also phylogeny in attempting to understand the evolution of primate social organization. The interrelations of these two classes of variables determines the expression of the character, in this case social structure. In some cases, the immediate ecological variables may limit the expression or development of social structure and, with other species and circumstances, variables of phylogeny may be the limiting parameters. For example, heterosexual groups with only one adult male seem to be typical of most *Cercopithecus* spp. and the closely related *Erythrocebus patas*. The only notable exception is *C. aethiops*. They are the only member of the Cercopithecini Tribe that have multi-male heterosexual groups, no solitary males, and no all-male groups. One wonders why they developed the multi-male heterosexual group rather than the all-male group as did *patas* in adapting to savanna life. In contrast, drills, baboons, and *Cercocebus* spp. typically have several adult males in their heterosexual groups regardless of gross ecological differences. *Gelada* and *hamadryas* are rather intermediate between these two extremes. This appears to be a case in which phylogenies are at least as important as ecology, if not more so, in understanding an aspect of the social structure. EISENBERG [1963] in his detailed behavioral study of heteromyid rodents reached a similar conclusion: "... it appears that the phylogenetic background has been very important in the expression of the social organization in the family (Heteromyidae)."

When one considers the distribution of African Cercopithecinae species in the various kinds of habitats, two trends become readily apparent. In terms of numbers of species and numbers of individuals the *Ceropithecus* species are more successful in forest habitats than are baboons (*Papio* spp.) who in contrast are more successful in savanna and arid-country than are the *Ceropithecus* species. This would imply that these two genera are and have been exposed to different selective pressures. Consequently, one would expect to find differences between these two groups in behavior having adaptive advantage such as social organization. Apparently, social organization among African Ceropithecinae is a relative inert or stable character.

I would certainly agree that more detailed ecological information is needed before the causal and casual factors of social organization of primates can be fully understood. But I would stress with equal emphasis that a better understanding of phylogenies can also contribute greatly to our comprehension of the evolution of primate social structure.

Summary

New data are presented on the social organization of several species of rain-forest cercopithecines from Cameroon, West Africa. These data are compared with studies on savanna and arid-country cercopithecines and discussed in view of earlier hypotheses that have attempted to relate ecology and social organization. Many observations of the rain-forest species are inconsistent with these earlier hypotheses and call for revision of them. It is concluded that, although some aspects of social organization are clearly related to ecology, others are not and are perhaps best understood from the standpoint of phylogenetic affinities.

APPENDIX

Critique of A. P. Buxton's [1952] paper on *Cercopithecus ascanius*.

In 1952 A. P. Buxton published a lengthy paper on a small and isolated population of *Cercopithecus ascanius* in Uganda. This population lived in a forest described by Buxton as a mixture of permanent swamp, raised wet forest, and raised seasonal forest. *C. ascanius* is typified as a rain-forest monkey and because Buxton's study is the only one published that presents data on the social structure of an African forest *Cercopithecus* sp. it is of direct relevance to this discussion. Buxton came to the following conclusions regarding the social structure of *C. ascanius*: (1) a typical family group consists of four monkeys (one adult male, one or two adult females, and one or two subadults or young); (2) the family groups sleep alone at night; (3) two or more family groups aggregate around 07.45 h; (4) between 09.00 and 10.30 h these aggregations break-up into family groups; (5) similar aggregations form again between 13.30 and 14.30 h and even larger bands form between 15.00 and 16.00 h; (6) after 16.00 h these bands split up into family groups that remain apart until the following morning. These conclusions are not in harmony with those presented in this paper on rain-forest cercopithecines in Cameroon. I have no evidence that group size is correlated with the time of day as is presumably demonstrated by Buxton (see his Tab. 5 and Fig. 9).

Re-analysis of Buxton's data reveal that his study suffers from certain major technical errors that invalidate his analysis of group size and consequently leave his conclusions on this subject as unsubstantiated opinions. His most serious error is that he assumed his data were statistically independent. But they obviously are not. He states: 'Each 5 min. period of observation from each post has been treated below as one unit observation period. Each monkey seen per unit observation period is treated as one monkey-observation.' The 'monkey-observations' were the data analyzed by Buxton with the chi-square test to form the basis of his conclusions on social structure (see his Tab. 5). If a particular monkey was observed for fifteen minutes one presumes that it constituted three monkey-observations and by definition these would not be independent observations. Furthermore, Buxton estimates that the total population studied was more than thirty but less than forty (p. 45). However, most of his observations are of more than forty (see his Tab. 5) which can only mean that some monkeys and groups were counted more than once for a given case or observation. Siegel [1956] in his discussion of the applicability of the chi-square test states: '... each observation must be independent of every other; thus one may not make several observations on the same person and count each as independent. To do so produces an inflated N.' Buxton wrongly assumed independence and was therefore not justified in applying the chi-square test to his data (p. 42). Consequently, he is not justified in concluding that the group size of *C. ascanius* fluctuates according to the time of day.

Even if his data were statistically independent, Buxton's analysis can be criticized on other grounds. He fails to describe how his African observers distinguished between the case of one large troop and the case of two small troops that were adjacent to one another, i. e., he gives no operational definition of a troop. In essence he admits this on page 42: 'After trying a large number of different groupings it was found convenient to classify the different sizes of groups arbitrarily into: solitary monkeys; 'small troops' of two, three and four individuals; 'large troops' of five to ten; and 'bands' of over ten monkeys. Groups of these sizes were chosen because clearer results were obtained with them than with any other grouping tried, though all groupings indicated the same general trends.' One is tempted to question whether 'clearer results' is not synonymous with Buxton's bias.

If one is attempting to understand the central tendency in group size of any species, the frequency distribution of the sizes of a representative sample of groups must be available. Buxton does not present such a frequency distribution. Instead he gives the percentage of monkeys seen in the four classes of group-size at thirteen different time periods between 05.35 and 18.30 h. This does not permit an adequate analysis of the problem of social organization of *C. ascanius*. With Buxton's method the larger troops and bands are given disproportionately more weight than smaller groupings. For example, in a hypothetical case there may be 100 monkeys, 25 (25%) of which are in one band, 25 (25%) in five large troops (5 in each), 25 (25%) in 12 small troops, and 25 (25%) solitaries. In this example the band is the least common social unit and yet considering only percentages, as Buxton did, it is given equal weight with the other classes.

Differences in the distribution of monkeys among the four classes of group-size could be due to differences in the size of the groups or to differences in the number of such groups. Buxton's data don't permit an evaluation of this problem. Assuming statistical independence and using Buxton's data in his Table 5, I estimated the number of groups seen in the four classes of social groups at each of the thirteen time periods. The number of monkey-observations for each group class at each time period was divided by its mean group-size, i. e., solitaries (1); small troops (3); large troops (7.5); and bands (18). The number of groups estimated in this manner for each of the four group classes at each of the thirteen time periods were then tested by the chi-square test. Contrary to Buxton's analysis, none of the classes showed any significant difference in their frequency throughout the thirteen time periods. The chi-square values ranged from 10.48 to 22.33 with 12 degrees of freedom in all cases. These values are clearly not significant and thus one concludes that the number of groups in each size class did not vary with the time of day.

ACKNOWLEDGEMENTS

I thank Dr. BRUCE A. BARRON for statistical consultation and Drs. P. MAR-LER, F. NOTTEBOHM, J. BRADBURY, J. S. GARTLAN, J. H. CROOK, and J. ELLEFSON for constructive criticism of the manuscript. This study was financed by a NSF grant (GB 5792) and sponsored by The New York Zoological Society and The Rockefeller University.

REFERENCES

ALTMANN, S. A. (ed.): Social communication among primates (Univ. of Chicago Press, Chicago 1967).

BOLWIG, N.: A study of the Chacma baboon, Papio ursinus. Behaviour *14:* 136 (1959).

BUXTON, A. P.: Observations on the diurnal behaviour of the redtail monkey *(Cercopithecus ascanius schmidti matschie)* in a small forest in Uganda. J. anim. Ecol. *21:* 25–58 (1952).

CHALMERS, N. R.: Group composition, ecology and daily activities of free-living mangabeys in Uganda. Folia primat. *8:* 247–262 (1968a). – The social behaviour of free-living mangabeys in Uganda. Folia primat. *8:* 263–281 (1968b).

CROOK, J. H.: Gelada baboon herd structure and movement – A comparative report. Symp. zool. Soc. Lond. *18:* 237–258 (1966).

CROOK, J. H. and GARTLAN, J. S.: Evolution of primate societies. Nature, Lond. *210:* 1200–1203 (1966).

DEVORE, I.: A comparison of the ecology and behavior of monkeys and apes. In: WASHBURN, S. L., Classification and human evolution, pp. 301–319 (Aldine Chicago 1963). – Primate behavior. Field studies of monkeys and apes (Holt, Rinehart and Winston, New York 1965).

DEVORE, I. and HALL, K. R. L.: Baboon ecology. In: DEVORE, I., Primate behavior: Field studies of monkeys and apes, pp. 20–52 (Holt, Rinehart and Winston, New York 1965).

DEVORE, I. and WASHBURN, S. L.: Baboon ecology and human evolution. In: HOWELL, F. and BOURLIÈRE, F., African ecology and human evolution, pp. 335–367 (Aldine, Chicago 1963).

EISENBERG, J. F.: The behavior of heteromyid rodents. Univ. of Calif. Publ. Zool. 69 (1963).

HADDOW, A. J.: Studies on *Cercopithecus ascanius schmidti.* Proc. zool. Soc. Lond. *122:* 337–373 (1952).

HALL, K. R. L.: Social organization of the old-world monkeys and apes. Symp. zool. Soc. Lond. *14:* 265–289 (1965a). – Behaviour and ecology of the wild patas monkey, *Erythrocebus patas,* in Uganda. J. Zool., Lond. *148:* 15–87 (1965b). – Distribution and adaptations of baboons. Some recent developments in comparative medicine. Symp. zool. Soc. Lond. *17* (1966).

HALL, K. R. L. and DEVORE, I.: Baboon social behaviour. In: DEVORE, I., Primate behavior: Field studies of monkeys and apes, pp. 53–110 (Holt, Rinehart and Winston, New York 1965).

HALL, K. R. L. and GARTLAN, J. R.: Ecology and behaviour of the vervet monkey, *Cercopithecus aethiops*, Lolui Island, Lake Victoria. Proc. zool. Soc. Lond. *145:* 37–56 (1965).

JOLLY, C. J.: Introduction to the Cercopithecoidea with notes on their use as laboratory animals. Symp. zool. Soc. Lond. *17:* 427–457 (1966).

JONES, C. and SABATER PI, J.: Comparative ecology of *Cercocebus albigena* (GRAY) and *Cercocebus torquatus* (KERR) in Rio Muni, West Africa. Folia primat. *9:* 99–113 (1968).

KUMMER, H.: Social organization of hamadryas baboons: A field study (Univ. of Chicago Press, Chicago 1968).

ROWELL, T. E.: Forest-living baboons in Uganda. J. Zool., Lond. *149:* 344–364 (1966).

SIEGEL, S.: Nonparametric statistics (McGraw-Hill, New York 1956).

STEVENSON-HAMILTON, J.: Wild life in South Africa (Cassell, London 1947).

STRUHSAKER, T. T.: Auditory communication among vervet monkeys, *(Cercopithecus aethiops)*. In: ALTMANN, S. A., Social communication among primates, pp. (Univ. of Chicago Press, Chicago 1967a). – Social structure among vervet monkeys *(Cercopithecus aethiops)*. Behaviour *29:* 2–4 (1967b). – Ecology of vervet monkeys *(Cercopithecus aethiops)* in the Masai-Amboseli Game Reserve, Kenya. Ecology *48:* 6 (1967c). – Behavior of vervet monkeys *(Cercopithecus aethiops)*. Univ. of Calif. Publ. Zool. 82 (1967d).

Part IV

GENETICS AND SOCIAL STRUCTURE

Editor's Comments
on Papers 19 and 20

19   HAMILTON
     The Genetical Evolution of Social Behaviour. I

20   DeFRIES and McCLEARN
     Social Dominance and Darwinian Fitness in the Laboratory Mouse

The quantitative genetic model presented in Paper 19, the first of two papers by W. D. Hamilton of the Galton Laboratory, University College, London, formed the basis of a paradigm shift, a scientific revolution, in our thinking about the evolution of social behavior. The model is concerned with the interactions between relatives and how these interactions may affect each other's inclusive fitness (Hamilton, 1964). Substance is given to the model in the second paper, in which more general applications are examined. The main principle emerging from this discussion is that "The social behaviour of a species evolves in such a way that in each distinct behavior-evoking situation the individual will seem to value his neighbor's fitness against his own according to the coefficients of relationship appropriate to that situation" (Hamilton, 1964, p. 19). Thus, such currently important concepts as altruism, parent-offspring conflict, group selection, kin selection, social parasitism, and the like are brought into analytic focus. An update of Hamiltons's concepts appears in his 1972 review article.

This theoretical population–genetic approach forms one of the main themes in Wilson's recently published *magnum opus, Sociobiology: The New Synthesis* (1975). Other specific papers dealing with aspects of this subject are Trivers (1974) and Alexander (1975).

The relationship between high rank in a dominance order and fitness has been a matter of interest and speculation for many years. Although intuitively plausible, hard evidence in support of the idea that high rankers outreproduce low rankers in a group organized in a dominance hierarchy is scanty. Paper 20 describes a laboratory study in which such evidence is presented. Although the study was conducted under highly confined conditions and made use of inbred strains of house mouse, our understanding of the natural distribution of wild house mouse populations into quite small demes provides ample credibility for the model used by DeFries and McClearn, both of the Institute of Behavioral Genetics, University of Colorado.

REFERENCES

Alexander, R. D. 1975. The evolution of social behavior. *Ann. Rev. Ecol. Syst.* 5: 325-383.

Hamilton, W. D. 1964. The genetical evolution of social behaviour. II. *J. Theor. Biol.* 7:17-52.

Hamilton, W. D. 1972. Altruism and related phenomena, mainly in social insects. *Ann. Rev. Ecol. Syst.* 3: 193-232.

Trivers, R. L. 1974. Parent-offspring conflict. *Amer. Zool.* 14: 249-264.

Wilson, E. O. 1975. Sociobiology: The New Synthesis. Belknap Press of Harvard University Press. Cambridge, Mass. 698 pp.

19

Reprinted from *J. Theor. Biol.,* 7:1–16 (1964)

The Genetical Evolution of Social Behaviour. I

W. D. HAMILTON

The Galton Laboratory, University College, London, W.C.2

(*Received* 13 *May* 1963, *and in revised form* 24 *February* 1964)

A genetical mathematical model is described which allows for inter-actions between relatives on one another's fitness. Making use of Wright's Coefficient of Relationship as the measure of the proportion of replica genes in a relative, a quantity is found which incorporates the maximizing property of Darwinian fitness. This quantity is named "inclusive fitness". Species following the model should tend to evolve behaviour such that each organism appears to be attempting to maximize its inclusive fitness. This implies a limited restraint on selfish competitive behaviour and possibility of limited self-sacrifices.

Special cases of the model are used to show (a) that selection in the social situations newly covered tends to be slower than classical selection, (b) how in populations of rather non-dispersive organisms the model may apply to genes affecting dispersion, and (c) how it may apply approximately to competition between relatives, for example, within sibships. Some artificialities of the model are discussed.

1. Introduction

With very few exceptions, the only parts of the theory of natural selection which have been supported by mathematical models admit no possibility of the evolution of any characters which are on average to the disadvantage of the individuals possessing them. If natural selection followed the classical models exclusively, species would not show any behaviour more positively social than the coming together of the sexes and parental care.

Sacrifices involved in parental care are a possibility implicit in any model in which the definition of fitness is based, as it should be, on the number of adult offspring. In certain circumstances an individual may leave more adult offspring by expending care and materials on its offspring already born than by reserving them for its own survival and further fecundity. A gene causing its possessor to give parental care will then leave more replica genes in the next generation than an allele having the opposite tendency. The selective advantage may be seen to lie through benefits conferred in-differently on a set of relatives each of which has a half chance of carrying the gene in question.

From this point of view it is also seen, however, that there is nothing special about the parent-offspring relationship except its close degree and a certain fundamental asymmetry. The full-sib relationship is just as close. If an individual carries a certain gene the expectation that a random sib will carry a replica of it is again one-half. Similarly, the half-sib relationship is equivalent to that of grandparent and grandchild with the expectation of replica genes, or genes "identical by descent" as they are usually called, standing at one quarter; and so on.

Although it does not seem to have received very detailed attention the possibility of the evolution of characters benefitting descendants more remote than immediate offspring has often been noticed. Opportunities for benefitting relatives, remote or not, in the same or an adjacent generation (i.e. relatives like cousins and nephews) must be much more common than opportunities for benefitting grandchildren and further descendants. As a first step towards a general theory that would take into account all kinds of relatives this paper will describe a model which is particularly adapted to deal with interactions between relatives of the same generation. The model includes the classical model for "non-overlapping generations" as a special case. An excellent summary of the general properties of this classical model has been given by Kingman (1961b). It is quite beyond the author's power to give an equally extensive survey of the properties of the present model but certain approximate deterministic implications of biological interest will be pointed out.

As is already evident the essential idea which the model is going to use is quite simple. Thus although the following account is necessarily somewhat mathematical it is not surprising that eventually, allowing certain lapses from mathematical rigour, we are able to arrive at approximate principles which can also be expressed quite simply and in non-mathematical form. The most important principle, as it arises directly from the model, is outlined in the last section of this paper, but a fuller discussion together with some attempt to evaluate the theory as a whole in the light of biological evidence will be given in the sequel.

2. The Model

The model is restricted to the case of an organism which reproduces once and for all at the end of a fixed period. Survivorship and reproduction can both vary but it is only the consequent variations in their product, net reproduction, that are of concern here. All genotypic effects are conceived as increments and decrements to a basic unit of reproduction which, if possessed by all the individuals alike, would render the population both stationary and non-evolutionary. Thus the fitness $a^\bullet$ of an individual is treated as the sum of his basic unit, the effect δa of his personal genotype and the total e° of effects on him due to his neighbours which will depend on their genotypes:

$$a^\bullet = 1 + \delta a + e^\circ. \tag{1}$$

The index symbol $\bullet$ in contrast to $\circ$ will be used consistently to denote the inclusion of the personal effect δa in the aggregate in question. Thus equation (1) could be rewritten

$$a^\bullet = 1 + e^\bullet.$$

In equation (1), however, the symbol $\bullet$ also serves to distinguish this neighbour modulated kind of fitness from the part of it

$$a = 1 + \delta a$$

which is equivalent to fitness in the classical sense of individual fitness.

The symbol δ preceding a letter will be used to indicate an effect or total of effects due to an individual treated as an addition to the basic unit, as typified in

$$a = 1 + \delta a.$$

The neighbours of an individual are considered to be affected differently according to their relationship with him.

Genetically two related persons differ from two unrelated members of the population in their tendency to carry replica genes which they have both inherited from the one or more ancestors they have in common. If we consider an autosomal locus, not subject to selection, in relative B with respect to the same locus in the other relative A, it is apparent that there are just three possible conditions of this locus in B, namely that both, one only, or neither of his genes are identical by descent with genes in A. We denote the respective probabilities of these conditions by c_2, c_1 and c_0. They are independent of the locus considered; and since

$$c_2 + c_1 + c_0 = 1,$$

the relationship is completely specified by giving any two of them. Li & Sacks (1954) have described methods of calculating these probabilities adequate for any relationship that does not involve inbreeding. The mean number of genes per locus i.b.d. (as from now on we abbreviate the phrase "identical by descent") with genes at the same locus in A for a hypothetical population of relatives like B is clearly $2c_2 + c_1$. One half of this number, $c_2 + \frac{1}{2}c_1$, may therefore be called the expected fraction of genes i.b.d. in a relative. It can be shown that it is equal to Sewall Wright's Coefficient of Relationship r (in a non-inbred population). The standard methods of calculating r without obtaining the complete distribution can be found in Kempthorne (1957). Tables of

$$f = \tfrac{1}{2}r = \tfrac{1}{2}(c_2 + \tfrac{1}{2}c_1) \quad \text{and} \quad F = c_2$$

for a large class of relationships can be found in Haldane & Jayakar (1962).

Strictly, a more complicated metric of relationship taking into account the parameters of selection is necessary for a locus undergoing selection, but the following account based on use of the above coefficients must give a good approximation to the truth when selection is slow and may be hoped to give some guidance even when it is not.

Consider now how the effects which an arbitrary individual distributes to the population can be summarized. For convenience and generality we will include at this stage certain effects (such as effects on parents' fitness) which must be zero under the restrictions of this particular model, and also others (such as effects on offspring) which although not necessarily zero we will not attempt to treat accurately in the subsequent analysis.

The effect of A on specified B can be a variate. In the present deterministic treatment, however, we are concerned only with the means of such variates. Thus the effect which we may write $(\delta a_{\text{father}})_A$ is really the expectation of the effect of A upon his father but for brevity we will refer to it as the effect on the father.

The full array of effects like $(\delta a_{\text{father}})_A$, $(\delta a_{\text{specified sister}})_A$, etc., we will denote

$$\{\delta a_{\text{rel.}}\}_A.$$

From this array we can construct the simpler array

$$\{\delta a_{r, c_2}\}_A$$

by adding together all effects to relatives who have the same values for the pair of coefficients (r, c_2). For example, the combined effect $\delta a_{\frac{1}{4}, 0}$ might contain effects actually occurring to grandparents, grandchildren, uncles, nephews and half-brothers. From what has been said above it is clear that as regards changes in autosomal gene-frequency by natural selection all the consequences of the full array are implied by this reduced array—at least, provided we ignore (a) the effect of previous generations of selection on the expected constitution of relatives, and (b) the one or more generations that must really occur before effects to children, nephews, grandchildren, etc., are manifested.

From this array we can construct a yet simpler array, or vector,

$$\{\delta a_r\}_A,$$

by adding together all effects with common r. Thus $\delta a_{\frac{1}{4}}$ would bring together effects to the above-mentioned set of relatives and effects to double-first cousins, for whom the pair of coefficients is $(\frac{1}{4}, \frac{1}{16})$.

Corresponding to the effect which A causes to B there will be an effect of similar type on A. This will either come from B himself or from a person who stands to A in the same relationship as A stands to B. Thus corresponding to an effect by A on his nephew there will be an effect on A by his uncle. The similarity between the effect which A dispenses and that which he receives is clearly an aspect of the problem of the correlation between relatives. Thus the term e° in equation (1) is not a constant for any given genotype of A since it will depend on the genotypes of neighbours and therefore on the gene-frequencies and the mating system.

Consider a single locus. Let the series of allelomorphs be $G_1, G_2, G_3, ..., G_n$, and their gene-frequencies $p_1, p_2, p_3, ..., p_n$. With the genotype G_iG_j associate the array $\{\delta a_{rel.}\}_{ij}$; within the limits of the above-mentioned approximations natural selection in the model is then defined.

If we were to follow the usual approach to the formulation of the progress due to natural selection in a generation, we should attempt to give formulae for the neighbour modulated fitnesses $a_{ij}^\bullet$. In order to formulate the expectation of that element of e_{ij}° which was due to the return effect of a relative B we would need to know the distribution of possible genotypes of B, and to obtain this we must use the double measure of B's relationship and the gene-frequencies just as in the problem of the correlation between relatives. Thus the formula for e_{ij}° will involve all the arrays $\{\delta a_{r,c_2}\}_{ij}$ and will be rather unwieldy (see Section 4).

An alternative approach, however, shows that the arrays $\{\delta a_r\}_{ij}$ are sufficient to define the selective effects. Every effect on reproduction which is due to A can be thought of as made up of two parts: an effect on the reproduction of genes i.b.d. with genes in A, and an effect on the reproduction of unrelated genes. Since the coefficient r measures the expected fraction of genes i.b.d. in a relative, for any particular degree of relationship this breakdown may be written quantitatively:

$$(\delta a_{rel.})_A = r(\delta a_{rel.})_A + (1-r)(\delta a_{rel.})_A.$$

The total of effects on reproduction which are due to A may be treated similarly:

$$\sum_{rel.} (\delta a_{rel.})_A = \sum_{rel.} r(\delta a_{rel.})_A + \sum_{rel.} (1-r)(\delta a_{rel.})_A,$$

or

$$\sum_r (\delta a_r)_A = \sum_r r(\delta a_r)_A + \sum_r (1-r)(\delta a_r)_A,$$

which we rewrite briefly as

$$\delta T_A^\bullet = \delta R_A^\bullet + \delta S_A,$$

where $\delta R_A^\bullet$ is accordingly the total effect on genes i.b.d. in relatives of A, and δS_A is the total effect on their other genes. The reason for the omission of an index symbol from the last term is that here there is, in effect, no question of whether or not the self-effect is to be in the summation, for if it is included it has to be multiplied by zero. If index symbols were used we should have $\delta S_A^\bullet = \delta S_A^\circ$, whatever the subscript; it therefore seems more explicit to omit them throughout.

If, therefore, all effects are accounted to the individuals that cause them, of the total effect $\delta T_{ij}^\bullet$ due to an individual of genotype G_iG_j a part $\delta R_{ij}^\bullet$ will involve a specific contribution to the gene-pool by this genotype, while the remaining part δS_{ij} will involve an unspecific contribution consisting of genes in the ratio in which the gene-pool already possesses them. It is clear that it is the matrix of effects $\delta R_{ij}^\bullet$ which determines the direction of selection progress in gene-frequencies; δS_{ij} only influences its magnitude. In view of this importance of the $\delta R_{ij}^\bullet$ it is convenient to give some name to the concept with which they are associated.

In accordance with our convention let

$$R_{ij}^\bullet = 1 + \delta R_{ij}^\bullet;$$

then $R_{ij}^\bullet$ will be called the *inclusive fitness*, $\delta R_{ij}^\bullet$ the *inclusive fitness effect* and δS_{ij} the *diluting effect*, of the genotype G_iG_j.

Let

$$T_{ij}^\bullet = 1 + \delta T_{ij}^\bullet.$$

So far our discussion is valid for non-random mating but from now on for simplicity we assume that it is random. Using a prime to distinguish the new gene-frequencies after one generation of selection we have

$$p_i' = \frac{\sum_j p_i p_j R_{ij}^\bullet + p_i \sum_{j,k} p_j p_k \delta S_{jk}}{\sum_{j,k} p_j p_k T_{jk}^\bullet} = p_i \frac{\sum_j p_j R_{ij}^\bullet + \sum_{j,k} p_j p_k \delta S_{jk}}{\sum_{j,k} p_j p_k T_{jk}^\bullet}.$$

The terms of this expression are clearly of the nature of averages over a part (genotypes containing G_i, homozygotes G_iG_i counted twice) and the whole of the existing set of genotypes in the population. Thus using a well known subscript notation we may rewrite the equation term by term as

$$p_i' = p_i \frac{R_{i.}^\bullet + \delta S_{..}}{T_{..}^\bullet}$$

$$\therefore \quad p_i' - p_i = \Delta p_i = \frac{p_i}{T_{..}^\bullet}(R_{i.}^\bullet + \delta S_{..} - T_{..}^\bullet)$$

or

$$\Delta p_i = \frac{p_i}{R_{..}^\bullet + \delta S_{..}^\bullet}(R_{i.}^\bullet - R_{..}^\bullet). \tag{2}$$

This form clearly differentiates the roles of the $R_{ij}^\bullet$ and $\delta S_{ij}^\bullet$ in selective progress and shows the appropriateness of calling the latter diluting effects.

For comparison with the account of the classical case given by Moran (1962), equation (2) may be put in the form

$$\Delta p_i = \frac{p_i}{T_{..}^{\bullet}} \left(\frac{1}{2} \frac{\partial R_{..}^{\bullet}}{\partial p_i} - R_{...}^{\bullet} \right)$$

where $\partial/\partial p_i$ denotes the usual partial derivative, written d/dp_i by Moran.

Whether the selective effect is reckoned by means of the $a_{ij}^{\bullet}$ or according to the method above, the denominator expression must take in all effects occurring during the generation. Hence $a_{..}^{\bullet} = T_{...}^{\bullet}$.

As might be expected from the greater generality of the present model the extension of the theorem of the increase of mean fitness (Scheuer & Mandel, 1959; Mulholland & Smith, 1959; a much shorter proof by Kingman, 1961a) presents certain difficulties. However, from the above equations it is clear that the quantity that will tend to maximize, if any, is $R_{..}^{\bullet}$, the mean inclusive fitness. The following brief discussion uses Kingman's approach.

The mean inclusive fitness in the succeeding generation is given by

$$R_{..}^{\bullet\prime} = \sum_{i,j} p_i' p_j' R_{ij}^{\bullet} = \frac{1}{T_{..}^{\bullet 2}} \sum_{i,j} p_i p_j R_{ij}^{\bullet} (R_{i.}^{\bullet} + \delta S_{..})(R_{.j}^{\bullet} + \delta S_{..}).$$

$$\therefore \quad R_{..}^{\bullet\prime} - R_{..}^{\bullet} = \Delta R_{..}^{\bullet} = \frac{1}{T_{..}^{\bullet 2}} \left\{ \sum_{i,j} p_i p_j R_{ij}^{\bullet} R_{i.}^{\bullet} R_{.j}^{\bullet} + 2\delta S_{..} \sum_{i,j} p_i p_j R_{ij}^{\bullet} R_{i.}^{\bullet} + \right.$$
$$\left. + R_{..}^{\bullet} \delta S_{..}^2 - R_{..}^{\bullet} T_{..}^{\bullet 2} \right\}.$$

Substituting $R_{..}^{\bullet} + \delta S_{..}$ for $T_{..}^{\bullet}$ in the numerator expression, expanding and rearranging:

$$\Delta R^{\bullet} = \frac{1}{T_{..}^{\bullet 2}} \left\{ \left(\sum_{i,j} p_i p_j R_{ij}^{\bullet} R_{i.}^{\bullet} R_{.j}^{\bullet} - R_{...}^{\bullet 3} \right) + \right.$$
$$\left. + 2\delta S_{..} \left(\sum_{i,j} p_i p_j R_{ij}^{\bullet} R_{i.}^{\bullet} - R_{...}^{\bullet 2} \right) \right\}.$$

We have ( ) $\geqslant 0$ in both cases. The first is the proven inequality of the classical model. The second follows from

$$\sum_{i,j} p_i p_j R_{ij}^{\bullet} R_{i.}^{\bullet} = \sum_i p_i R_{i.}^{\bullet 2} \geqslant \left(\sum_i p_i R_{i.}^{\bullet} \right)^2 = R_{...}^{\bullet 2}.$$

Thus a sufficient condition for $\Delta R_{..}^{\bullet} \geqslant 0$ is $\delta S_{..} \geqslant 0$. That $\Delta R_{..}^{\bullet} \geqslant 0$ for positive dilution is almost obvious if we compare the actual selective changes with those which would occur if $\{R_{ij}^{\bullet}\}$ were the fitness matrix in the classical model.

It follows that $R_{..}^{\bullet}$ certainly maximizes (in the sense of reaching a local maximum of $R_{..}^{\bullet}$) if it never occurs in the course of selective changes that $\delta S_{..} < 0$. Thus $R_{..}^{\bullet}$ certainly maximizes if all $\delta S_{ij} \geqslant 0$ and therefore also if all $(\delta a_{\mathrm{rel.}})_{ij} \geqslant 0$. It still does so even if some or all δa_{ij} are negative, for, as we have seen δS_{ij} is independant of δa_{ij}.

Here then we have discovered a quantity, inclusive fitness, which under the conditions of the model tends to maximize in much the same way that fitness tends to maximize in the simpler classical model. For an important class of genetic effects where the individual is supposed to dispense benefits to his neighbours, we have formally proved that the average inclusive fitness in the population will always increase. For cases where individuals may dispense harm to their neighbours we merely know, roughly speaking, that the change in gene frequency in each generation is aimed somewhere in the direction of a local maximum of average inclusive fitness,† but may, for all the present analysis has told us, overshoot it in such a way as to produce a lower value.

As to the nature of inclusive fitness it may perhaps help to clarify the notion if we now give a slightly different verbal presentation. Inclusive fitness may be imagined as the personal fitness which an individual actually expresses in its production of adult offspring as it becomes after it has been first stripped and then augmented in a certain way. It is stripped of all components which can be considered as due to the individual's social environment, leaving the fitness which he would express if not exposed to any of the harms or benefits of that environment. This quantity is then augmented by certain fractions of the quantities of harm and benefit which the individual himself causes to the fitnesses of his neighbours. The fractions in question are simply the coefficients of relationship appropriate to the neighbours whom he affects: unity for clonal individuals, one-half for sibs, one-quarter for half-sibs, one-eighth for cousins, ... and finally zero for all neighbours whose relationship can be considered negligibly small.

Actually, in the preceding mathematical account we were not concerned with the inclusive fitness of individuals as described here but rather with certain averages of them which we call the inclusive fitnesses of types. But the idea of the inclusive fitness of an individual is nevertheless a useful one. Just as in the sense of classical selection we may consider whether a given character expressed in an individual is adaptive in the sense of being in the interest of his personal fitness or not, so in the present sense of selection we may consider whether the character or trait of behaviour is or is not adaptive in the sense of being in the interest of his inclusive fitness.

3. Three Special Cases

Equation (2) may be written

$$\Delta p_i = p_i \frac{\delta R_{i.}^\bullet - \delta R_{..}^\bullet}{1 + \delta T_{..}^\bullet}. \tag{3}$$

† That is, it is aimed "uphill": that it need not be at all directly towards the local maximum is well shown in the classical example illustrated by Mulholland & Smith (1959).

Now $\delta T_{ij}^{\bullet} = \sum_r (\delta a_r)_{ij}$ is the sum and $\delta R^{\bullet} = \sum_r r(\delta a_r)_{ij}$ is the first moment about $r = 0$ of the array of effects $\{\delta a_{rel.}\}_{ij}$ cause by the genotype $G_i G_j$; it appears that these two parameters are sufficient to fix the progress of the system under natural selection within our general approximation.

Let

$$r_{ij}^{\bullet} = \frac{\delta R_{ij}^{\bullet}}{\delta T_{ij}^{\bullet}}, \qquad (\delta T_{ij}^{\bullet} \neq 0); \tag{4}$$

and let

$$r_{ij}^{\circ} = \frac{\delta R_{ij}^{\circ}}{\delta T_{ij}^{\circ}}, \qquad (\delta T_{ij}^{\circ} \neq 0). \tag{5}$$

These quantities can be regarded as average relationships or as the first moments of reduced arrays, similar to the first moments of probability distributions.

We now consider three special cases which serve to bring out certain important features of selection in the model.

(a) The sums $\delta T_{ij}^{\bullet}$ differ between genotypes, the reduced first moment $r^{\bullet}$ being common to all. If all higher moments are equal between genotypes, that is, if all arrays are of the same "shape", this corresponds to the case where a stereotyped social action is performed with differing intensity or frequency according to genotype.

Whether or not this is so, we may, from equation (4), substitute $r^{\bullet} \delta T_{ij}^{\bullet}$ for $\delta R_{ij}^{\bullet}$ in equation (3) and have

$$\Delta p_i = p_i r^{\bullet} \frac{\delta T_{i.}^{\bullet} - \delta T_{..}^{\bullet}}{1 + \delta T_{..}^{\bullet}}.$$

Comparing this with the corresponding equation of the classical model,

$$\Delta p_i = p_i \frac{\delta a_{i.} - \delta a_{..}}{1 + \delta a_{..}}. \tag{6}$$

we see that placing genotypic effects on a relative of degree $r^{\bullet}$ instead of reserving them for personal fitness results in a slowing of selection progress according to the fractional factor $r^{\bullet}$.

If, for example, the advantages conferred by a "classical" gene to its carriers are such that the gene spreads at a certain rate the present result tells us that in exactly similar circumstances another gene which conferred similar advantages to the sibs of the carriers would progress at exactly half this rate.

In trying to imagine a realistic situation to fit this sort of case some concern may be felt about the occasions where through the probabilistic nature of things the gene-carrier happens not to have a sib, or not to have one suitably placed to receive the benefit. Such possibilities and their frequencies of realization must, however, all be taken into account as the effects $(\delta a_{sibs})_A$, etc., are being evaluated for the model, very much as if in a classical case allowance were being made for some degree of failure of penetrance of a gene.

(b) The reduced first moments $r_{ij}^{\bullet}$ differ between genotypes, the sum $\delta T^{\bullet}$ being common to all. From equation (4), substituting $r_{ij}^{\bullet}\delta T^{\bullet}$ for $\delta R_{ij}^{\bullet}$ in equation (3) we have

$$\Delta p_i = p_i \frac{\delta T^{\bullet}}{T^{\bullet}}(r_{i.}^{\bullet} - r_{..}^{\bullet}).$$

But it is more interesting to assume δa is also common to all genotypes. If so it follows that we can replace $^{\bullet}$ by $^{\circ}$ in the numerator expression of equation (3). Then, from equation (5). substituting $r_{ij}^{\circ}\delta T^{\circ}$ for δR_{ij}°, we have

$$\Delta p_i = p_i \frac{\delta T^{\circ}}{T^{\bullet}}(r_{i.}^{\circ} - r_{..}^{\circ}).$$

Hence, if a giving-trait is in question (δT° positive), genes which restrict giving to the nearest relative ($r_{i.}^{\circ}$ greatest) tend to be favoured; if a taking-trait (δT° negative), genes which cause taking from the most distant relatives tend to be favoured.

If all higher reduced moments about $r = r_{ij}^{\circ}$ are equal between genotypes it is implied that the genotype merely determines whereabouts in the field of relationship that centres on an individual a stereotyped array of effects is placed.

With many natural populations it must happen that an individual forms the centre of an actual local concentration of his relatives which is due to a general inability or disinclination of the organisms to move far from their places of birth. In such a population, which we may provisionally term "viscous", the present form of selection may apply fairly accurately to genes which affect vagrancy. It follows from the statements of the last paragraph but one that over a range of different species we would expect to find giving-traits commonest and most highly developed in the species with the most viscous populations whereas uninhibited competition should characterize species with the most freely mixing populations.

In the viscous population, however, the assumption of random mating is very unlikely to hold perfectly, so that these indications are of a rough qualitative nature only.

(c) $\delta T_{ij}^{\bullet} = 0$ for all genotypes.

$$\therefore \quad \delta T_{ij}^{\circ} = -\delta a_{ij}$$

for all genotypes, and from equation (5)

$$\delta R_{ij}^{\circ} = -\delta a_{ij} r_{ij}^{\circ}.$$

Then, from equation (3), we have

$$\Delta p_i = p_i(\delta R_{i.}^{\bullet} - \delta R_{..}^{\bullet}) = p_i\{(\delta a_{i.} + \delta R_{i.}^{\circ}) - (\delta a_{..} + \delta R_{..}^{\circ})\}$$
$$= p_i\{\delta a_{i.}(1 - r_{i.}^{\circ}) - \delta a_{..}(1 - r_{..}^{\circ})\}.$$

Such cases may be described as involving transfers of reproductive potential. They are especially relevant to competition, in which the individual can be considered as endeavouring to transfer prerequisites of survival and reproduction from his competitors to himself. In particular, if $r_{ij}^{\circ} = r^{\circ}$ for all genotypes we have

$$\Delta p_i = p_i(1 - r^{\circ})(\delta a_{i.} - \delta a_{..}).$$

Comparing this to the corresponding equation of the classical model (equation (6)) we see that there is a reduction in the rate of progress when transfers are from a relative.

It is relevant to note that Haldane (1923) in his first paper on the mathematical theory of selection pointed out the special circumstances of competition in the cases of mammalian embryos in a single uterus and of seeds both while still being nourished by a single parent plant and after their germination if they were not very thoroughly dispersed. He gave a numerical example of competition between sibs showing that the progress of gene-frequency would be slower than normal.

In such situations as this, however, where the population may be considered as subdivided into more or less standard-sized batches each of which is allotted a local standard-sized pool of reproductive potential (which in Haldane's case would consist almost entirely of prerequisites for pre-adult survival), there is, in addition to a small correcting term which we mention in the short general discussion of competition in the next section, an extra overall slowing in selection progress. This may be thought of as due to the wasting of the powers of the more fit and the protection of the less fit when these types chance to occur positively assorted (beyond any mere effect of relationship) in a locality; its importance may be judged from the fact that it ranges from zero when the batches are indefinitely large to a halving of the rate of progress for competition in pairs.

4. Artificialities of the Model

When any of the effects is negative the restrictions laid upon the model hitherto do not preclude certain situations which are clearly impossible from the biological point of view. It is clearly absurd if for any possible set of gene-frequencies any $a_{ij}^\bullet$ turns out negative; and even if the magnitude of δa_{ij} is sufficient to make $a_{ij}^\bullet$ positive while $1 + e_{ij}^\circ$ is negative the situation is still highly artificial, since it implies the possibility of a sort of overdraft on the basic unit of an individual which has to be made good from his own takings. If we call this situation "improbable" we may specify two restrictions: a weaker, $e_{ij}^\circ > -1$, which precludes "improbable" situations; and a stronger, $e_{ij}^\bullet > -1$, which precludes even the impossible situations, both being required over the whole range of possible gene-frequencies as well as the whole range of genotypes.

As has been pointed out, a formula for $e_{ij}^\bullet$ can only be given if we have the arrays of effects according to a double coefficient of relationship. Choosing the double coefficient (c_2, c_1) such a formula is

$$e_{ij}^\bullet = \sum_{c_2, c_1}^\bullet [c_2 \operatorname{Dev}(\delta a_{c_2, c_1})_{ij} + \tfrac{1}{2}c_1 \{\operatorname{Dev}(\delta a_{c_2, c_1})_{i.} + \operatorname{Dev}(\delta a_{c_2, c_1})_{.j}\}] + \delta T_{..}^\circ$$

where

$$\operatorname{Dev}(\delta a_{c_2, c_1})_{ij} = (\delta a_{c_2, c_1})_{ij} - (\delta a_{c_2, c_1})_{..} \quad \text{etc.}$$

Similarly

$$e_{ij}^\circ = \sum^\circ ['']+ \delta T_{..}^\circ,$$

the self-effect $(\delta a_{1, 0})_{ij}$ being in this case omitted from the summations.

The following discussion is in terms of the stronger restriction but the argument holds also for the weaker; we need only replace $\bullet$ by $\circ$ throughout.

If there are no dominance deviations, i.e. if

$$(\delta a_{\text{rel.}})_{ij} = \tfrac{1}{2}\{(\delta a_{\text{rel.}})_{ii} + (\delta a_{\text{rel.}})_{jj}\} \quad \text{for all } ij \text{ and rel.,}$$

it follows that each ij deviation is the sum of the $i.$ and the $j.$ deviations. In this case we have

$$e_{ij}^\bullet = \sum^\bullet r \operatorname{Dev}(\delta a_r)_{ij} + \delta T_{..}^\bullet.$$

Since we must have $e_{..}^\bullet = \delta T_{..}^\bullet$, it is obvious that some of the deviations must be negative.

Therefore $\delta T_{..}^\bullet > -1$ is a necessary condition for $e_{ij}^\bullet > -1$. This is, in fact, obvious when we consider that $\delta T_{..}^\bullet = -1$ would mean that the aggregate of individual takings was just sufficient to eat up all basic units exactly. Considering that the present use of the coefficients of relationships is only valid when selection is slow, there seems little point in attempting to derive mathematically sufficient conditions for the restriction to hold;

intuitively however it would seem that if we exclude over- and under-dominance it should be sufficient to have no homozygote with a net taking greater than unity.

Even if we could ignore the breakdown of our use of the coefficient of relationship it is clear enough that if $\delta T^{\bullet}_{\ldots}$ approaches anywhere near -1 the model is highly artificial and implies a population in a state of catastrophic decline. This does not mean, of course, that mutations causing large selfish effects cannot receive positive selection; it means that their expression must moderate with increasing gene-frequency in a way that is inconsistent with our model. The "killer" trait of *Paramoecium* might be regarded as an example of a selfish trait with potentially large effects, but with its only partially genetic mode of inheritance and inevitable density dependance it obviously requires a selection model tailored to the case, and the same is doubtless true of most "social" traits which are as extreme as this.

Really the class of model situations with negative neighbour effects which are artificial according to a strict interpretation of the assumptions must be much wider than the class which we have chosen to call "improbable". The model assumes that the magnitude of an effect does not depend either on the genotype of the effectee or on his current state with respect to the prerequisites of fitness at the time when the effect is caused. Where taking-traits are concerned it is just possible to imagine that this is true of some kinds of surreptitious theft but in general it is more reasonable to suppose that following some sort of an encounter the limited prerequisite is divided in the ratio of the competitive abilities. Provided competitive differentials are small however, the model will not be far from the truth; the correcting term that should be added to the expression for Δp_i can be shown to be small to the third order. With giving-traits it is more reasonable to suppose that if it is the nature of the prerequisite to be transferable the individual can give away whatever fraction of his own property that his instincts incline him to. The model was designed to illuminate altruistic behaviour; the classes of selfish and competitive behaviour which it can also usefully illuminate are more restricted, especially where selective differentials are potentially large.

For loci under selection the only relatives to which our metric of relationship is strictly applicable are ancestors. Thus the chance that an arbitrary parent carries a gene picked in an offspring is $\frac{1}{2}$, the chance that an arbitrary grandparent carries it is $\frac{1}{4}$, and so on. As regards descendants, it seems intuitively plausible that for a gene which is making steady progress in gene-frequency the true expectation of genes i.b.d. in a n-th generation descendant will exceed $\frac{1}{2}^n$, and similarly that for a gene that is steadily declining in frequency the reverse will hold. Since the path of genetic connection with a simple same-generation relative like a half-sib includes an "ascending part" and a "descending part" it is tempting to imagine that the ascending part can be treated with multipliers of exactly $\frac{1}{2}$ and the descending part by multipliers consistently more or less than $\frac{1}{2}$ according to which type of selection is in progress. However, a more rigorous attack on the problem shows that it is more difficult than the corresponding one for simple descendants, where the formulation of the factor which actually replaces $\frac{1}{2}$ is quite easy at least in the case of classical selection, and the author has so far failed to reach any definite general conclusions as to the nature and extent of the error in the foregoing account which his use of the ordinary coefficients of relationship has actually involved.

Finally, it must be pointed out that the model is not applicable to the selection of new mutations. Sibs might or might not carry the mutation depending on the point in the germ-line of the parent at which it had occurred, but for relatives in general a definite number of generations must pass before the coefficients give the true—or, under selection, the approximate—expectations of replicas. This point is favourable to the establishment of taking-traits and slightly against giving-traits. A mutation can, however, be expected to overcome any such slight initial barrier before it has recurred many times.

5. The Model Limits to the Evolution of Altruistic and Selfish Behaviour

With classical selection a genotype may be regarded as positively selected if its fitness is above the average and as counter-selected if it is below. The environment usually forces the average fitness $a..$ towards unity; thus for an arbitrary genotype the sign of δa_{ij} is an indication of the kind of selection. In the present case although it is $T_{..}^{\bullet}$ and not $R^{\bullet}$ that is forced towards unity, the analogous indication is given by the inclusive fitness effect $\delta R_{ij}^{\bullet}$, for the remaining part, the diluting effect δS_{ij}, of the total genotypic effect $\delta T_{ij}^{\bullet}$ has no influence on the kind of selection. In other words the kind of selection may be considered determined by whether the inclusive fitness of a genotype is above or below average.

We proceed, therefore, to consider certain elementary criteria which determine the sign of the inclusive fitness effect. The argument applies to any genotype and subscripts can be left out.

Let

$$\delta T^\circ = k \, \delta a. \tag{7}$$

According to the signs of δa and δT° we have four types of behaviour as set out in the following diagram:

| | Neighbours | |
| | gain;
δT° +ve | lose;
δT° −ve |

		Neighbours gain; δT° +ve	Neighbours lose; δT° −ve
Individual	gains; δa +ve	k +ve *Selected*	k −ve Selfish behaviour ?
	loses; δa −ve	k −ve Altruistic behaviour ?	k +ve *Counter- selected*

The classes for which k is negative are of the greatest interest, since for these it is less obvious what will happen under selection. Also, if we regard fitness as like a substance and tending to be conserved, which must be the case in so far as it depends on the possession of material prerequisites of survival and reproduction, k −ve is the more likely situation. Perfect conservation occurs if $k = -1$. Then $\delta T^\bullet = 0$ and $T^\bullet = 1$: the gene-pool maintains constant "volume" from generation to generation. This case has been discussed in Case (c) of section 3. In general the value of k indicates the nature of the departure from conservation. For instance, in the case of an altruistic action $|k|$ might be called the ratio of gain involved in the action: if its value is two, two units of fitness are received by neighbours for every one lost by an altruist. In the case of a selfish action, $|k|$ might be called the ratio of diminution: if its value is again two, two units of fitness are lost by neighbours for one unit gained by the taker.

The alarm call of a bird probably involves a small extra risk to the individual making it by rendering it more noticeable to the approaching predator but the consequent reduction of risk to a nearby bird previously unaware of danger must be much greater.† We need not discuss here just how risks are to be reckoned in terms of fitness: for the present illustration it is reasonable to guess that for the generality of alarm calls k is negative but $|k| > 1$. How large must $|k|$ be for the benefit to others to outweigh the risk to self in terms of inclusive fitness?

† The alarm call often warns more than one nearby bird of course—hundreds in the case of a flock—but since the predator would hardly succeed in surprising more than one in any case the total number warned must be comparatively unimportant.

$$\delta R^{\bullet} = \delta R^{\circ} + \delta a$$
$$= r^{\circ}\,\delta T^{\circ} + \delta a \qquad\qquad \text{from (5)}$$
$$= \delta a(kr^{\circ} + 1) \qquad\qquad \text{from (7)}.$$

Thus of actions which are detrimental to individual fitness (δa −ve) only those for which $-k > \dfrac{1}{r^{\circ}}$ will be beneficial to inclusive fitness ($\delta R^{\bullet}$ +ve).

This means that for a hereditary tendency to perform an action of this kind to evolve the benefit to a sib must average at least twice the loss to the individual, the benefit to a half-sib must be at least four times the loss, to a cousin eight times and so on. To express the matter more vividly, in the world of our model organisms, whose behaviour is determined strictly by genotype, we expect to find that no one is prepared to sacrifice his life for any single person but that everyone will sacrifice it when he can thereby save more than two brothers, or four half-brothers, or eight first cousins . . . Although according to the model a tendency to simple altruistic transfers ($k = -1$) will never be evolved by natural selection, such a tendency would, in fact, receive zero counter-selection when it concerned transfers between clonal individuals. Conversely selfish transfers are always selected except when from clonal individuals.

As regards selfish traits in general (δa +ve, k −ve) the condition for a benefit to inclusive fitness is $-k < \dfrac{1}{r^{\circ}}$. Behaviour that involves taking too much from close relatives will not evolve. In the model world of genetically controlled behaviour we expect to find that sibs deprive one another of reproductive prerequisites provided they can themselves make use of at least one half of what they take; individuals deprive half-sibs of four units of reproductive potential if they can get personal use of at least one of them; and so on. Clearly from a gene's point of view it is worthwhile to deprive a large number of distant relatives in order to extract a small reproductive advantage.

REFERENCES

HALDANE, J. B. S. (1923). *Trans. Camb. phil. Soc.* **23**, 19.

HALDANE, J. B. S. & JAYAKAR, S. D. (1962). *J. Genet.* **58**, 81.

KEMPTHORNE, O. (1957). "An Introduction to Genetical Statistics". New York: John Wiley & Sons, Inc.

KINGMAN, J. F. C. (1961a). *Quart. J. Math.* **12**, 78.

KINGMAN, J. F. C. (1961b). *Proc. Camb. phil. Soc.* **57**, 574.

LI, C. C. & SACKS, L. (1954). *Biometrics*, **10**, 347.

MORAN, P. A. P. (1962). *In* "The Statistical Processes of Evolutionary Theory", p. 54. Oxford: Clarendon Press.

MULHOLLAND, H. P. & SMITH, C. A. B. (1959). *Amer. math. Mon.* **66**, 673.

SCHEUER, P. A. G. & MANDEL, S. P. H. (1959). *Heredity*, **31**, 519.

20

Reprinted from *Amer. Naturalist,* 104(938):408–411 (1970)

SOCIAL DOMINANCE AND DARWINIAN FITNESS IN THE LABORATORY MOUSE

Levin, Petras, and Rasmussen (1969) have recently discussed the maintenance of the t-allele polymorphism in natural populations of house mice (*Mus musculus*). Although t alleles are recessive lethal or male sterile, most natural populations are polymorphic for alleles at the T locus. An abnormal segregation or transmission ratio of about 0.95 in heterozygous males is apparently responsible for the maintenance of this polymorphism. However, the observed t-allele frequency in natural populations is approximately half that expected on the basis of deterministic models which assume infinite population size and random mating. Lewontin and Dunn (1960) have suggested that this discrepancy may be due to the effects of random genetic drift. They postulated that mice live in small family units (demes), a large fraction of which would become fixed for the normal allele; thus, when a geographical population composed of demes is sampled, the observed frequency of the t allele will be less than that calculated from deterministic models.

Levin, Petras, and Rasmussen (1969) employed a Monte Carlo simulation of a geographical population of house mice to examine the range of breeding-unit sizes and interdemic migration rates which were compatible with the random-drift hypothesis. Assuming a transmission ratio of 0.95 for lethal t alleles, it was found that random drift would have an important effect on t-allele frequencies in populations with migration between demes approaching 0% only when the effective breeding-unit size was less than 12. For populations with interdemic migration rates of 1% or 3%, the effective population size would have to be less than eight or four, respectively.

In a study of the deme size of house mice released in a population cage, Reimer and Petras (1967) previously observed that a deme is generally composed of a dominant male, several females, and several subordinate males. A valid estimate of the effective population size was not available from this study however, since no definite information about the subordinate males' contribution to the gene pool was obtained. The authors also observed that interdemic migration was rare and appeared possible only by females. This low rate of migration is in accord with the earlier findings by Young, Strecker, and Emlen (1950) of rather localized activity in indoor populations of wild house mice.

The above observations suggest that the effective population size of house mice is indeed small. However, the actual breeding-unit size will be greatly affected by the role of the subordinate males. If subordinate males sire relatively few offspring, the effective population size could be less than four. Therefore, it seems important to assess the contribution of subordinate males to the gene pool. The present communication reports the results of two experiments which suggest that the Darwinian fitness of subordinate males is very low; thus, the effective population size appears to be sufficiently small to permit random drift to play an important role in natural populations of house mice.

In order to facilitate study of the genetics of social behavior in laboratory mice, standardized social living units have been developed. Each "triad" is constructed from three standard living cages ($9 \times 4 \times 4$ inches high). An opening is cut into one end of each of the three cages, which are then connected with a Y-shaped plastic manifold constructed of $1\frac{3}{4}$-inch clear Plexiglas tubing, thus permitting free movement of animals from one cage to another. In the first experiment, three males (60–80 days of age), each from a different inbred strain (A/Ibg, BALB/cIbg, C57BL/Ibg, DBA/2Ibg, I/Ibg, or C3H/Ibg, symbolized henceforth as A, B, C, D, I, and 3, respectively), were placed in triads with three B females (over 50 days of age). Combination of males was dictated solely by the possibility of ascertaining paternity from the coat color of subsequent litters. B females were chosen because of their availability and because they are recessive for two coat color loci (*bbcc*), thus facilitating identification of paternity. Male combinations and number of replications were as follows: ACD, 10; BCD, 7; BD3, 2; AD3, 2; and ACI, 1. Males were examined daily for number of tail wounds (skin punctures) and then weighed. Location of the animals within the triad was recorded three times daily. One cage of the triad was changed each day, and food and water were available ad libitum. After 2 weeks, males were discarded and females retained until litters were born and old enough to permit classification by coat color.

A·typical pattern of social interaction has emerged in the triad situation: Males begin to fight within a few minutes of introduction. Within 1 or 2 days, one male is established as the clearly dominant individual. Wounds are usually found on the tails of subordinate males, but wounding of hind quarters may also result. A criterion of least number of tail wounds appears to be an almost infallible index of social dominance. In all cases observed, this criterion agreed with evaluations of dominance based upon observation of behavioral interactions. The two subordinate males tend to stay in one cage. Out of 623 observations in which all three males were still alive in the triad, both subordinate males were in one cage 522 times (84%). If the subordinate males venture from their cage, they are driven back by the dominant male. The dominant male may enter the subordinate males' cage at any time, eat, drink, and attack them with no obvious provocation, and then leave. Although females are rarely if ever attacked, a male may briefly harass a female during the early hours of establishment of the population when fighting among males is intense. Fighting among females is rare but has been observed. Females are free to come and go in the triad and do enter the cage of the subordinate males. However, when a female leaves this cage, the subordinate males do not follow. In about half of the triads, one or both of the subordinate males died before the 2 weeks of observations were completed. In most cases, the wounds did not seem sufficiently serious to be fatal, suggesting that death may have been due to stress.

Of the 10 replications which employed the ACD combination of inbred males, A males were dominant in six, and C and D males were each dominant in two. In eight triads, the D male was most subordinate, as indicated by the greatest number of tail wounds or earliest death. Of the seven BCD triads, B males were dominant in four, and C males were dominant in three; D males were most subordinate in six. These results suggest that A and B males tend to be most dominant and that D males are most subordinate in triad social situations. The number of replications of the other male combinations was too small to suggest reliable strain differences. Social dominance in this experiment was closely related to body weight: A highly significant correlation ($r = -.45$, df $= 63$, $P < .01$) was observed between initial body weight and maximum number of tail wounds during the first 3 days in the triad (animals that died during the first 3 days were assigned a score of 20).

The relationship between social dominance and Darwinian fitness as measured by relative number of litters sired in the triad situation was particularly striking. In 18 of the 22 triads, the dominant male sired all of the litters. In all 22 triads, the dominant male sired at least one litter. In three triads a subordinate male sired one litter, and in only one case did a subordinate male sire two litters. Of the 61 litters obtained, 56 (92%) were sired by the dominant male ($\chi^2 = 94.1$, df $= 1$, $P < .0001$). In order to assess the possibility that this relationship was due in part to the killing of subordinate males by the dominant male, litters were excluded which were conceived when only the dominant male was alive, that is, those litters which were born more than 21 days after the death of the second subordinate male. Of the remaining 58 litters, 53 (91%) were sired by the dominant male. Thus, social dominance appears to be an overwhelming component of Darwinian fitness in the triad situation and does not depend upon death of the subordinate animals.

In order to rule out the possibility that this relationship was due to the use of only inbred B females in the triads, the experiment was repeated with outbred females derived from an initial cross of eight inbred strains. Because of the greater difficulty in ascertaining paternity with these females, only those which were homozygous recessive for at least one of several coat color loci were utilized (albino, brown, or dilute). Pairs of males were placed with either two or three females in these triads, combinations again being determined on the basis of possible identification of paternity by coat color of pups. Male combinations and number of replicates in this experiment were as follows: BD, 10; CD, 3; BI, 3; AB, 2; DI, 1; and BC, 1. In the 10 triads in which BD males were employed, eight were dominated by B males, again demonstrating the social dominance of B males relative to D males. In the remaining 10 triads, replications were insufficient to indicate reliable strain differences. However, of the 42 litters produced in this experiment, 40 (95%) were sired by the dominant male ($\chi^2 = 34.4$, df $= 1$, $P < .0001$). Of the 32 litters which were conceived when the subordinate male was still alive, 30 (94%) were sired by the dominant male. Thus, social dominance appears to be a major component of Darwinian fitness in the triad situation and to be independent of the genotype of the female mice employed.

If the subordinate males in natural populations contribute as little to the gene pool as those in the triad experiments, the effective breeding-unit size is indeed small. From the expression used to calculate effective population size (N_e), $N_e = 4N_mN_f/(N_m + N_f)$, it may be seen that when the number of male parents (N_m) is equal to one, N_e in finite populations is always less than four. Thus, if the dominant male sires 90% or more of the offspring in a deme, the effective population size will be sufficiently small that random drift will have an important effect, even when the migration rate is as high as 3%. In a natural environment, of course, out-migration of subordinate animals may occur. However, the net effect on reproductive success within the deme would be essentially the same. In view of the competitive disadvantage of a strange male in another male's territory (Reimer and Petras 1967) it would seem that the *effective* rate of migration among established demes may be less than 3%.

ACKNOWLEDGMENT

This work was supported in part by Research grant GM-14547 from the National Institute of General Medical Sciences. We thank Mr. Jerry Howard for his assistance during the course of this study.

LITERATURE CITED

Levin, B. R., M. L. Petras, and D. I. Rasmussen. 1969. The effect of migration on the maintenance of a lethal polymorphism in the house mouse. Amer. Natur. 103: 647–661.

Lewontin, R. C., and L. C. Dunn. 1960. The evolutionary dynamics of a polymorphism in the house mouse. Genetics 45:705–722.

Reimer, J. D., and M. L. Petras. 1967. Breeding structure of the house mouse, *Mus musculus*, in a population cage. J. Mammal. 48:88–99.

Young, H., R. L. Strecker, and J. T. Emlen, Jr. 1950. Localization of activity in two indoor populations of house mice, *Mus musculus*. J. Mammal. 31:403–410.

J. C. DeFries
G. E. McClearn

Institute for Behavioral Genetics
University of Colorado
Boulder, Colorado 80302
March 9, 1970

AUTHOR CITATION INDEX

SUBJECT INDEX

409

411

About the Editor

EDWIN M. BANKS is Professor of Ethology, Animal Science, and Psychology at the University of Illinois, Urbana. He received the Ph.D. in 1955 from the University of Florida under the direction of the late W. C. Allee. He served as President of the Animal Behavior Society in 1972 and was Assistant Editor of Animal Behavior, 1968–1972. Professor Banks is the current Secretary of the U. S. International Ethological Conference Committee and has published extensively in the areas of vertebrate social behavior and organization.